Statistics for Biology and Health

Series Editors
M. Gail, K. Krickeberg, J. Samet, A. Tsiatis, W. Wong

Statistics for Biology and Health

Bacchieri/Cioppa: Fundamentals of Clinical Research
Borchers/Buckland/Zucchini: Estimating Animal Abundance: Closed Populations
Burzykowski/Molenberghs/Buyse: The Evaluation of Surrogate Endpoints
Cook/Lawless: The Statistical Analysis of Recurrent Events
Everitt/Rabe-Hesketh: Analyzing Medical Data Using S-PLUS
Ewens/Grant: Statistical Methods in Bioinformatics: An Introduction, 2nd ed.
Gentleman/Carey/Huber/Irizarry/Dudoit: Bioinformatics and Computational Biology Solutions Using R and Bioconductor
Hougaard: Analysis of Multivariate Survival Data
Keyfitz/Caswell: Applied Mathematical Demography, 3rd ed.
Klein/Moeschberger: Survival Analysis: Techniques for Censored and Truncated Data, 2nd ed.
Kleinbaum/Klein: Survival Analysis: A Self-Learning Text, 2nd ed.
Kleinbaum/Klein: Logistic Regression: A Self-Learning Text, 2nd ed.
Lange: Mathematical and Statistical Methods for Genetic Analysis, 2nd ed.
Manton/Singer/Suzman: Forecasting the Health of Elderly Populations
Martinussen/Scheike: Dynamic Regression Models for Survival Data
Moyé: Multiple Analyses in Clinical Trials: Fundamentals for Investigators
Nielsen: Statistical Methods in Molecular Evolution
O'Quigley: Proportional Hazards Regression
Parmigiani/Garrett/Irizarry/Zeger: The Analysis of Gene Expression Data: Methods and Software
Proschan/LanWittes: Statistical Monitoring of Clinical Trials: A Unified Approach
Siegmund/Yakir: The Statistics of Gene Mapping
Simon/Korn/McShane/Radmacher/Wright/Zhao: Design and Analysis of DNA Microarray Investigations
Sorensen/Gianola: Likelihood, Bayesian, and MCMC Methods in Quantitative Genetics
Stallard/Manton/Cohen: Forecasting Product Liability Claims: Epidemiology and Modeling in the Manville Asbestos Case
Sun: The Statistical Analysis of Interval-censored Failure Time Data
Therneau/Grambsch: Modeling Survival Data: Extending the Cox Model
Ting: Dose Finding in Drug Development
Vittinghoff/Glidden/Shiboski/McCulloch: Regression Methods in Biostatistics: Linear, Logistic, Survival, and Repeated Measures Models
Wu/Ma/Casella: Statistical Genetics of Quantitative Traits: Linkage, Maps, and QTL
Zhang/Singer: Recursive Partitioning in the Health Sciences
Zuur/Ieno/Smith: Analyzing Ecological Data

Rongling Wu
Chang-Xing Ma
George Casella

Statistical Genetics of Quantitative Traits

Linkage, Maps, and QTL

Springer

Rongling Wu
Department of Statistics
University of Florida
Gainesville, FL 32611
rwu@stat.ufl.edu

Chang-Xing Ma
Department of Biostatistics
State University of New York at Buffalo
Buffalo, NY 14214
cxma@buffalo.edu

George Casella
Department of Statistics
University of Florida
Gainesville, FL 32611
casella@stat.ufl.edu

ISBN 978-0-387-20334-8 e-ISBN 978-0-387-68154-2

Library of Congress Control Number: 2006938665

Printed on acid-free paper.

9 8 7 6 5 4 3 2 1

springer.com

To my parents, wife, and son
RW

To my parents and Yuehua, David, and Eric
CM

To my family, and Lulu, too
GC

Preface

Most traits in nature and of importance to agriculture are quantitatively inherited. These traits are difficult to study due to the complex nature of their inheritance. However, recent developments of genomic technologies provide a revolutionary means for unraveling the secrets of genetic variation in quantitative traits. Genomic technologies allow the molecular characterization of polymorphic markers throughout the entire genome that are then used to identify and map the genes or quantitative trait loci (QTLs) underlying a quantitative trait based on linkage analysis.

Statistical analysis is a crucial tool for analyzing genome data, which are now becoming increasingly available for a variety of species, and for giving precise explanations regarding genetic variation in quantitative traits occurring among species, populations, families, and individuals. In 1989, Lander and Botstein published a hallmark methodological paper for interval mapping that enables geneticists to detect and estimate individual QTL that control the phenotype of a trait. Today, interval mapping is an important statistical tool for studying the genetics of quantitative traits at the molecular level, and has led to the discovery of thousands of QTLs responsible for a variety of traits in plants, animals, and humans. In a recent study published in *Science*, Li, Zhou, and Sang (2006, *311*, 1936–1939) were able to characterize the molecular basis of the reduction of grain shattering – a fundamental selection process for rice domestication – at the detected QTL by interval mapping. Among many other examples of the success of interval mapping are the positional cloning of QTLs responsible for fruit size and shape in tomato (Frary et al. 2000, *Science 289*, 85–88) and for branch, florescence, and grain architecture in maize (Doebley et al. 1997, *Nature 386*, 485–488; Gallavotti et al. 2004, *Nature 432*, 630–635; Wang et al. 2005, *Nature 436*, 714–719).

To make it suitable for various practical applications, interval mapping has been extensively modified and extended during the past 15 years. A host of useful statistical methods for QTL mapping have been produced through the collective efforts of statistical geneticists. However, these methods generally have various objectives and utilities and are sporadically distributed in a massive amount of literature. A single volume synthesizing statistical developments for genetic mapping may be helpful for many researchers, especially those with a keen interest in building a bridge between genetics and statistics, to acquaint themselves with this expanding area as quickly as possible.

This book intends to provide geneticists with the tools needed to understand and model the genetic variation for quantitative traits based on genomic data collected in mapping research and equip statisticians with the uniqueness and ideas in relation to the exploration of genetic secrets using their computational skills. This book also intends to attract researchers toward multidisciplinary research and to introduce them to new paradigms in genomic science. In this book, the statistical and computational theories applied to genetic mapping are developed hand in hand and a number of examples displaying the implications of statistical genomics are introduced.

This book contains 14 chapters, broadly divided into three parts. Part 1, including Chapters 1 and 2, provides introductory genetics and statistics at the level appropriate for understanding general genetic concepts and statistical models for genetic mapping. Part 2, composed of Chapters 3–7, attempts to provide a thorough and comprehensive coverage of linkage analysis with molecular markers. Models and methods for linkage analysis and map construction are systematically introduced for different designs, such as the backcross/F_2 (Chapter 3), outbred crosses (Chapter 4), recombinant inbred lines (Chapter 5) and structured pedigrees (Chapter 7), and for special marker types including distorted and misclassified markers (Chapter 6) and dominant markers (Chapters 4 and 7). Part 3, composed of Chapters 8–14, covers statistical models and algorithms of QTL mapping. The topics include simple marker-phenotype association analyses (Chapter 8), the statistical structure of interval mapping (Chapter 9), regression- (Chapter 10) and maximum likelihood-based analysis of interval mapping (Chapter 11), threshold and confidence interval determination (Chapter 12), composite interval mapping using multiple markers as cofactors (Chapter 13), and interval mapping for outbred mapping populations (Chapter 14). In the Appendices, we provide general statistical theories directly related to the genetic mapping approaches introduced and R programs for some of the examples used in the book. A webpage (http://www.buffalo.edu/∼cxma/book/) was constructed for this book, which includes a complete list of programs and algorithms written in MatLab or R for all the examples.

Writing a book in such a rapidly developing and changing field is a pain but, more precisely speaking, full of excitement. In the summer of 1997, Wu delivered a series of lectures on statistical methods for QTL mapping to graduate students and faculty at Nanjing Forestry University, China. In the spring semester of 2002, Wu taught a statistical genetics course at the master's level at the University of Florida and then was joined for coteaching by Casella in the spring of 2003 and Ma in the spring of 2005. This course is now taught by Wu at the University of Florida and by Ma at the State University of New York at Buffalo on the regular basis. We all gave many lectures or short courses related to statistical genetics at other places and times. At each place and time, we were heavily impressed by the enthusiasm of students and other audiences to learn this fascinating area. All these encouraged us to write a book that can cover basic methods for statistical genetics research. The concepts, models and algorithms related to genetic mapping have been published in a variety of statistics and genetics journals by a large number of authors, but part of the material contained in this book comes from our collaborative research program in the past five years. In particular, we apologize for those authors whose work was not mentioned in this book because of limited space.

During the writing of this book, many of our colleagues and friends both at the University of Florida and outside provided valuable help from different perspective. Wu is warmly grateful to his postdoctoral advisor, Dr. Zhao-Bang Zeng at North Carolina State University, for tremendous guidance and for leading him to the field of statistical genetics. Dr. Bruce Walsh at the University of Arizona provided insightful reviews of the book manuscript in different stages. Several anonymous reviewers gave constructive comments that significantly improve the presentation of the book. Students or postdocs who attended our lectures and classes or are working with us on statistical genetics in different places have provided many insightful suggestions to improve our presentation of the book. The following students or postdocs in our group, former or current, deserve special thanks: Yuehua Cui, Wei Hou, Hongying Li, Min Lin, Tian Liu, Fei Long, Xiang-Yang Lou, Qing Lu, Damaris Santana, Zhaojie Wang, Zuoheng Wang, Jiasheng Wu, Song Wu, Jie Yang, John Yap, Li Zhang, Wei Zhao, and Yun Zhu. The data used for examples in the book were kindly supplied by Dr. James Cheverud at Washington University (mouse), Dr. Junyi Gai at Nanjing Agricultural University (soybean), Rory Todhunter at Cornell University (dog), Drs. Stan Wullschleger and Tongming Yin at Oak Ridge National Laboratory (poplar), and Dr. Jun Zhu at Zhejiang University (rice).

We are grateful to the Department of Statistics and the Institute of Food and Agricultural Sciences at the University of Florida for writing this book and performing our research program. Finally, we are greatly indebted to our respective families for their continuous support of our research activities over the years. This work is partially supported by NSF grant 0540745.

Gainesville, FL *Rongling Wu*
Buffalo, NY *Chang-Xing Ma*
December 2006 *George Casella*

Contents

1

Basic Genetics

1.1 Introduction

There have been enormous advances in the science of genetics. A huge amount of information regarding the precise molecular mechanisms of genetic transmission from parent to offspring is becoming increasingly available. In this chapter, we briefly review basic terminology and principles of genetics from Mendelian, population, quantitative and molecular perspectives at a level appropriate for understanding the research methods to be described in this book. Much of the description for classic Mendelian genetics is adapted from Bailey's (1961) book. To learn more about modern genetics, please look into the more general genetics textbooks that are listed at the end of this chapter.

1.2 Genes and Chromosomes

Genes are discrete units in which biological characteristics are inherited from parents to offspring. Genes are normally transmitted unchanged from generation to generation, and they usually occur in pairs. If a given pair consists of similar genes, the individual is said to be *homozygous* for the gene in question, while if the genes are dissimilar, the individual is said to be *heterozygous*. For example, if we have two alternative genes, say A and a, there are two kinds of homozygotes, namely AA and aa, and one kind of heterozygote, namely Aa. These alternative genes are called *alleles*. With a single pair of alleles, there are three different kinds of possible organisms represented by the three *genotypes* AA, Aa, and aa.

Genes are generally very numerous, and situated within the cell nucleus, where they lie in linear order along microscopic bodies called *chromosomes*. The chromosomes occur in similar, or *homologous*, pairs, where the number of pairs is constant for each species. For example, *Drosophila* has 4 pairs of chromosomes, pine has 12, the house mouse has 20, humans have 23, etc. The totality of these pairs constitutes the *genome* of a particular organism. One of the chromosome pairs in the genome are the sex chromosomes (typically denoted by **X** and **Y**) that determine genetic sex. The other pairs are *autosomes* which guide the expression of most other traits.

Each gene pair has a certain place or *locus* on a particular chromosome. Since the chromosomes occur in pairs, the loci and the genes occupying them also occur in pairs. Therefore, it is the loci that have the fixed linear order, although a given locus may be occupied by any gene from the series of alleles (more than two alleles or *multialleles*) determining a particular trait. The most important purpose of a genome mapping project is to locate the genes affecting trait expressions on chromosomes.

1.3 Meiosis

When ordinary body cells divide and multiply, the cell nucleus undergoes a process of division called *mitosis*, which results in the two daughter cells, each having a full set of paired chromosomes exactly like the parent cell. But in the production of reproductive cells or *gametes* (ova and spermatozoa), we have a different mechanism, called *meiosis*. This ensures that only one chromosome from each homologous pair passes into each gamete. It follows that gametes also possess only one gene from each gene pair. The number of chromosomes in a gamete is referred to as the *haploid* number, in contrast to the full complement possessed by a fertilized egg, or *zygote*, which is *diploid*.

A diagram is drawn to illustrate the biological process of meiosis (Fig. 1.1). The chromosomes are already duplicated by the time they become visible at the start of the first meiotic division. Each pair of duplicates is joined at the *centromere*, a small particle at which two arms of the chromosome are connected. The duplicated pairs remain joined throughout the first anaphase. The paternal *homolog* (a duplicated pair) moves to one pole; the maternal homolog (another duplicated pair) moves to the other. The immediate products of the first meiotic division are two cells, each containing a diploid chromosome set. However, each homologous pair of chromosomes in one of these cells is a pair of maternally originated chromosomes or a pair of paternally originated chromosomes. The assortment between the two cells is random, with each resulting cell normally containing some chromosome pairs of maternal origin and others of paternal origin. In the second meiotic division, the number of chromosomes is halved and each of the two products of the first division produces identical daughter cells with half the usual number of chromosomes.

The significance of reduction division in meiosis is that it can maintain a diploid (double) chromosome set after fertilization, the fusion of a male gamete (sperm) with a female gamete (egg). A second essential characteristic of meiosis is that there is an interchange of genetic material between the two chromosomes of a homologous pair. Thus, the haploid gamete chromosome set contains a mixture of chromosomes, some derived from the father and some from the mother.

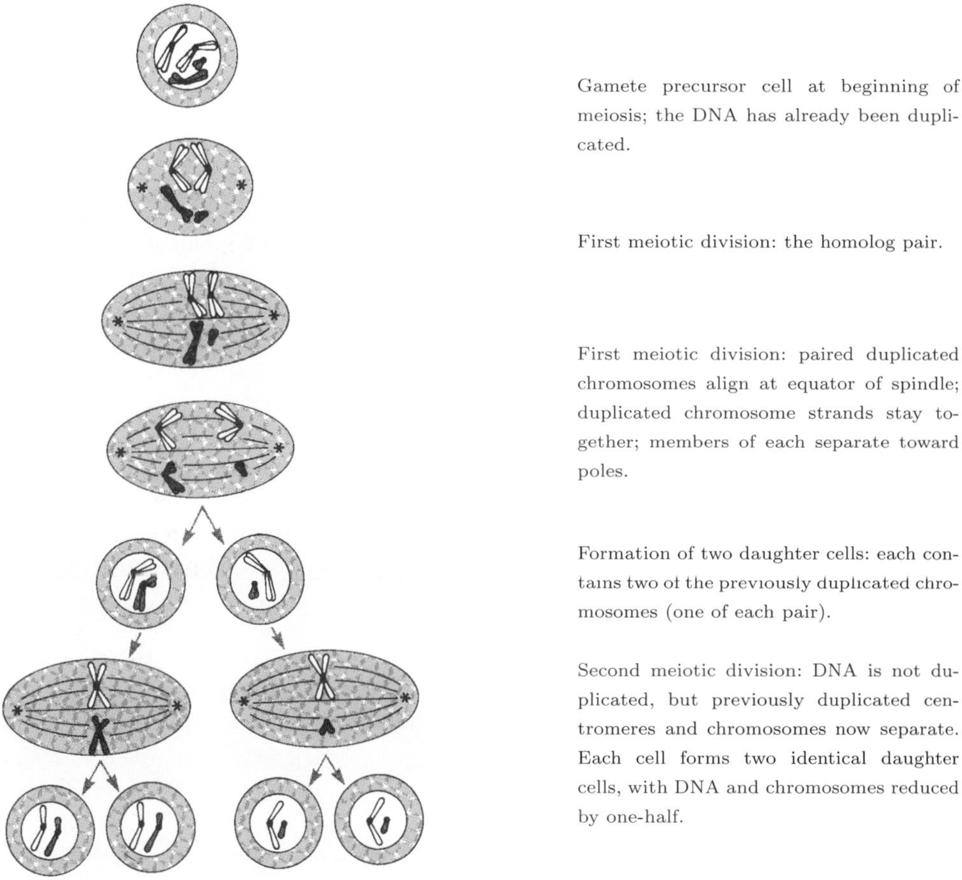

Fig. 1.1. Schematic diagram of meiosis in a hypothetical male who has one pair of identical autosomes (white) and one dissimilar XY pair (shaded). Adapted from Cavalli-Sforza and Bodmer (1971).

1.4 Mendel's Laws

1.4.1 Mendel's First Law

Genes are present in pairs in all cells of an adult organism, except for gametes. The gametes have only one gene from any given pair. Thus if an adult has genotype AA, all the gametes produced are of type A. But if the genotype is Aa, two types of gametes are possible, A and a, and these are normally produced in equal numbers. When fertilization occurs, a sperm carrying one gene from the male parent is united with an ovum carrying one gene from the female parent, thus making up a complete pair. The fertilized egg, or *zygote*, then develops to produce an organism in each body cell, of which one gene is derived from one parent and one from the other. The new individual produces its own reproductive cells, and so the process can continue.

The considerations above constitute Mendel's first law, the *Law of Segregation.* This states that characteristics are controlled by pairs of genes that segregate or separate during the formation of the reproductive cells, thus passing into different gametes. The pairs are restored when fertilization occurs, and this leads to the production of different types of offspring in certain definite proportions. In effect, segregation shuffles the genes and redeals them to the next generation. Characters themselves may also be said to show segregation, but the precise manner in which this happens depends on the nature of the genes involved and their dominant and recessive relationships.

Suppose we cross two individuals, represented by AA and aa. All gametes from the first will be A and all from the second will be a. Thus, all zygotes F_1 will be of the heterozygous type Aa. We now cross two individuals from the F_1 to form a new F_2 generation. Each F_1 heterozygous Aa produces two kinds of gametes, A and a, in equal numbers. At fertilization, there are four ways in which a zygote can be formed: one A gene from each parent; one a from each parent; A from the male and a from the female; or A from the female and a from the male. We therefore expect the three types of offspring AA, Aa, and aa in the ratios of 1:2:1 in the F_2 generation. But, if A is dominant, the first two classes will be phenotypically indistinguishable, giving the characters A and a in a 3:1 ratio.

If one of the heterozygous F_1 offspring is mated back to the homozygous parent, a *backcross* population is generated. The genotype of an individual in the backcross depends only on the heterozygous F_1 in which two kinds of gametes, A and a, are formed in equal numbers. Thus, the segregation ratio of the genotypes in the backcross follows a 1:1 ratio.

1.4.2 Mendel's Second Law

Mendel's second law says that when two or more pairs of genes segregate simultaneously, they do so independently. This is the *Law of Independent Assortment.* In some cases, this law is adequate, but it is subject to certain very important exceptions. These arise because of the phenomenon of linkage, a main topic of this book.

Suppose we have two pairs of genes represented by $\mathbf{A}$, with two alleles A and a, and $\mathbf{B}$ with two alleles B and b. If we cross two individuals, one homozygous for both A and B and the other homozygous for both a and b (i.e., the mating $AABB \times aabb$), it is obvious that all offspring will be $AaBb$. This is because the first parent must produce gametes that are all AB, and the second parent must produce gametes which are all ab. We now consider the intercross $AaBb \times AaBb$. If the segregation is to be independent then each of these individuals will produce four kinds of gametes, namely AB, Ab, aB and ab, in equal numbers. Combining the four alternative types of gametes from one parent with the four alternatives from the other leads to 16 combinations, which are not, however, all different. The various possibilities are most easily presented as shown in the diagram of Fig. 1.2. It will be seen from the diagram that there are in fact nine distinct genotypes, $AABB$ (1), $AABb$ (2), $AAbb$ (1), $AaBB$ (2), $AaBb$ (4), $Aabb$ (2), $aaBB$ (1), $aaBb$ (2), and $aabb$ (1), where the number given in parentheses is the forming number of each genotype. But if each gene pair exhibits a dominant/recessive relationship, there will be only four separate phenotypic classes, AB, Ab, aB, and ab occurring in the ratio 9:3:3:1.

		Gametes			
		AB	*Ab*	*aB*	*ab*
Gametes	*AB*	*AABB* (*AB*)	*AABb* (*AB*)	*AaBB* (*AB*)	*AaBb* (*AB*)
	Ab	*AABb* (*AB*)	*AAbb* (*Ab*)	*AaBb* (*AB*)	*Aabb* (*Ab*)
	aB	*AaBB* (*AB*)	*AaBb* (*AB*)	*aaBB* (*aB*)	*aaBb* (*aB*)
	ab	*AaBb* (*AB*)	*Aabb* (*Ab*)	*aaBb* (*aB*)	*aabb* (*ab*)

Fig. 1.2. Gene segregation of an intercross, $AaBb \times AaBb$, involving two gene pairs. When each pair exhibits dominance, the resultant phenotypes are given in brackets. The degree of dominance is roughly described by different darknesses of the cells.

1.5 Linkage and Mapping

Mendel's second law applies to genes whose loci lie on different chromosomes. Genes whose loci lie on the same chromosome will tend to remain together. Loci on the same chromosome are said to be *syntenic*, and those on different chromosomes are said to be *nonsyntenic*. The extent to which syntenic loci remain together depends on their closeness. We are thus led to consider the phenomenon of *linkage*.

In order to see what essentially is involved in linkage, let us consider the formation of gametes by a heterozygote $AaBb$. If the loci for the gene pairs A, a and B, b lie on the same kind of chromosome, we can specify more exactly the composition of the homologous pair of chromosomes. Thus, one chromosome may contain A and B, the other a and b; i.e.,

$$\begin{array}{c|c|c} A & & a \\ B & & b \end{array} \,, \tag{1.1}$$

where the two vertical lines stand for the two homologous chromosomes. Or, alternatively, A and b may lie on one chromosome, while the other contains a and B; i.e.,

$$(1.2) \qquad \begin{array}{c|c|c} A & & a \\ b & & B \end{array} \; .$$

Definition 1.1. [Some Basic Terms] For alleles A and B, the arrangement displayed in diagram (1.1) is termed *coupling* and is written AB/ab; the arrangement in diagram (1.2) is called *repulsion* and is indicated by Ab/aB. The relative arrangement of *nonalleles* (i.e., A vs. B, A vs. b, a vs. B, or a vs. b) at different loci along a chromosome is called the *linkage phase.*

At an early stage of meiosis, the two chromosomes 1 and 2 lie side by side with corresponding loci aligned. If the parental genotype is AB/ab, we can represent the alignment as in Fig. 1.3A. Each of the paired chromosomes is then duplicated to form two sister strands (*chromatids*) connected to each other at a region called the *centromere.* The homologous chromosomes form pairs, so that each resulting complex consists of four chromatids known as a *tetrad* (Fig. 1.3B). At this stage, the non-sister chromatids adhere to each other in a semi-random fashion at regions called *chiasmata.* Each chiasma represents a point where *crossing over* between two non-sister chromatids can occur (Fig. 1.3C). Chiasmata do not occur entirely at random, as they are more likely farther away from the centromere, and it is unusual to find two chiasmata in very close proximity to each other.

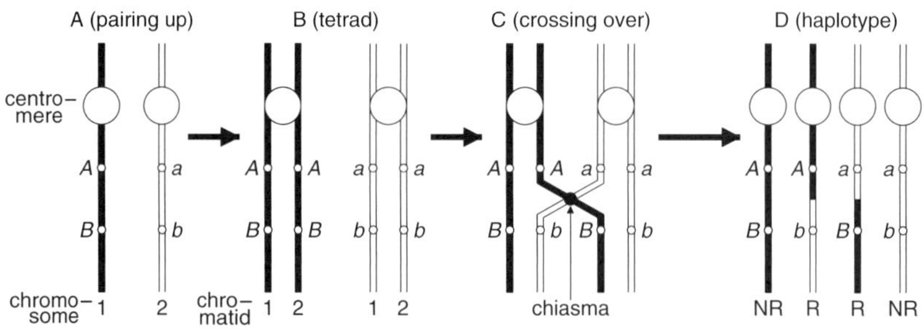

Fig. 1.3. Diagram for crossing–over between linked loci **A** and **B**.

Each gamete receives one chromatid from a tetrad to make up the haploid complement (Fig. 1.3D). Since it is possible that more than one crossover occurs on the chromosomes, some chromosomes in the haploid complement consist of a number of segments from the two parental chromosomes. The number of segments is determined by the number of crossovers that occurred in the formation of the chromatid that became the chromosome. If no crossovers occur, then the chromosome will be a replicate of an entire parental chromosome. If one crossover occurs between two loci $\mathcal{A}$ and $\mathcal{B}$, then the chromosome will consist of two segments, one from each parental chromosome. In the former case, the resultant gametes must be AB or ab, just like the *parental* chromosomes. In the latter case, where there is one point of exchange, we have the new combinations Ab and aB, called *recombinant* types. In general, if there are an even number of points of exchange between the two loci, the final result will be indistinguishable from AB or ab. But if there are an odd number of points of exchange, the result will be like Ab or aB.

The existence of linkage means that there will be more gametes like AB and ab, and fewer like Ab and aB. Let us suppose that the proportion of recombinant gametes is r, which we call the *recombination fraction*, and that the proportion of parental type is $1 - r$. The recombination fraction can be estimated on the basis of the expected number of recombinants in a segregating progeny (see Chapter 3). In general, we should not expect to find recombination fractions greater than one-half, though in certain unusual circumstances there may be a tendency for chromosomes inherited from one parent or from particular stocks to associate nonrandomly.

From the definition of the recombination fraction, it follows that the special case $r = 1/2$ is equivalent to independent segregation or no linkage. Actually, if two loci on one chromosome are a long way apart, odd and even numbers of points of exchange will be about equally frequent (i.e., 50 percent each), so this case will not be immediately distinguishable from the case where the loci are on different chromosomes. Alternatively, if two loci are close together, the frequency of points of exchange will be low, and the corresponding recombination fraction will be small. To some extent, we can use the latter as a measure of the distance between any two loci.

A better scale of measurement is that afforded by the density of points of exchange.

Definition 1.2. [Map Distance] The *map distance* between any two loci is the average number of points of exchange occurring in the segment.

The map distance is a quantity that is automatically additive. There is a very simple relationship between the recombination fraction and the map distance for a pair of loci in the simplest case of no interference. Such a relationship is called a *map function* and will be discussed in Section 3.10. When the recombination fractions between pairs of loci on a single chromosome have been determined from an appropriate linkage experiment, it is a simple matter to transform them into map distances and hence construct a chromosome map. Since there is no reason to suppose that chromosomes are homogeneous along their lengths with regard to the frequency of crossing–over, we cannot assume that there is necessarily a very close correspondence between genetic map distance and the actual physical distance between the corresponding genes.

When many genes are considered, an issue arises about their linear arrangement within each chromosome. The loci of any organism fall into linkage groups, where any locus in one group is unlinked to any locus in a different group. Within any group, however, the loci can be arranged in a linear order. For sufficiently close loci, the recombination fraction between any pair may, in an elementary analysis, be used as a direct measure of the distance between the loci. To retain additivity at greater separations, we must work in terms of the average number of crossovers rather than the recombination fraction (which only measures the frequency of an odd number of crossovers). We thus need to know how the recombination fractions observed between many pairs of loci lying on a single chromosome can be fitted into a unifying picture based on the notion of a chromosome *map*. This will critically rely upon the development of theoretical models and statistical algorithms for constructing genetic linkage maps, which is one of the major themes of this book.

1.6 Interference

In the simplest case, we assume that the points of exchange occur at random, so that the pattern of crossing–over in any segment of a chromosome is independent of the pattern in any other segment. In practice, however, nonrandomness is common and was named *interference* by H. J. Muller (1916). When, as usual, this is positive, the occurrence of a point of exchange tends to inhibit the formation of other such points in its neighborhood. Various models are available for describing the phenomenon of interference, and some of these entail the occurrence of recombination fractions greater than one-half in sufficiently long chromosomes.

As mentioned earlier, each chromosome splits longitudinally into a pair of identical daughter–chromosomes (chromatids) during the relevant part of a meiotic division. The two chromatids are initially held together by the centromere (Fig. 1.3B). Crossing–over always occurs between chromatids from different chromosomes of a homologous pair, as shown in Fig. 1.3C. Thus, the phenomenon of crossing–over actually involves all four chromatids, or strands, of any pair of homologous chromosomes. A pair of homologous chromosomes united by crossing–over is often called *bivalent*.

We may envision the occurrence of several points of exchange or chiasma, each of which now entails the X-like arrangement of chromatids shown in Fig. 1.3C. We can distinguish between two kinds of interference.

Definition 1.3. [Kinds of Interference] One type of interference is *chiasma interference*, in which the occurrence of one chiasma influences the chance of another occurring in its neighborhood, and another is *chromatid interference*, which is a nonrandom relationship between the pair of strands involved in one chiasma and the pair involved in the next chiasma.

Chiasma interference is common, and some distributions have been observed in which the variance of interference was as low as a quarter of its mean. Chromatid interference, on the other hand, is much more difficult to detect, and evidence for its existence is more scant. It has been proven that chiasma interference alone is incapable of causing recombination fractions of more than 50 percent (Mather 1938).

1.7 Quantitative Genetics

1.7.1 Population Properties of Genes

Mendelian segregation leads to simple and predictable segregation ratios in the offspring of specific mating types but only applies to a progeny population derived from two parents of known genotype. However, different mating types can occur simultaneously to generate the offspring in a natural or experimental population in which the ratios of the different genotypes are weighted averages of the segregation ratios of all the possible mating types, the weights being the relative frequencies of the different mating types. The population properties of genes can be described by the allele frequencies, genotype frequencies, and Hardy-Weinberg law.

Consider a gene with two alleles, A and a, with respective frequencies p_1 and p_0, in a population. Let P_2, P_1, and P_0 be the population frequencies of three genotypes, AA, Aa and aa, respectively. When the mating type frequencies arise from random mating, the ratios of the different genotypes follow a mathematical model established independently by the English mathematician Hardy (1908) and the German physician Weinberg (1908). This well-known model, today called the Hardy-Weinberg Law, states that, if individuals in the population mated with each other at random, these frequencies would satisfy the relationship

$$P_1^2 = 4P_2P_0, \tag{1.3}$$

and each of these frequencies is kept unchanged from generation to generation. The population that follows equation (1.3) is said to be at Hardy-Weinberg equilibrium, in which the genotype frequencies can be expressed as $P_2 = p_1^2$, $P_1 = 2p_1p_0$, and $P_0 = p_0^2$, respectively. Approaches exist to test whether or not a population is at Hardy-Weinberg equilibrium (Falconer and Mackay 1996; Lynch and Walsh 1998).

1.7.2 A General Quantitative Genetic Model

A gene that is segregating in a population may affect the phenotype of a trait. For a complex or quantitatively inherited trait, the genes that determine it may be numerous and their relationships with the environment may be complicated. The study of the genetic basis of a quantitative trait is the theme of quantitative genetics.

Consider a quantitative trait with phenotypic value P, which is determined by the genetic (G) and environmental factors (E) and their interaction (G $\times$ E), expressed as

$$\mathrm{P} = \mathrm{G} + \mathrm{E} + \mathrm{G} \times \mathrm{E}. \tag{1.4}$$

Assuming that all terms in equation (1.4) are independent of one another, we partition the phenotypic variance of the trait into the corresponding genetic, environmental, and genotype $\times$ environment interaction variance components:

$$V_{\mathrm{P}} = V_{\mathrm{G}} + V_{\mathrm{E}} + V_{\mathrm{G}\times\mathrm{E}}. \tag{1.5}$$

In statistics, the variance is generally symbolized by V or σ^2. The genetic variance, V_{G} or σ_G^2, is due to the effects of all genes that determine the trait. Consider a gene with genotypes AA, Aa, and aa whose genotypic values and frequencies in a population at Hardy-Weinberg equilibrium are expressed as follows:

Genotype	Genotypic Value	Frequency
AA	$\mu_2 = \mu + a$	$P_2 = p_1^2$
Aa	$\mu_1 = \mu + d$	$P_1 = 2p_1p_0$
aa	$\mu_0 = \mu - a$	$P_0 = p_0^2$

The three different genotypes are symbolized by j ($j = 2$ for AA, 1 for Aa, and 0 for aa). Genotypic values are composed of the overall mean of the trait (μ), the *additive effect* (a) of the gene due to the substitution of alleles from A to a, or the *dominance effect* (d) due to the interaction effect of different alleles A and a at the gene. If there is no dominance, $d = 0$; if allele A is dominant over a, d is positive; and if allele a is dominant over A, d is negative. Dominance is complete if d is equal to $+a$ or $-a$, and there is overdominance if d is greater than $+a$ or less than $-a$. The degree of dominance is described by the ratio d/a.

The population mean of the three genotypes with different frequencies is calculated as

$$\bar{\mu} = \sum_{j=0}^{2} P_j \mu_j = (p_1 - p_0)a + 2p_1p_0d,$$

and we have the genetic variance for this gene,

$$\begin{aligned}\sigma_g^2 &= \sum_{j=0}^{2} P_j \left(\mu_j - \bar{\mu}\right)^2 \\ &= 2p_1p_0[a + (p_1 - p_0)d]^2 + 4p_1^2p_0^2d^2 \\ &= 2p_1p_0\alpha^2 + 4p_1^2p_0^2d^2 \\ &\stackrel{def}{=} \sigma_a^2 + \sigma_d^2,\end{aligned}$$

where $\alpha = a + (p_1 - p_0)d$ is the *average effect* due to the substitution of alleles from A to a (Falconer and Mackay 1996). The first term of the genetic variance, σ_A^2, is the *additive genetic variance* component, and the second term, σ_D^2, is the *dominance genetic variance* component. These two expressions can be readily extended to include the effects of all underlying genes for a trait. If gene interactions are ignored, the variances contributed by all the genes are expressed as $\sigma_G^2 = \sum \sigma_g^2$, $\sigma_A^2 = \sum \sigma_a^2$, and $\sigma_D^2 = \sum \sigma_d^2$.

1.7.3 Genetic Models for the Backcross and F$_2$ Design

The partitioning of the genetic variance can be made for different genetic settings. Consider two parental populations, P$_1$ and P$_2$, fixed with favorable alleles $A_1, ..., A_m$ and unfavorable alleles $a_1, ..., a_m$, respectively, for all m loci. The two parents are crossed to generate an F$_1$. The F$_1$ is backcrossed to one of the parents to form a backcross or self-crossed to form an F$_2$.

Let a_k and d_k be the additive and dominance effects of gene k, respectively, and r_{kl} be the recombination fraction between any two genes k and l. Consider a pair of genes, $\mathbf{A}_k$ and $\mathbf{A}_l$, whose genotypic values (upper) and frequencies (lower) in the F_2 population are expressed as

(1.6)

	A_lA_l	A_la_l	a_la_l
A_kA_k	$\mu + a_k + a_l$	$\mu + a_k + d_l$	$\mu + a_k - a_l$
	$\frac{1}{4}(1 - r_{kl})^2$	$\frac{1}{2}r_{kl}(1 - r_{kl})$	$\frac{1}{4}r_{kl}^2$
A_ka_k	$\mu + d_k + a_l$	$\mu + d_1 + d_2$	$\mu + d_k - a_l$
	$\frac{1}{2}r_{kl}(1 - r_{kl})$	$\frac{1}{2}[r_{kl}^2 + (1 - r_{kl})^2]$	$\frac{1}{2}r_{kl}(1 - r_{kl})^2$
a_ka_k	$\mu - a_k + a_l$	$\mu - a_k + d_l$	$\mu - a_k - a_l$
	$\frac{1}{4}r_{kl}^2$	$\frac{1}{2}r_{kl}(1 - r_{kl})$	$\frac{1}{4}(1 - r_{kl})^2$

where the genotypic values are composed of the additive and dominance effects at the two genes because gene interactions are ignored, and the derivation of the genotype frequencies in the F_2, expressed in terms of the recombination fraction between two genes, needs knowledge of linkage analysis, described in Section 3.5. From display 1.6, we can derive the genetic variance of the trait as

$$
\begin{aligned}
\sigma_G^2 &= \frac{1}{2}\sum_{k=1}^{m} a_k^2 + \frac{1}{4}\sum_{k=1}^{m} d_k^2 \\
&+\frac{1}{2}\sum_{k=1}^{m}\sum_{l=1,k\neq l}^{m}(1-2r_{kl})a_ka_l + \frac{1}{4}\sum_{k=1}^{m}\sum_{l=1,k\neq l}^{m}(1-2r_{kl})^2 d_kd_l. \qquad (1.7)
\end{aligned}
$$

The first term on the right side of equation (1.7) for the F_2 is the additive variance within loci, the second is the dominance variance within loci, the third is the additive covariance between different loci, and the fourth is the dominance covariance between different loci.

For the backcross, in which the dominance effect cannot be defined due to inadequate degrees of freedom, we can derive a similar but simpler genetic variance, expressed as

$$
\sigma_G^2 = \frac{1}{4}\sum_{k=1}^{m} a_k^2 + \frac{1}{4}\sum_{k\neq l}^{m}(1-2r_{kl})a_ka_l. \qquad (1.8)
$$

From equation (1.8), the genetic variance in a backcross consists of the additive genetic variance and additive covariance between different loci.

1.7.4 Epistatic Model

Genes may affect quantitative traits in an interactive way. The effect due to gene interaction was coined as *epistasis* by W. Bateson (1902). From a physiological perspective, epistasis describes the dependence of gene effects at one locus upon those at the other locus. Fisher (1918) first partitioned the genetic variance into additive, dominance, and epistatic components using the least squares principle. Cockerham (1954) further partitioned the two-gene epistatic variance into the additive × additive, additive × dominance, dominance × additive, and dominance × dominance interaction components. There are many approaches for specifying epistasis, but we will model epistasis using Mather and Jinks' (1982) approach.

Consider two genes, one denoted by **A**, with three genotypes, AA, Aa, and aa, and the second denoted by **B**, with three genotypes, BB, Bb, and bb. These two genes form nine two-locus genotypes, whose genotypic values, denoted by $\mu_{j_1 j_2}$, can be partitioned into different components

$$
\begin{aligned}
\mu_{j_1 j_2} = \quad & \mu && \text{overall mean} \\
& + (j_1 - 1)a_1 + (j_2 - 1)a_2 && \text{additive effects} \\
& + j_1(2 - j_1)d_1 + j_2(2 - j_2)d_2 && \text{dominance effects} \\
& + (j_1 - 1)(j_2 - 1)i_{aa} && \text{additive } \times \text{ additive effect} \\
& + (j_1 - 1)j_2(2 - j_2)i_{ad} && \text{additive } \times \text{ dominance effect} \\
& + j_1(2 - j_1)(j_2 - 1)i_{da} && \text{dominance } \times \text{ additive effect} \\
& + j_1(2 - j_1)j_2(2 - j_2)i_{dd} && \text{dominance } \times \text{ dominance effect,}
\end{aligned} \tag{1.9}
$$

where

$$
j_1, j_2 = \begin{cases} 2 & \text{for } AA \text{ or } BB \\ 1 & \text{for } Aa \text{ or } Bb \\ 0 & \text{for } aa \text{ or } bb \end{cases} .
$$

The second line of equation (1.9) is the additive effects of single genes, the third line is the dominance effects of single genes, and the fourth, fifth, sixth, and seventh lines are the epistatic effects between the two genes, additive × additive (i_{aa}), additive × dominance (i_{ad}), dominance × additive (i_{da}), and dominance × dominance (i_{dd}), respectively.

For the two genes that are cosegregating with the recombination fraction of r in an F_2 population, the genotypic values and frequencies are expressed in Table 1.1. Note that the genotype frequencies are calculated in terms of r. Based on Table 1.1, the genetic variance of a trait can be derived.

1.7.5 Heritability and Its Estimation

According to equation (1.5), the total phenotypic variance of a quantitative trait is decomposed into its genetic, environment and genotype × environment interaction

Table 1.1. Genotypic values (upper) and frequencies (lower) of the nine genotypes at two genes, **A** and **B**.

	BB	*Bb*	*bb*
AA	$\mu + a_1 + a_2 + i_{aa}$	$\mu + a_1 + d_2 + i_{ad}$	$\mu + a_1 - a_2 - i_{aa}$
	$\frac{1}{4}(1-r)^2$	$\frac{1}{2}r(1-r)$	$\frac{1}{4}r^2$
Aa	$\mu + d_1 + a_2 + i_{da}$	$\mu + d_1 + d_2 + i_{dd}$	$\mu + d_1 - a_2 - i_{da}$
	$\frac{1}{2}r(1-r)$	$\frac{1}{2}[r^2 + (1-r)^2]$	$\frac{1}{4}r(1-r)$
aa	$\mu - a_1 + a_2 - i_{aa}$	$\mu - a_1 + d_2 - i_{ad}$	$\mu - a_1 - a_2 + i_{aa}$
	$\frac{1}{4}r^2$	$\frac{1}{2}r(1-r)$	$\frac{1}{4}(1-r)^2$

variance components. The ratio of the genetic variance over the phenotypic variance is defined as *broad-sense heritability*, i.e.,

$$H^2 = \frac{\mathrm{V_G}}{\mathrm{V_G} + \mathrm{V_E} + \mathrm{V_{G\times E}}}. \tag{1.10}$$

As shown above, the genetic effect or variance can be partitioned into additive (A) and nonadditive (NA) effects or variances. Thus, we have

$$\begin{aligned}\mathrm{P} &= \mathrm{G} + \mathrm{E} + \mathrm{G} \times \mathrm{E} \\ &= \mathrm{A} + \mathrm{NA} + \mathrm{E} + \mathrm{A} \times \mathrm{E} + \mathrm{NA} \times \mathrm{E},\end{aligned}$$

and

$$\begin{aligned}V_\mathrm{P} &= V_\mathrm{G} + V_\mathrm{E} + V_\mathrm{G\times E} \\ &= V_\mathrm{A} + V_\mathrm{NA} + V_\mathrm{E} + V_\mathrm{A\times E} + V_\mathrm{NA\times E},\end{aligned}$$

if all the effects terms are independent of each other.

The nonadditive effect or variance is the summation of dominance and epistatic effect or variance. Because the additive effect can be inherited from the parents to offspring whereas the nonadditive effect cannot, we use the ratio of the additive variance over the total phenotypic variance, define as the *narrow-sense heritability*, i.e.,

$$h^2 = \frac{\mathrm{V_A}}{\mathrm{V_A} + \mathrm{V_{NA}} + \mathrm{V_E} + \mathrm{V_{A\times E}} + \mathrm{V_{NA\times E}}}, \tag{1.11}$$

to quantify the degree with which the phenotypic value of a quantitative trait is unchanged from one generation to next. The two heritability parameters (1.10) and (1.11) are traditionally used to describe the degree of overall genetic control for a trait, including the contributions of all the underlying genes (Lynch and Walsh 1998). These two parameters are now commonly used to describe the contributions of individual genes if these genes can be detected by an approach like genetic mapping, described in Chapters 8–14.

In practice, genetic variances can be estimated on the basis of a quantitative genetic theory founded by Cockerham (1954, 1963). According to this theory, a set of parents is crossed to generate multiple crosses in a mating design. The progeny from the mating design is then grown in a particular experimental design, from which the phenotypic data collected are analyzed by statistical approaches, such as analysis of variance, to obtain various experimental variances. Based on the resemblance between relatives, the estimated experimental variances are used to estimate the additive and dominance genetic variances and, therefore, the broad- and narrow-sense heritabilities.

Comparable to Cockerham's models, Mather and Jinks (1982) proposed a different approach based on generation differences to estimate genetic effect or variance components. Consider study material composed of three generations, inbred parents P_1 and P_2, the non-segregating F_1 and the segregating F_2, which are grown under the same condition. The phenotypic variance of a trait for the two pure parent lines ($\mathrm{V}_{\mathrm{P}_1}$ and $\mathrm{V}_{\mathrm{P}_2}$) and F_1 progeny ($\mathrm{V}_{\mathrm{F}_1}$) is purely due to environmental factors, whereas the phenotypic variance of the same trait in the F_2 ($\mathrm{V}_{\mathrm{F}_2}$) includes a sum of genetic, environmental and genotype $\times$ environmental variance. Thus, the genetic variance of the trait can be estimated by

$$\mathrm{V}_{\mathrm{G}} = \mathrm{V}_{\mathrm{F}_2} - \mathrm{V}_{\mathrm{F}_1}, \tag{1.12}$$

or

$$\mathrm{V}_{\mathrm{G}} = \mathrm{V}_{\mathrm{F}_2} - \frac{1}{4}\left(\mathrm{V}_{\mathrm{P}_1} + \mathrm{V}_{\mathrm{P}_2} + 2\mathrm{V}_{\mathrm{F}_1}\right). \tag{1.13}$$

The estimates of individual genetic variance components can be obtained by the inclusion of more generations (Mather and Jinks 1982).

1.7.6 Genetic Architecture

Most quantitative traits are determined by a web of many interacting loci and by an array of environmental factors (Falconer and Mackay 1996). The traditional *polygenic* theory of quantitative traits (Mather 1943) envisaged a fairly large number of loci, each with relatively small and equal effects, acting in a largely additive way. Over the years it has indeed been observed that a quantitative trait may display complicated genetic architecture (Mackay 1996, 2001), expressed as

(1) It may be controlled by a fairly large number of loci; for example, of the order of 50, according to the work of Shrimpton and Robertson (1988a,b);
(2) Genes act in ways which may be additive, dominance, epistatic with other genes, and interactive with environmental factors;
(3) The magnitude of the effect produced by each locus can vary considerably;
(4) The same genes may affect different phenotypic traits through pleiotropic effects;

(5) The genes affecting the trait may be distributed over the genome at random or in a certain pattern.

With a deep use of genetic mapping to analyze quantitative traits, increasing evidence has been observed for the third point, which suggests that typically a small number of loci account for a very large fraction of the variation in the trait. For this reason, the traditional polygenic model may be replaced by a new *oligogenic* model in which a small number of major genes each with a large effect, combined with many minor genes each with a small effect, determine the genetic variation of a quantitative trait (see Mackay 1996 for an excellent review).

1.7.7 The Estimation of Gene Number

The actual number of genes that control a quantitative trait is one of the most important elements for the genetic architecture of the trait. Gene number can be estimated by a biometrical approach, although it depends on some critical assumptions (Lande 1981; Lynch and Walsh 1998). The number of genes estimated by this approach basically reflects the effective number of genes that contribute a major part of genetic variation of a trait. The most widely used approach for estimating gene number is based on the phenotypic means and variances of two parental lines and their hybrids, i.e., F_1, F_2 and backcrosses. The biometrical approach for the enumeration of effective genes was first proposed by Castle (1921).

Suppose there are two contrasting parental lines, one (P_1) being homozygous for all increasing alleles and the second (P_2) being homozygous for all decreasing alleles. These two lines are crossed to generate the F_1 and F_2. There are a total of unlinked m_e effective genes each with the same effect (a) that is purely additive. The mean phenotype of the P_1 and P_2 line can be written, respectively, as

$$\mu_{P_1} = \mu + \sum_{i=1}^{m_e} a = \mu + m_e a,$$

$$\mu_{P_1} = \mu - \sum_{i=1}^{m_e} a = \mu - m_e a,$$

whose difference is

$$\Delta = \mu_{P_1} - \mu_{P_2} = 2m_e a, \tag{1.14}$$

with the overall mean μ being canceled. Based on equation (1.7), the genetic variance of the F_2 is rewritten as

$$V_G = \frac{1}{2}\sum_{i=1}^{m_e} a^2 = \frac{1}{2} m_e a^2, \tag{1.15}$$

under the assumptions as mentioned above. Combining equations (1.14) and (1.15), we obtain the Castle-Wright estimator of gene number as

$$\hat{m}_e = \frac{\Delta^2}{8\mathrm{V_G}}, \tag{1.16}$$

where $\mathrm{V_G}$ is estimated by equation (1.12) or (1.13). The sampling variance of $\hat{n}_e$ can be approximated by

$$\mathrm{Var}(\hat{m}_e) = \hat{m}_e^2 \left[\frac{4(\mathrm{V_{P_1}} + \mathrm{V_{P_2}})}{\Delta^2} + \frac{\mathrm{Var}(\mathrm{V_G})}{\mathrm{V_G^2}} \right], \tag{1.17}$$

where

$$\mathrm{Var}(\mathrm{V_G}) = \frac{2\mathrm{V_{F_2}^2}}{n_{\mathrm{F_2}} + 2} + \frac{2\mathrm{V_{F_1}^2}}{n_{\mathrm{F_1}} + 2}$$

with $n_{\mathrm{F_2}}$ and $n_{\mathrm{F_1}}$ being the sample sizes, if equation (1.12) is used.

After the Castle-Wright estimator, several studies were pursued to improve the estimation of gene number. Lande (1981) generalized the Castle-Wright estimator for use with outcrossing populations. Zeng et al. (1990) and Zeng (1992) relaxed some of the critical assumptions, including unlinkage and equal additive effect, used for the Castle-Wright estimator. Epistatic effects between different genes were considered in Wu (1996) who extended gene enumeration to estimate a more complete picture of genetic architecture. In particular, Wu's model allows for the estimation of more genetic parameters by including multiple generations, $\mathrm{P_1}$, $\mathrm{P_2}$, $\mathrm{F_1}$, $\mathrm{F_2}$ and backcrosses, in the same experiment. Generally speaking, use of biometrical approaches for the estimation of gene number has been limited in practice, despite their significance in helping to understand general quantitative genetic theory. A more precise approach for gene enumeration is based on genetic mapping with molecular markers in which the association between markers and phenotypic variation is analyzed and tested by statistical models (Lander and Botstein 1989).

1.8 Molecular Genetics

Molecular genetics applied to linkage analysis is concerned with genetic marker technologies. Molecular genetic markers are readily assayed phenotypes that have a direct 1:1 correspondence with DNA sequence variation at a specific location in the genome. In principle, the assay for a genetic marker is not affected by environmental factors. Genetic markers are DNA sequence polymorphisms that show Mendelian inheritance. For genome mapping, the ideal genetic marker is codominant, multiallelic, and hypervariable (i.e., segregates in almost every family). However, some dominant markers are also very useful and powerful in particular situations.

Molecular markers have many different types. Restriction fragment length polymorphisms (RFLPs) were the first genetic markers that were widely used for genomic mapping and population studies. RFLP markers are obtained by using restriction endonucleases to precisely cleave a genomic DNA fragment containing a particular gene sequence. If two organisms differ in the distance between sites of cleavage of a particular restriction endonuclease, they will produce different lengths of the fragments when the DNA is digested with a restriction enzyme. The fragments can then be separated by gel electrophoresis. The diagram in Fig. 1.4 illustrates the segregation of an RFLP marker among eight different organisms. Individuals 1 and 5 are the homozygote AA for the larger RFLP alleles, individuals 4 and 7 are the homozygotes for the smaller RFLP allele, and the others are heterozygous. The mode of inheritance of RFLP markers is codominant, allowing for all three genotypes at a single locus to be scored.

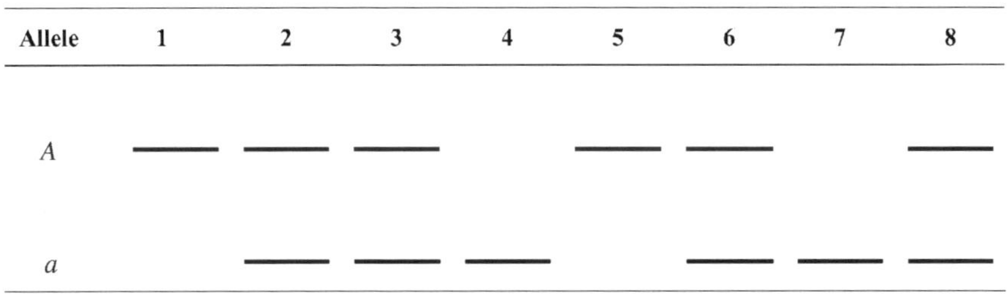

Fig. 1.4. Diagram for the segregating pattern of an RFLP marker detected by gel electrophoresis.

The detection of RFLPs requires the hybridization of a labeled DNA probe to denatured single-strand DNA fragments. The probe can be cloned DNA with known or unknown sequences. Most RFLP variation is due to insertion/deletion differences that are located between restriction enzyme recognition sites, and a small portion of the variation can be due to sequence variation within restriction sites. RFLPs require large amounts of genomic DNA and are laborious to carry out compared with polymerase chain reaction methods.

The polymerase chain reaction (PCR) provides a useful way to obtain genetic markers based on amplification of specific DNA fragments from small quantities of genomic DNA templates. PCR-amplified markers can be based on anonymous genomic DNA fragments that vary in size (codominant inheritance), can be amplified from some individuals but not others (dominant), or can be cleaved differentially by restriction enzymes. The PCR can also readily provide many genetic markers based on amplification from genomic DNA templates using a single short primer (randomly amplified polymorphic DNA, or RAPD). These markers are anonymous DNA sequences flanked by the primer sequence in opposite orientation. The mode of inheritance for RAPDs is usually dominant; the sequence either amplifies or not, and one copy cannot be distinguished readily from two copies. Another PCR-based anonymous marker system is amplified fragment length polymorphisms (AFLPs). AFLPs are produced by cleaving genomic DNA using a pair of restriction endonucleases and then ligating adapters to the ends of the DNA fragments. A subset of the fragment is selectively amplified from these ligated fragments. These DNA fragments can be resolved on DNA sequencing gels and have a higher multiplex ratio than RAPDs, often 20–40 markers per gel lane. Figure 1.5 is a diagram for the segregation of two dominant

(RAPD or AFLP) markers, **A** and **B**, for eight individuals. At marker **A**, individuals 1, 2, 3, 5, 6, and 8 show a band, suggesting that they are either the homozygote for the dominant allele or the heterozygote. For this marker, individuals 4 and 7 with no band should be homozygous for the recessive allele. A similar inference can be made about the segregating patterns of markers **B** and **C**.

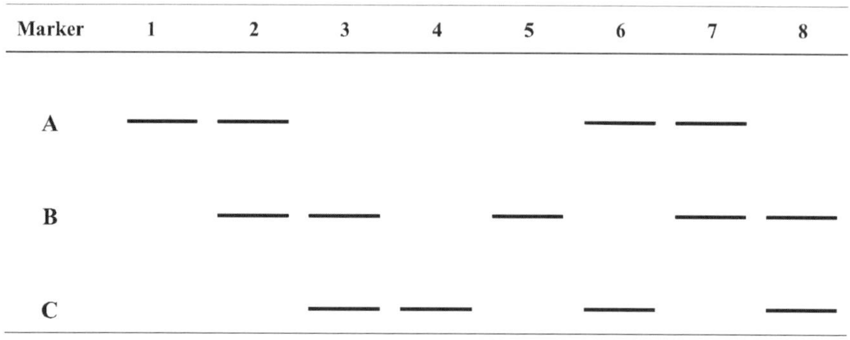

Fig. 1.5. Diagram for the segregating pattern of three dominant markers detected by gel electrophoresis.

Microsatellite markers, also called simple sequence repeats (SSRs), are based on the PCR–amplification of a genomic region containing a simple sequence (mono-, di- or trinucleotide) that is repeated. The repeat number of microsatellites can be highly variable, so that most individuals are heterozygous and the proportion of individuals that have the same marker genotype is small. Microsatellites may have many alleles (up to 70 or 80) at a single SSR locus and are inherited as codominant markers. In Fig. 1.6, eight hypothesized individuals are segregating for a triallelic microsatellite marker. Individual 1 is homozygous for the largest allele, individual 2 is heterozygous for the largest and second largest alleles, and others can be observed accordingly.

In practice, two ways are commonly used to identify microsatellite loci suitable for use as genetic markers. For some species, such as the human, mouse, *Arabidopsis*, and rice, in which a large number of DNA sequences are already available, microsatellites may be identified by searching from the DNA sequence databases for sequences containing simple repeats. However, for most plant and animal species, a large effort in hybridization and sequencing is needed to identify microsatellites suitable for use as genetic markers.

1.9 SNP

A single nucleotide polymorphism (SNP) is a site in the genome where the DNA sequences of many individuals differ by a single A, T, C, or G. For example, two

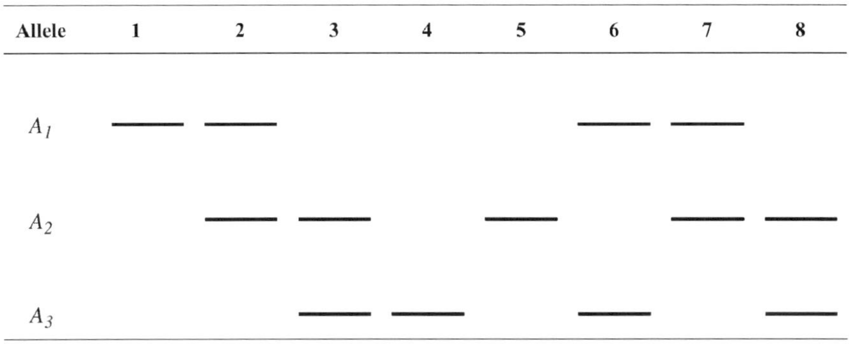

Fig. 1.6. Diagram for the segregating pattern of a multiallelic microsatellite marker detected by gel electrophoresis.

sequenced DNA fragments from different individuals, AAGCCTA and AAGCTTA, contain a difference in a single nucleotide. In this case, two nucleotides that make the two individuals different are C and T, which are called two alleles. SNPs, as the newest markers, have been the focus of much attention in genetics because they are extremely abundant and well-suited for automated large-scale genotyping. A dense set of SNP markers opens up the possibility of studying the genetic basis of complex diseases by population approaches.

SNPs can be detected in either a sequence-specific or sequence-nonspecific way. Sequence-nonspecific detection is based on the capture, cleavage, or mobility change during electrophoresis or liquid chromatography of mismatched heteroduplexes formed between allelic DNA molecules or single-stranded DNA molecules that assume slightly different conformations under nondenaturing conditions. Although sequence nonspecific detection of polymorphisms is the mainstay in polymorphism/mutation discovery, it is not an acceptable approach to genotyping because one is never certain if the inferred genotyping is the true genotype. Sequence-specific detection relies on four general mechanisms for allelic discrimination: allele-specific hybridization, allele-specific nucleotide incorporation, allele-specific oligonucleotide ligation, and allele-specific invasive cleavage.

General Genetics Textbooks for Further Reading

[1] Anthony J. F. Griffiths, William M. Gelbart, Richard C. Lewontin and Jeffrey H. Miller (2002) *Modern Genetic Analysis: Integrating Genes and Genomes.* 2nd edition. W. H. Freeman, New York.

[2] Daniel L. Hartl and Elizabeth W. Jones (2001) *Genetics: Analysis of Genes and Genomes.* 6th edition. Jones and Bartlett Publishers, Sudbury, Massachusetts.

[3] Benjamin Lewin (2005) *Essential Genes.* Prentice-Hall, Engewood Cliff, New Jersey.

1.10 Exercises

1.1 Equations (1.7) and (1.8) describe the genetic variance σ_G^2 due to m genes for the F_2 and backcross, respectively. Show how these two equations are derived.

Hint: Consider two linked genes, **A** and **B**, with recombination fraction r, in a backcross. The additive effects of these two genes are denoted by a_1 and a_2, respectively. The genetic values of the four backcross genotypes at the two genes are defined, along with their frequencies in terms of r, as follows:

Genotype	Code	Genetic Value μ_j	Frequency P_j
$AaBb$	1	$\mu + a_1 + a_2$	$\frac{1}{2}(1-r)$
$Aabb$	2	$\mu + a_1$	$\frac{1}{2}r$
$aaBb$	3	$\mu + a_2$	$\frac{1}{2}r$
$aabb$	4	μ	$\frac{1}{2}(1-r)$

The genetic variance due to these two genes is calculated as

$$\begin{aligned}\sigma_G^2 &= \sum_{j=1}^{4} P_j\mu_j^2 - \left(\sum_{j=1}^{4} P_j\mu_j\right)^2 \\ &= \frac{1}{4}(a_1^2 + a_2^2) + \frac{1}{4}(1-2r)a_1a_2.\end{aligned}$$

This can be extended to include m genes and derived similarly for the F_2 progeny.

1.2 Equation (1.16) is the Castle-Wright estimator for gene number. Lists the assumptions used for its derivation.

1.3 Modify the Castle-Wright estimator to suit the following situations:

(a) Each gene displays both the additive and dominance effects.

(b) Some genes are of increasing effect in the P_1, with the rest of decreasing effect, and vice versa for the P_2.

(c) Different genes are linked on the same chromosome.

(d) Different genes interact to display epistatic effects.

You may consider including more generations in your model, which empowers you to estimate more genetic parameters. Read Zeng et al. (1990), Zeng (1992), and Wu (1996) for some thorough discussions on each issue above.

1.11 Note

In what follows, we present an approach to model unequal genetic effects across different loci for a quantitative trait. Genetic effects are found to vary among different genes in a particular oligogenic pattern, as shown in Fig. 1.11.1, i.e., typically a small number of loci account for a large proportion of the genetic variation, with many others contributing a small proportion (Shrimpton and Robertson 1988a,b). This pattern allows the modeling of genetic effects across loci by a Gamma function or geometric series.

1.11.1 Modeling Unequal Genetic Effects by the Gamma Function

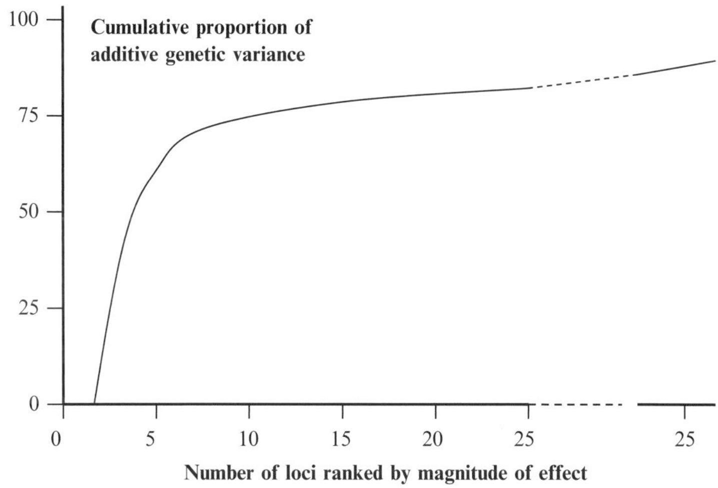

Fig. 1.7. Graphical theoretical representation of the relationship between the number of loci determining a typical character and the cumulative proportion of the additive genetic variance account for by such loci.

Suppose there are n_e genes responsible for a trait whose additive and dominance genetic effects are different. A three-generation pedigree including the P_1, P_2, F_1, F_2, and backcrosses (B_1 and B_2) is considered. Assume that the additive effect (a) at a locus follows a gamma distribution (Kimura 1979; Hill and Rasbash 1986), whose density function, respectively, as

$$f(a) = \frac{\alpha^{\beta} e^{-\alpha a} a^{\beta - 1}}{\Gamma(\beta)}, \ 0 \le a < \infty, 0 < \alpha, \beta < \infty,$$

where α is the scale parameter of the gamma distribution of the additive effect and β is the corresponding shape parameter. The moments for this distribution are

$$\mathcal{E}(a) = \frac{\beta}{\alpha}, \ \mathcal{E}(a^2) = \frac{\beta(1+\beta)}{\alpha^2}, \ \mathrm{V}(a) = \frac{\beta}{\alpha^2}, \text{ for } a$$

where $\mathcal{E}$ denotes expectation. The parameter β can be used to measure the equality of additive and dominance effects at various genes. When $\beta \to \infty$, the distribution converges to the case of equal genetic effect. According to the above moments, we obtain the following expression

$$\frac{\sum_{i=1}^{m_e} a_i^2}{m^*} = \overline{a^2} = \mathcal{E}(a^2) = \frac{1+\beta}{\beta}[\mathcal{E}(a)]^2,$$

where $\overline{a}$ and $\overline{a^2}$ are the average values of additive effects and of squared additive effects across all relevant genes, respectively. Ignoring the dominance effect, the genetic variance of the F_2 due to m^* unlinked unequally-sized genes according to 1.7 can be expressed as

$$V_G = \frac{1}{2}\sum_{i=1}^{m^*} a_i^2 = \frac{1}{2}m^*\mathcal{E}(a^2) = \frac{1}{2}m^*\left(\frac{1+\beta}{\beta}\right)(\overline{a})^2$$

The difference between the two parent lines is shown as

$$\Delta = \mu_{P_1} - \mu_{P_2} = 2m^*\overline{a}.$$

Combining the above two equations leads to the estimate of n^* as

$$\hat{m}^* = \frac{\Delta^2}{8V_G}\left(\frac{1+\beta}{\beta}\right) = \hat{m}_e\left(\frac{1+\beta}{\beta}\right), \tag{1.18}$$

where $\hat{m}_e$ is the estimate of gene number when all the genes are assumed to have an equal effect (equation (1.16)). Parameter $(1+\beta)/\beta$ can be used to describe the pattern of distribution of genetic effects. If individual effects (a_i) are normally distributed, then we have $(1+\beta)/\beta = \pi/2 = 1.57$. However, a highly leptokurtic distribution can lead to a $(1+\beta)/\beta$ value larger than $\pi/2$ (Mackay et al. 1992). Therefore, by assuming the same genetic effects the gene number is always underestimated when genetic effects are virtually unequal among genes.

1.11.2 Modeling Unequal Genetic Effects by a Geometric Series

The oligogenic model stating the genetic control of a quantitative trait by a few genes of large effects and many genes of small effects implies that the distribution of additive genetic effects (a) may be approximated by a geometric series (Lande and Thompson 1990), i.e.,

$$a, ar, ar^2, ar^3, \ldots, ar^m, \quad r \neq 1,$$

where r is the ratio determining the relative magnitude of the additive effect of each of m genes. In this case, the difference between the two parents and the genetic variance of the F_2 can be written as

$$\Delta = 2a\left(\frac{1-r^m}{1-r}\right),$$

$$V_G = \frac{a^2}{2}\left(\frac{1-r^{2m}}{1-r^2}\right),$$

which lead to

$$\widehat{m}_e = \frac{A^2}{8V_G} = \frac{1-r^m}{1+r^m}\left(\frac{1+r}{1-r}\right). \tag{1.19}$$

We thus have

$$r^m = \frac{1 - \widehat{m}_e\left(\frac{1-r}{1+r}\right)}{1 + \widehat{m}_e\left(\frac{1-r}{1+r}\right)},$$

and

$$\hat{m} = \log_r\left[1 - \widehat{m}_e\left(\frac{1-r}{1+r}\right)\right] - \log_r\left[1 + \widehat{m}_e\left(\frac{1-r}{1+r}\right)\right]. \tag{1.20}$$

Letting $u = \frac{1+r^m}{1-r^m}$, we derive the expression of r, based on equation (1.19), as

$$r = \frac{\widehat{m}_e u - 1}{\widehat{m}_e u + 1}.$$

For the oligogenic control model for which m is large and r is small, it is not unreasonable to assume $u \simeq 1$, which leads to

$$r^* = \frac{\widehat{m}_e - 1}{\widehat{m}_e + 1}.$$

Therefore, the number of genes can be approximately estimated by substituting r^* into equation (1.20).

2

Basic Statistics

2.1 Introduction

Now that we have seen the basics of genetics, we turn to an introduction to the statistical methodologies that we will use throughout this book. Most of the statistical inferences that we will make will be based on likelihood analysis, and we will be concerned not only with constructing the appropriate likelihood function for a given model but also with methods for computing and optimizing the likelihood. We start here with some basics and work our way toward likelihood analysis of a genetic linkage model.

2.1.1 Populations and Models

The goal of a statistical analysis is to draw conclusions about a *population*, a collection of objects (typically infinite) not all of which can be measured, based on examination of a *sample*, a smaller collection of objects drawn from the population, all of which can be measured. To connect the sample to the population and have the ability to make an inference, we use a *model*.

Example 2.1 (Tomato Plant Heights). Suppose that we have the following data $\mathbf{y}$ on heights (in cm) of 12 tomato plants of a particular species:

$$\mathbf{y} = (y_1, y_2, \ldots, y_{12}) = (79, 82, 85, 87, 100, 101, 102, 103, 124, 125, 126, 127).$$

We may take the following simple model. The true mean height of the population is μ, and we observe data Y_i according to

$$Y_i = \mu + \varepsilon_i, \tag{2.1}$$

where ε_i is an error term, typically taken to have a normal distribution with mean 0 and variance σ^2.

For the model (2.1), it is often assumed that the ε_i follow a normal distribution $N(0, \sigma^2)$, where the $N(\mu, \sigma^2)$ probability density function is given by

$$\phi(y|\mu, \sigma) = \frac{1}{\sqrt{2\pi\sigma^2}} \exp\left\{-\frac{1}{2\sigma^2}(y-\mu)^2\right\}.$$

Thus, under the normality assumption, we can also write the model (2.1) as $Y_i \sim \phi(y|\mu, \sigma)$.

As the process that we are trying to describe gets more complicated, so will the model. In this book, we will examine a range of models, with forms that are often dictated by the biology of the problem. Here are some examples of other models:

(a) *Linear Regression.* To describe a linear relationship between a dependent variable Y and an independent variable (or *covariate*) x, we could use the model

$$Y_i = a + bx_i + \varepsilon_i.$$

If we have many different covariates, we could use a *multiple regression model* $Y_i = a + \sum_j b_j x_{ij} + \varepsilon_i$.

(b) To describe a relationship between a covariate and a Bernoulli random variable (a variable Y that only takes the values 0 and 1, with $P(Y = 1) = p$), a *logistic regression model* is often used. This has the form

$$\text{logit}(p(x)) = a + bx,$$

where the *logit* is defined as $\text{logit}(p) = \log\left(\frac{p}{1-p}\right)$, and $P(Y = 1|x) = p(x)$.

(c) There are many models used to describe the growth process as a function of time. One popular model is the *logistic growth curve*, given by

$$g(t) = \frac{a}{1 + be^{-rt}},$$

where t typically is a time measurement. If we observe actual growth, we would use this model in the form $Y_i = \frac{a}{1+be^{-rt_i}} + \varepsilon_i$.

(d) The last model that we will describe here will find much use in QTL mapping (see Chapter 9). It is a *mixture model*, given by

$$Y_i = \begin{cases} \mu_1 + \varepsilon_i & \text{with probability} \quad p \\ \mu_2 + \varepsilon_i & \text{with probability} \quad 1-p \end{cases},$$

or, equivalently,

$$Y_i \sim \begin{cases} N(\mu_1, \sigma^2) & \text{with probability} \quad p \\ N(\mu_2, \sigma^2) & \text{with probability} \quad 1-p \end{cases}.$$

We can also write the mixture model as

$$Y_i = \mu_1 I(X = 1) + \mu_2 I(X = 0) + \varepsilon_i, \quad \varepsilon_i \sim N(0, \sigma^2),$$

where $P(X = 1) = p$, and I is the *indicator function*, which is equal to 1 if the argument is true and equal to 0 otherwise.

2.1.2 Samples

To begin an investigation, we would typically draw a *random sample* from the population of interest, a small collection of objects that are representative of the population. A random sample is, formally, a sample of n objects obtained in such a way that all sets of n objects have the same chance of being the sample. In practice, however, we hope to have a sample that is *independent* and *identically distributed*, or iid.

To be specific, if we are sampling from a population with probability density function $f(y|\theta)$, where θ contains the unknown parameters, and we draw an iid sample $Y_1, Y_2, \ldots, Y_n$, we want each variable to satisfy $Y_i \sim f(y|\theta)$ (identical) and for the variable to be independent. If we obtain an iid sample $y_1, y_2, \ldots, y_n$, the density function of the sample is

$$f(y_1, y_2, \ldots, y_n|\theta) = \prod_{i=1}^{n} f(y_i|\theta).$$

Example 2.2 (Normal Sample Density). Suppose that $Y_1, Y_2, \ldots, Y_n$ are an iid sample from an $N(\mu, \sigma^2)$ population. The sample density is

$$\begin{aligned}
\phi(y_1, y_2, \ldots, y_n|\mu, \sigma^2) &= \prod_{i=1}^{n} \phi(y_i|\mu, \sigma^2) \\
&= \prod_{i=1}^{n} \frac{1}{\sqrt{2\pi\sigma^2}} \exp\left\{-\frac{1}{2\sigma^2}(y_i - \mu)^2\right\} \\
&= \left(\frac{1}{\sqrt{2\pi\sigma^2}}\right)^{n/2} \exp\left\{-\frac{1}{2\sigma^2}\sum_{i=1}^{n}(y_i - \mu)^2\right\}. \qquad (2.2)
\end{aligned}$$

The sample density defines the relationship between the sample and the model and is thus the only means we have of estimating the unknown parameters. Actually, there are other methods, but they all lack efficiency when compared with estimation based on the sample density (see Casella and Berger 2001). Moreover, all of the information that the sample has about the parameters is contained in the sample density function. A consequence of this is that we should base our parameter estimation method on the sample density.

2.2 Likelihood Estimation

The most common estimation method, which is typically a very good method, is based on the sample density. We define the *likelihood function* as

$$L(\theta|y_1, y_2, \ldots, y_n) = f(y_1, y_2, \ldots, y_n|\theta),$$

which is merely treating the sample density f as a function of the parameter, holding the data fixed. The method of maximum likelihood estimation takes as the estimate of θ the value that maximizes the likelihood.

Example 2.3. (**Finding a Maximum Likelihood Estimator (MLE)**). To find the MLEs in Example 2.2, we need to find the values of μ and σ^2 that maximize equation (2.2). Since likelihood functions from iid samples are products (which are nasty to maximize), it is often easier to work with the logarithm of the likelihood (known as the log-likelihood). The maxima are the same as the original function, and the calculations are quite a bit easier.

From (2.2), the normal log-likelihood is

$$logL(\mu, \sigma^2) = -\frac{n}{2}\log 2\pi - \frac{n}{2}\log\sigma^2 - \frac{1}{2\sigma^2}\sum_{i=1}^{n}(y_i - \mu)^2.$$

Differentiating and setting equal to zero

$$\frac{\partial}{\partial\mu} = \frac{1}{\sigma^2}\sum_{i=1}^{n}(y_i - \mu) = 0,$$

$$\frac{\partial}{\partial\sigma^2} = -\frac{n}{2\sigma^2} + \frac{1}{2(\sigma^2)^2}\sum_{i=1}^{n}(y_i - \mu)^2 = 0,$$

gives the MLEs $\hat{\mu} = \bar{y}$ and $\hat{\sigma}^2 = \sum_{i=1}^{n}(y_i - \bar{y})^2/n$. For the data of Example 2.1, we have $\hat{\mu} = 103.41$ and $\hat{\sigma}^2 = 303.24$.

The rationale behind maximum likelihood estimation is the following. If the sample $y_1, y_2, \ldots, y_n$ is representative, then the values of y_i should come from the regions of f that have high probability; that is, the sample should "sit" under the mode of f. We do not know where this mode is because we do not know the value of θ. However, for the given sample, we can find the value of θ that makes $f(y_1, y_2, \ldots, y_n|\theta)$, or equivalently $L(\theta|y_1, y_2, \ldots, y_n)$, the highest. This puts the sample in the highest probability region, and the resulting estimator is the maximum likelihood estimator (MLE).

Example 2.4. (**Tomato Data Revisited**). As a slightly more realistic example, we return to the tomato heights of Example 2.1, but we now ask if there may be evidence of genetic control of height. Specifically, if a "height" gene is segregating in an F_2 progeny according to Mendel's first law, then we would expect to see the genotype AA:Aa:aa segregating in the ratio 1:2:1. We hypothesize that there is a gene associated with height, and it is segregating in the ratio p:q:$1-p-q$ for genotypes AA:Aa:aa. As the values of p and q are unknown, and as we do not know the genotype of the plant that we observed, the model of Example 2.1 becomes the mixture model

$$Y_i = \begin{cases} \mu_{AA} + \varepsilon_i & \text{with probability} \quad p \\ \mu_{Aa} + \varepsilon_i & \text{with probability} \quad q \\ \mu_{aa} + \varepsilon_i & \text{with probability} \quad 1-p-q \end{cases},$$

where $\varepsilon_i \sim N(0, \sigma^2)$. If we let ϕ_{AA} denote a normal density with mean μ_{AA} and variance σ^2, and if we define ϕ_{Aa} and ϕ_{aa} similarly, then, writing $\mathbf{y} = (y_1, y_2, \ldots, y_n)$, the likelihood and log-likelihood functions are

$$L(\mu_{AA}, \mu_{Aa}, \mu_{aa}, \sigma^2|\mathbf{y}) = \prod_{i=1}^{n} \left[p\phi_{AA}(y_i) + q\phi_{Aa}(y_i) + (1-p-q)\phi_{aa}(y_i) \right],$$

$$\log L(\mu_{AA}, \mu_{Aa}, \mu_{aa}, \sigma^2|\mathbf{y}) = \sum_{i=1}^{n} \log \left[p\phi_{AA}(y_i) + q\phi_{Aa}(y_i) + (1-p-q)\phi_{aa}(y_i) \right].$$

If the gene were segregating in the ratio 1:2:1 then we would have an Mendelian F_2 population.

We can find the MLEs through differentiation. Although we will see that there is no explicit solution, the equations will lead to a nice iterative solution. This solution will have more general applicability, so to address that we will solve the likelihood equations for a general mixture model and then return to the example.

Given iid observations $\mathbf{y} = (y_1, y_2, \ldots, y_n)$ from the general mixture model

$$Y_i \sim \sum_{j=1}^{k} p_j f(y_i|\theta_j), \quad \sum_j p_j = 1,$$

the log likelihood is

$$\log L(\boldsymbol{\theta}|\mathbf{y}) = \sum_{i=1}^{n} \log \left(\sum_{j=1}^{k} p_j f(y_i|\theta_j) \right),$$

where we write $\boldsymbol{\theta} = (\theta_1, \ldots, \theta_k)$.

For now, we will assume that p and q are known, but we will return to this case later. Differentiating with respect to $\theta_{j'}$ gives

$$\frac{\partial}{\partial \theta_{j'}} \log L(\boldsymbol{\theta}|\mathbf{y}) = \sum_{i=1}^{n} \frac{\sum_{j=1}^{k} p_j \frac{\partial}{\partial \theta_{j'}} f(y_i|\theta_j)}{\sum_{j=1}^{k} p_j f(y_i|\theta_j)}.$$

Notice that if $\theta_{j'}$ does not contain any of the same components as θ_j, the derivative in the numerator will be zero unless $j = j'$. We will see this in detail in Example 2.5.

For convenience, we now write

$$\frac{\partial}{\partial \theta_{j'}} f(y_i|\theta_j) = f(y_i|\theta_j) \frac{\partial}{\partial \theta_{j'}} \log \left(f(y_i|\theta_j) \right)$$

$$P_j(y_i) = \frac{p_j f(y_i|\theta_j)}{\sum_{j=1}^{k} p_j f(y_i|\theta_j)},$$

and then we have

$$\frac{\partial}{\partial \theta_{j'}} \log L(\boldsymbol{\theta}|\mathbf{y}) = \sum_{i=1}^{n} \sum_{j=1}^{n} P_j(y_i) \frac{\partial}{\partial \theta_{j'}} \log \left(f(y_i|\theta_j) \right). \tag{2.3}$$

We can solve for the MLEs with the following iterative algorithm. Start with initial values $\theta_j^{(0)}, P_j^{(0)}(y_i)$ for $j = 1, \ldots, k$. For $t = 0, 1, \ldots$:

1. For $j = 1, \ldots, k$, set $P_j^{(t+1)}(y_i) = \frac{p_j f(y_i|\theta_j^{(t)})}{\sum_{j=1}^k p_j f(y_i|\theta_j^{(t)})}$.
2. For $j' = 1, \ldots, k$, solve for $\theta_{j'}^{(t+1)}$ in

$$\sum_{i=1}^{n} \sum_{j=1}^{k} P_j^{(t+1)}(y_i) \frac{\partial}{\partial \theta_{j'}} \log\left(f(y_i|\theta_j)\right) = 0.$$

3. Increment t and return to 1. Repeat until convergence.

Example 2.5. (**Normal Mixture**). If $f(y|\theta_j)$ is $N(\mu_j, \sigma^2)$, when differentiating with respect to $\mu_{j'}$, the numerator in equation (2.3) contains only one term, the one with $j = j'$. However, the differentiation with respect to σ^2 contains the entire sum, as the parameter σ^2 is common to all densities in the mixture. We have

$$\sum_{i=1}^{n} P_j^{(t+1)}(y_i) \frac{\partial}{\partial \mu_j} \log\left(f(y_i|\mu_j, \sigma^2)\right) = \sum_{i=1}^{n} P_j^{(t+1)}(y_i) \frac{1}{2\sigma^2}(y_i - \mu_j),$$

$$\sum_{i=1}^{n} \sum_{j=1}^{k} P_j^{(t+1)}(y_i) \frac{\partial}{\partial \sigma^2} \log\left(f(y_i|\mu_j, \sigma^2)\right) = \sum_{i=1}^{n} \sum_{j=1}^{k} P_j^{(t+1)}(y_i) \times \left[-\frac{n}{2\sigma^2} + \frac{1}{2(\sigma^2)^2}(y_i - \mu_j)^2\right].$$

In setting these equations equal to zero and solving, we can treat each μ_j separately and get

$$\mu_j^{(t+1)} = \frac{\sum_{i=1}^n y_i P_j^{(t+1)}(y_i)}{\sum_{i=1}^n P_j^{(t+1)}(y_i)}.$$

For σ^2, after substituting $\mu_j^{(t+1)}$ for μ_j,we have

$$\sigma^{2(t+1)} = \frac{\sum_{i=1}^n \sum_{j=1}^k P_j^{(t+1)}(y_i) \left(y_i - \mu_j^{(t+1)}\right)^2}{n \sum_{j=1}^k P_j^{(t+1)}(y_i)}.$$

Example 2.6. (**Tomato Data Revisited, Completed**). If the gene were segregating in the ratio 1:2:1 then we would have a Mendelian F_2 population. We now estimate the normal parameters under this scenario.

Using the iterations described above, we estimate

$$\hat{\mu}_{AA} = 125.5, \quad \hat{\mu}_{Aa} = 101.5, \quad \hat{\mu}_{aa} = 83.25, \quad \hat{\sigma}^2 = 0.324.$$

The sequence of iterations is shown in Fig. 2.1, where the rapid convergence of the algorithm can be seen. An `R` program to reproduce Figure 2.1 is given in Appendix B.2.

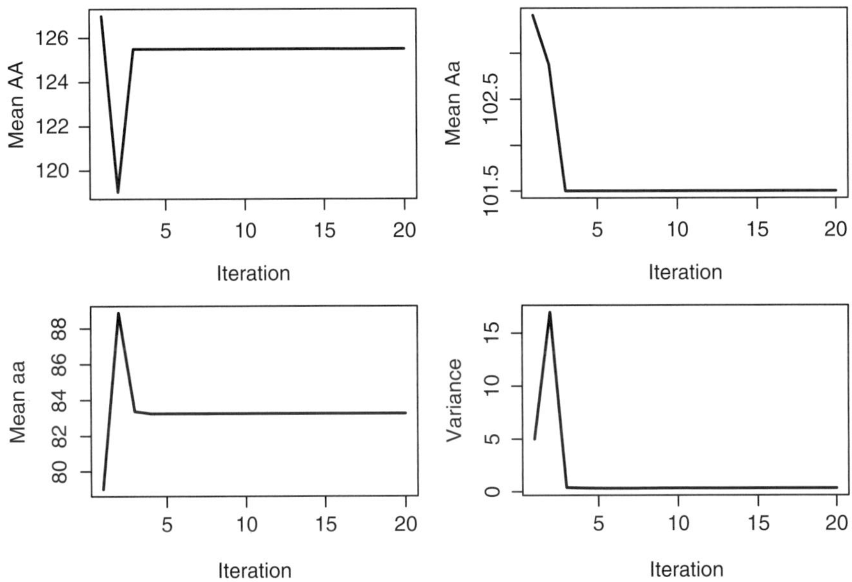

Fig. 2.1. Graphs of the iterations for the MLE of Example 2.6. The plots show the convergence of the estimates of the three means and the variance.

2.3 Hypothesis Testing

A major activity in statistical analysis is the testing of hypotheses. Here we review a number of approaches, both classical and more recent.

2.3.1 The Pearson Chi-Squared Test

We first look at a situation where there are two classes of genotypes. Suppose that there are N individuals with probability p of being in the first class. If the individuals are independent, the probability of observing n_1 individuals in the first class (and $n_2 = N - n_1$ in the second class) is given by the binomial distribution

$$P(n_1, n_2) = \frac{N!}{n_1! n_2!} p^{n_1} (1-p)^{n_2}.$$

The mean of the distribution is Np and the variance is $Np(1-p)$. In large samples, it is typical to approximate the binomial distribution by a normal distribution with the same mean and variance. Thus

$$Z = \frac{n_1 - Np}{(Np(1-p))^{1/2}} \tag{2.4}$$

is approximately a standard normal random variable, and hence Z^2 is approximately a chi-squared random variable with one degree of freedom. It can be shown that

(2.5) $$Z^2 = \chi_1^2 = \frac{(n_1 - Np)^2}{Np} + \frac{(n_2 - N(1-p))^2}{N(1-p)},$$

which is the usual formula for the *Pearson chi-squared statistic*; that is, the sum of the squares of the differences between observed (O) and expected counts (E) divided by expected counts:

(2.6) $$\chi^2 = \sum \frac{(O - E)^2}{E}.$$

Note that in testing for the segregation ratio 1:1, equation (2.6) can be written $(n_1 - n_2)^2/N$. In fact, the segregation test can be based directly on the standard normal test using equation (2.4), which would produce the same result as the Pearson chi-squared test of equation (2.5). However, the Pearson chi-squared test has an advantage in that the test statistic (2.6) can be generalized to situations involving more than two categories of genotypes. In such cases, the multinomial distribution is used to model the observations, and the chi-squared approximation is given by equation (2.6).

In general, when there are m observed counts, the Pearson chi-squared statistic that arises from the calculation of m "expected counts" is asymptotically chi-squared with $m-k$ degrees of freedom. Here m is one less than the number of observed counts (cells) and k is the number of parameters to be estimated in the calculation of the expected counts (see Section 2.3.2).

For the backcross, in which there are two genotype classes, $m = 1$. This is because only one class can be filled arbitrarily, as once a number is assigned to the first class the number in the second class is determined. For the F_2, which has a total of three genotype classes at a codominant marker, $m = 2$.

If N is not very large, the binomial distribution may not be well-approximated by a normal. In such cases, it may be best to calculate an exact p-value (see Section 2.3.3). We can also use a continuity corrected χ^2 statistic:

(2.7) $$\chi^2 = \sum \frac{(O - E)^2 - |O - E| + \frac{1}{4}}{E},$$

which trades ease of use for some accuracy in the test.

Example 2.7. (**F_1 Hybrid Population**). We use an example from Yin et al. (2001), who constructed a genetic linkage map using molecular markers in an F_1 interspecific hybrid population between two different poplar species, Chinese quaking poplar and white poplar. The mapping population was comprised of 103 hybrid trees, each genotyped with RAPD markers. Some markers are heterozygous in one parent but null in the other, whereas other markers have an inverse pattern. Since these markers are segregating in the same pattern as two-way backcross markers do, such an F_1 population is called a *two-way pseudo-test backcross* (Grattapaglia and Sederoff 1994). In a two-way pseudo-test backcross population, two parent-specific maps can be constructed. In this example, six markers are chosen that formed linkage group 16 in the white poplar genetic linkage map (Yin et al. 2001).

Table 2.1. Pearson test statistic for testing Mendelian segregation 1:1 using RAPD markers in an interspecific poplar hybrid population.

Marker	n_1	n_2	Pearson χ^2: χ^2	Pearson χ^2: p-value
I18_1090	44	59	2.184	0.139
W2_1050	45	58	1.641	0.200
AK12_2700	51	52	0.010	0.920
N18_1605	49	54	0.243	0.622
AK17_1200	40	63	5.135	0.023
I13_1080	41	62	4.282	0.038

Three different methods were used to test whether these six testcross markers follow the Mendelian segregation 1:1 in the mapping population. According to the Pearson chi-squared test (2.6), assuming a large sample size, markers I18_1090, W2_1050, AK12_2700, and N18_1605 follow the 1:1 ratio ($p > .05$), but markers AK17_1200 and I13_1080 do not (Table 2.1). This suggests that the latter two markers display possible distorted segregation.

The sample size here is actually rather large, and the continuity-corrected statistic (2.7) gave exactly the same answers. We also note that when doing multiple tests, there is the danger of rejection by chance alone. Thus, although we found two significant markers, it is best to consider this only evidence of significance, and further investigation should be done.

2.3.2 Likelihood Ratio Tests

The testing of hypotheses can be carried out very efficiently using likelihood methodology. We first consider a general setup where we observe iid observations $\mathbf{Y} = (Y_1, \ldots, Y_n)$ from a population with density function $f(y|\boldsymbol{\theta})$, where $\boldsymbol{\theta}$ could be a vector of parameters. For example, $\boldsymbol{\theta} = (\mu, \sigma^2)$ if we are modeling normals, or $\boldsymbol{\theta} = (p_1, \ldots, p_k)$, $\sum_j p_j = 1$, if we are modeling probabilities.

Given a sample $\mathbf{y} = (y_1, \ldots, y_n)$, the likelihood function for these data is $L(\boldsymbol{\theta}|\mathbf{y}) = \prod_{i=1}^n f(y_i|\boldsymbol{\theta})$. To test a hypothesis of the form

$$H_0 : \boldsymbol{\theta} = \boldsymbol{\theta}_0 \text{ vs. } H_1 : \boldsymbol{\theta} \neq \boldsymbol{\theta}_0,$$

we evaluate the ratio of the maxima of the likelihood functions under both hypotheses:

$$\lambda(\mathbf{y}) = \frac{\max_{\boldsymbol{\theta}:\boldsymbol{\theta}=\boldsymbol{\theta}_0} L(\boldsymbol{\theta}|\mathbf{y})}{\max_{\boldsymbol{\theta}} L(\boldsymbol{\theta}|\mathbf{y})}.$$

The value of λ is less than 1 by construction since the denominator is calculating the maximum over a larger set and hence must be a bigger number. Moreover, values of λ close to 1 provide support for H_0, as then the restricted likelihood function is close to the unrestricted one, which in turn should be close to the truth. On the other hand, small values of λ lead to rejection of H_0.

To actually carry out the test, we could calculate a p-value as

$$P(\lambda(\mathbf{Y}) < \lambda(\mathbf{y})|H_0 \text{ is true}) = p(\mathbf{y}) \tag{2.8}$$

and reject H_0 for small values of $p(\mathbf{y})$, say $p(\mathbf{y}) < .01$. In general it is not an easy job to figure out the exact distribution of the probability in equation (2.8), but there is a famous approximate result that is often helpful (see Appendix A.1).

It is typical to transform λ to $-2\log\lambda$, and in this form we would now reject H_0 if $-2\log\lambda$ is large, the usual scenario. However, there is a second, more important consequence, described in the following result (see Appendix A.1 for technical details).

If $Y_1, \ldots, Y_n$ is a random sample from a density $f(x|\theta)$, then an approximate level α test of the hypothesis

$$H_0 : \theta \in \Theta_0 \text{ vs. } H_1 : \theta \notin \Theta_0$$

is to reject H_0 if

$$-2\log\lambda(\mathbf{X}) > \chi^2_{\nu,\alpha},$$

where $\chi^2_{\nu,\alpha}$ is the upper α cutoff from a chi-squared distribution with ν degrees of freedom, equal to the difference between the number of free parameters specified by H_0 and the number of free parameters specified by H_1.

So, for example, if f is $N(\mu, \sigma^2)$, the test of $H_0 : \mu - \mu_0, \quad \sigma^2 = \sigma_0^2$ vs. $H_0 : \mu \neq \mu 0, \sigma^2 \neq \sigma_0^2$ has two degrees of freedom, while the test of $H_0 : \mu = \mu_0$ vs. $H_0 : \mu \neq \mu 0$ since σ^2 is free in both hypotheses.

Example 2.8. (**First Testing Example**). As a first simple example, return to the situation of Example 2.2, where we have $Y_1, Y_2, \ldots, Y_n$ iid from an $N(\mu, \sigma^2)$ population. To test $H_0 : \mu = \mu_0$ vs. $H_0 : \mu \neq \mu 0$ we calculate

$$\begin{aligned}\lambda(\mathbf{y}) &= \frac{\max_{\mu=\mu_0} L(\mu, \sigma^2|\mathbf{y})}{\max_{\mu,\sigma^2} L(\mu, \sigma^2|\mathbf{y})} \\ &= \frac{\max_{\mu=\mu_0} L(\mu, \sigma^2|\mathbf{y})}{L(\hat{\mu}, \hat{\sigma}^2|\mathbf{y})},\end{aligned}$$

where we see that the denominator maximum is attained by substituting the MLEs for the parameter values (see Example 2.3). In maximizing the numerator, we set $\mu = \mu_0$ and maximize in σ^2, giving

$$\max_{\mu=\mu_0} L(\mu, \sigma^2|\mathbf{y}) = L(\mu_0, \hat{\sigma}_0^2|\mathbf{y}),$$

where $\hat{\sigma}_0^2 = (1/n)\sum_{i=1}^n (y_i - \mu_0)^2$. This gives the LR statistic

$$
\begin{aligned}
\lambda(\mathbf{y}) &= \frac{L(\mu_0, \hat{\sigma}_0^2|\mathbf{y})}{L(\hat{\mu}, \hat{\sigma}^2|\mathbf{y})} \\
&= \frac{\left(\frac{1}{2\pi\hat{\sigma}_0^2}\right)^{n/2} \exp\left\{\frac{1}{2\hat{\sigma}_0^2}\sum_{i=1}^n (y_i - \mu_0)^2\right\}}{\left(\frac{1}{2\pi\hat{\sigma}^2}\right)^{n/2} \exp\left\{\frac{1}{2\hat{\sigma}^2}\sum_{i=1}^n (y_i - \hat{\mu})^2\right\}} \\
&= \left(\frac{\hat{\sigma}^2}{\hat{\sigma}_0^2}\right)^{n/2},
\end{aligned}
$$

which is, in fact, equivalent to the usual t–test.

We next look at another model and likelihood ratio test, based on the *multinomial distribution*. A classic experiment can be analyzed with this distribution and method.

Example 2.9. (**Morgan (1909) Data**). In 1909, Morgan experimented on fruit flies, crossing two inbred lines with the following genotypic traits:

Eye color	A:red	a:purple
Wing length	B:normal	b:vestigial

He then obtained 2839 crosses of $AABB \times aabb$ and observed the four genotypes $AaBb$, $Aabb$, $aaBb$, and $aabb$. (Note that the middle two genotypes are recombinants.)

A model for the Morgan experiment can be based on the multinomial distribution, a discrete distribution that is used to model frequencies. Suppose that a random variable Y can take on one of k values, the integers $1, 2, \ldots, k$, each with probability $p_1, p_2, \ldots, p_k$. More precisely,

$$P(Y = j) = p_j, j = 1, \ldots k.$$

Note that if $k = 2$ we have the binomial distribution.

If we now have an iid sample $Y_1, \ldots, Y_n$, and we let $\mathbf{p} = (p_1, p_2, \ldots, p_k)$, where $\sum_j p_j = 1$, then the sample density (the likelihood function) is

$$L(\mathbf{p}|\mathbf{y}) = \prod_{i=1}^n f(y_i|\mathbf{p}) = p_1^{n_1} p_2^{n_2} \cdots p_k^{n_k},$$

where n_j = number of $y_1, \ldots, y_n$ equal to j.

Example 2.10. (**Morgan (1909) Data Continued**). For the Morgan experiment, we have $k = 4$ categories, the four genotypes $AaBb$, $Aabb$, $aaBb$, and $aabb$. There are $n = 2839$ observations, which were observed to be

$AaBb$	$Aabb$	$aaBb$	$aabb$
1339	151	154	1195

so $n_1 = 1339$, etc.

Typical null hypotheses specify patterns in the cell probabilities p_j, with the hypothesis of *equal* cell probabilities being a popular one. That is, test

$$H_0\colon p_1 = p_2 = \cdots = p_k \qquad \text{versus} \qquad H_1\colon H_0 \text{is not true.}$$

Of course, under this H_0, all of the $p_j's$ equal $1/k$. As a slightly more interesting example, and one that is applicable to the Morgan experiment, suppose that $k = 4$ and we want to test

$$H_0\colon p_1 = p_4, \quad p_2 = p_3 \text{ vs. } H_1\colon H_0 \text{ is nottrue.} \tag{2.9}$$

using a likelihood ratio test. We proceed as before and calculate

$$\lambda(\mathbf{y}) = \frac{\max_{p_1=p_4, \quad p_2=p_3} L(\mathbf{p}|\mathbf{y})}{\max_{\mathbf{p}} L(\mathbf{p}|\mathbf{y})} = \frac{\max_{p_1=p_4, \quad p_2=p_3} p_1^{n_1} p_2^{n_2} p_3^{n_3} p_4^{n_4}}{\max_{\mathbf{p}} p_1^{n_1} p_2^{n_2} p_3^{n_3} p_4^{n_4}}. \tag{2.10}$$

To maximize the numerator, recall that $\sum_j p_j = 1$ implies that $p_k = 1 - \sum_{j=1}^{k-1} p_j$ (so there are really only $k-1$ parameters in the general problem). If $p_1 = p_4 = p/2$, then $p_2 = p_3 = (1-p)/2$, where p is the only unknown parameter. The numerator of equation (2.10) becomes

$$\max_p \left(\frac{p}{2}\right)^{n_1} \left(\frac{1-p}{2}\right)^{n_2} \left(\frac{1-p}{2}\right)^{n_3} \left(\frac{p}{2}\right)^{n_4} = \max_p \left(\frac{p}{2}\right)^{n_1+n_4} \left(\frac{1-p}{2}\right)^{n_2+n_3}.$$

This is a binomial likelihood, and taking logs and differentiating will show that the MLE of p under H_0 is $\hat{p}_0 = (n_1+n_4)/(n_1+n_2+n_3+n_4) = (n_1+n_4)/n$. For the denominator, write $p_4 = 1 - p_1 - p_2 - p_3$ and take logs to get

$$L(\mathbf{p}|\mathbf{y}) = n_1 \log p_1 + n_2 \log p_2 + n_3 \log p_3 + n_4 \log(1 - p_1 - p_2 - p_3),$$

and differentiating with respect to p_1, p_2, p_3 shows that $\hat{p}_j = n_j/n, j = 1, \ldots, 4$. The likelihood test statistic then becomes

$$\lambda(\mathbf{y}) = \frac{\hat{p}_0^{n_1+n_4}(1-\hat{p}_0)^{n_2+n_3}}{\hat{p}_1^{n_1}\hat{p}_2^{n_2}\hat{p}_3^{n_3}\hat{p}_4^{n_4}} = \frac{\left(\frac{n_1+n_4}{2}\right)^{n_1+n_4}\left(\frac{n_2+n_3}{2}\right)^{n_2+n_3}}{n_1^{n_1} n_2^{n_2} n_3^{n_3} n_4^{n_4}}.$$

To perform the hypothesis test, we can use the approximation that $-2\log\lambda(\mathbf{y})$ has a chi-squared distribution. But first we must get the degrees of freedom correct.

Under H_1, there is no restriction on the parameters other than that they must sum to one, so there are three free parameters. Under H_0, we have placed an additional two restrictions, so there is one free parameter under H_0, and the chi square has $3 - 1 = 2$ degrees of freedom.

Example 2.11. (**Morgan (1909) Data–First Conclusion**). The four genotypes are nonrecombinant (*AaBb* and *aabb*) and recombinant (*Aabb* and *aaBb*), and the hypothesis (2.9) specifies that the nonrecombinant genotypes segregate with a common parameter, and the recombinant genotypes segregate with a common parameter, but these two common parameters need not be equal. The alternative hypothesis merely says that this is not so.

The observed value of $\lambda(\mathbf{y})$ is $\lambda(\mathbf{y}) = 0.0164$ with $-2\log\lambda(\mathbf{y}) = 8.217$, to be compared with a chi-squared distribution with two degrees of freedom. The .05 cutoff, $\chi^2_{2,.05} = 5.99$, leads us to reject the null hypothesis.

2.3.3 Simulation-Based Approach

The chi-squared approximation used in Section 2.3.2 relies on an asymptotic approximation – its validity is dependent on having large cell sizes. Moreover, the discrepancy in the size of the cells can also have an effect on the adequacy of the approximation.

If there is reason to suspect the adequacy of the approximation, or if the evidence in the data is difficult to interpret, it may be reasonable to try another approach to assessing the evidence against H_0. (Actually, this is a good idea in most cases.)

We describe in this section a simulation technique that is sometimes known as the *parametric bootstrap* (Efron and Tibshirani 1993); it is an all-purpose simulation-based technique. We first describe it in general, then apply it to the Morgan data.

Simulation-Based Hypothesis Assessment

Given an iid sample $\mathbf{Y} = (Y_1, \ldots, Y_n)$ and a density $f(x|\theta)$, we assess the hypotheses $H_0 : \theta \in \Theta_0$ vs. $H_1 : \theta \notin \Theta_0$ as follows.

(1) Estimate θ with the MLE $\hat{\theta}$, and calculate the observed likelihood statistic $-2 \log \lambda(\mathbf{y})$.
(2) Generate $t = 1, \ldots, M$ new iid samples $\mathbf{Y}^* = (Y_1^*, \ldots, Y_n^*)$, where $Y_i^* \sim f(x|\hat{\theta})$, and calculate $-2 \log \lambda(\mathbf{y}_t^*)$.
(3) A p-value for the test can be calculated as

$$\hat{p}(\mathbf{y}) = \frac{1}{M} \sum_{t=1}^{M} I\left(\lambda(\mathbf{y}_t^*) > \lambda(\mathbf{y})\right),$$

where $I(\cdot)$ is the indicator function, which is equal to 1 if the argument is true and 0 otherwise. A histogram of the $\lambda(\mathbf{y}_t^*)$ can also be drawn.

Example 2.12. (**Morgan (1909) Data–Second Conclusion**). To do a simulation-based assessment of the hypotheses (2.9) we would simulate random variables Y_1^*, Y_2^*, ... from a multinomial distribution with $n = 2839$ and probability vector

$$\mathbf{p} = \left(\frac{1339}{2839}, \frac{151}{2839}, \frac{154}{2839}, \frac{1195}{2839}\right) = (0.471, 0.054, 0.053, 0.421).$$

Figure 2.2 shows the results of the simulation of 10000 values of $-2 \log \lambda(\mathbf{y}^*)$, and they are quite different from the results of the chi-square test of Example 2.11. The observed value of $-2 \log \lambda(\mathbf{y}) = 8.217$ is now right in the middle of the distribution, with an estimated p-value of .5718, meaning that 5718 of the 10000 random variables simulated were larger than the observed value of the test statistic. This puts the statistic right in the middle of the distribution and thus leads us to accept the null hypothesis.

It should be mentioned that the overall conclusion is not crystal clear, but the evidence is certainly pointing toward the conclusion that H_0 is a tenable hypothesis, and the asymptotic approximation of the chi-square distribution is not the best in this case.

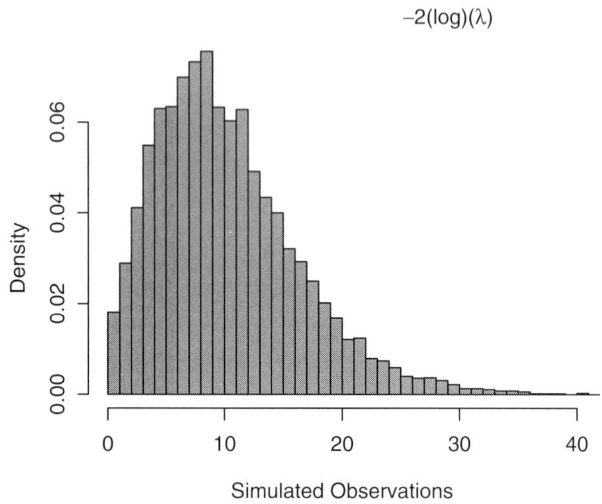

Fig. 2.2. Histogram of 10000 values of the likelihood ratio statistic $-2\log\lambda$ for the Morgan data.

2.3.4 Bayesian Estimation

Throughout this chapter, we have been describing the classical approach to statistics, where we base our evidence on a repeated-trials assessment of error. There is an alternative approach, the Bayesian approach, which is fundamentally different from the classical approach. Here the assessment is based on an experimenter's prior belief and how that belief is altered by the data.

It is not constructive to view the two approaches in opposition. Rather, it is better to examine each problem at hand, and choose the approach that will give the most useful answer.

In the classical approach, the parameter, say θ, is thought to be an unknown, but fixed, quantity. A random sample $X_1, \ldots, X_n$ is drawn from a population indexed by θ and, based on the observed values in the sample, knowledge about the value of θ is obtained. In the Bayesian approach, θ is considered to be a quantity whose variation can be described by a probability distribution (called the *prior distribution*). This is a subjective distribution, based on the experimenter's belief, and is formulated before the data are seen. A sample is then taken from a population indexed by θ, and the prior distribution is updated with this sample information. The updated prior distribution is called the *posterior distribution.*

If we denote the prior distribution by $\pi(\theta)$ and the sampling distribution by $f(\mathbf{x}|\theta)$, then the posterior distribution, the conditional distribution of θ given the sample, $\mathbf{x}$, is calculated using Bayes' Rule as

$$\pi(\theta|\mathbf{x}) = \frac{f(\mathbf{x}|\theta)\pi(\theta)}{\int f(\mathbf{x}|\theta)\pi(\theta)d\,\theta} = \frac{f(\mathbf{x}|\theta)\pi(\theta)}{m(\mathbf{x})}, \tag{2.11}$$

where $m(\mathbf{x}) = \int f(\mathbf{x}|\theta)\pi(\theta)d\theta$ is the marginal distribution of $\mathbf{X}$.

Once $\pi(\theta|\mathbf{x})$ is obtained, it contains all of the information about the parameter θ. We can plot this distribution to see the shape, and perhaps where the modes are, and whether it is symmetric or not. It is also typical to calculate the posterior mean and variance to get a point estimator and a measure of spread. These are given by

$$\mathrm{E}(\theta|\mathbf{x}) = \int \theta\pi(\theta|\mathbf{x})d\theta, \quad \mathrm{Var}(\theta|\mathbf{x}) = \int [\theta - \mathrm{E}(\theta|\mathbf{x})]^2\pi(\theta|\mathbf{x})d\theta.$$

We illustrate Bayesian estimation with some examples.

Example 2.13. (**Bayes Estimation in the Normal Distribution**). Let $X \sim n(\theta, \sigma^2)$, and suppose that the prior distribution on θ is $n(\mu, \tau^2)$. (Here we assume that σ^2, μ, and τ^2 are all known.) The posterior distribution of θ is

$$\begin{aligned}\pi(\theta|x) &\propto \frac{1}{\sqrt{2\pi\sigma^2}}e^{-\frac{1}{2\sigma^2}(x-\mu)^2}\frac{1}{\sqrt{2\pi\tau^2}}e^{-\frac{1}{2\tau^2}(\theta-\mu)^2} \\ &\propto \left[\frac{\sqrt{\sigma^2+\tau^2}}{\sqrt{2\pi\sigma^2\tau^2}}e^{-\frac{\sigma^2+\tau^2}{2\sigma^2\tau^2}(\theta-\mathrm{E}(\theta|x))^2}\right]\left[\frac{1}{\sqrt{2\pi(\sigma^2+\tau^2)}}e^{-\frac{1}{2\sigma^2\tau^2}(x-\mu)^2}\right],\end{aligned}$$

where the distribution in the first square brackets is the posterior and the second distribution is the marginal. The calculation is somewhat long and tedious, and depends on completing the square in the exponent.

Upon inspection, we see that the posterior distribution of θ is also normal, with mean and variance given by

$$\begin{aligned}\mathrm{E}(\theta|x) &= \frac{\tau^2}{\tau^2+\sigma^2}x + \frac{\sigma^2}{\sigma^2+\tau^2}\mu, \\ \mathrm{Var}\,(\theta|x) &= \frac{\sigma^2\tau^2}{\sigma^2+\tau^2}.\end{aligned} \tag{2.12}$$

The Bayes estimator is a linear combination of the prior and sample means. Notice also that as τ^2, the prior variance, is allowed to tend to infinity, the Bayes estimator tends toward the sample mean. We can interpret this as saying that as the prior information becomes more vague, the Bayes estimator tends to give more weight to the sample information. On the other hand, if the prior information is good, so that $\sigma^2 > \tau^2$, then more weight is given to the prior mean.

Example 2.14. (**Bayes Estimation in the Binomial**). Let $X_1, \ldots, X_n$ be iid Bernoulli (p). Then $Y = \sum X_i$ is binomial(n, p). We take the prior distribution on p to be

$$\mathrm{beta}(\alpha, \beta) = \frac{\Gamma(\alpha+\beta}{\Gamma(\alpha)\Gamma(\beta)}p^{\alpha-1}(1-p)^{\beta-1}.$$

The posterior distribution of p is

$$\pi(p|y) \propto \left[\binom{n}{y} p^y (1-p)^{n-y}\right]\left[\frac{\Gamma(\alpha+\beta)}{\Gamma(\alpha)\Gamma(\beta)} p^{\alpha-1}(1-p)^{\beta-1}\right]$$
$$\propto \frac{\Gamma(\alpha+\beta)}{\Gamma(\alpha)\Gamma(\beta)} p^{y+\alpha-1}(1-p)^{n-y+\beta-1},$$

which is again a beta distribution, now with parameters $y+\alpha$ and $n-y+\beta$. We can estimate p with the mean of the posterior distribution

$$\mathrm{E}(p|y) = \frac{y+\alpha}{\alpha+\beta+n}.$$

Consider how the Bayes estimate of p is formed. The prior distribution has mean $\alpha/(\alpha+\beta)$, which would be our best estimate of p without having seen the data. Ignoring the prior information, we would probably use the MLE $p = y/n$ as our estimate of p. The Bayes estimate of p combines all of this information. The manner in which this information is combined is made clear if we write $\mathrm{E}(p|y)$ as

$$\mathrm{E}(p|y) = \left(\frac{n}{\alpha+\beta+n}\right)\left(\frac{y}{n}\right) + \left(\frac{\alpha+\beta}{\alpha+\beta+n}\right)\left(\frac{\alpha}{\alpha+\beta}\right).$$

Thus p_{B} is a linear combination of the prior mean and the sample mean, with the weights being determined by α, β, and n.

Example 2.15. (**Bayes Estimation in the Binomial**). We revisit the data of Example 2.7 and now estimate p, the true proportion, using a Bayes estimator. Consider marker I18_1090 with counts of 44 and 59 in the two genotype classes. The MLE of the true proportion in the first genotype is $44/(44+59) = .427$.

Suppose we estimate p using a Bayes estimator. If the experimenter believes that the genes are segregating independently, then he believes that $p = 1/2$. We can reflect this belief with a beta distribution that is symmetric around 1/2. We choose a beta $(2, 2)$, which does not give much weight to the prior distribution. Under this prior distribution the Bayes estimator of p is $(44+2)/(44+59+2+2) = .430$, which is very similar to the MLE. Thus, for this choice of prior distribution, the information in the data is very strong and the Bayes estimator is virtually the same as the MLE.

If the experimenter is very certain that p is close to 1/2, this can be reflected in the prior distribution by increasing the parameter values. If we choose $\alpha = \beta = 100$, the prior mean remains 1/2 but the prior variance is decreased. The resulting Bayes estimator is $(44+100)/(44+59+200) = .475$, and the posterior distribution is now quite symmetric. This is illustrated in Fig. 2.3, where we see the symmetry of the prior distributions, but one is more concentrated. The resulting posterior distributions are close to the likelihood function and skewed when $\alpha = \beta = 2$ and symmetric and far from the likelihood function when $\alpha = \beta = 100$.

2.4 Exercises

2.1 (a) Illustrate the binomial approximation to the normal as described in Section 2.3.1. For $N = 5, 15, 50$ and $p = .25, .5$, draw the binomial histogram and the normal density. Calculate the .05 and .01 tail area in each case.

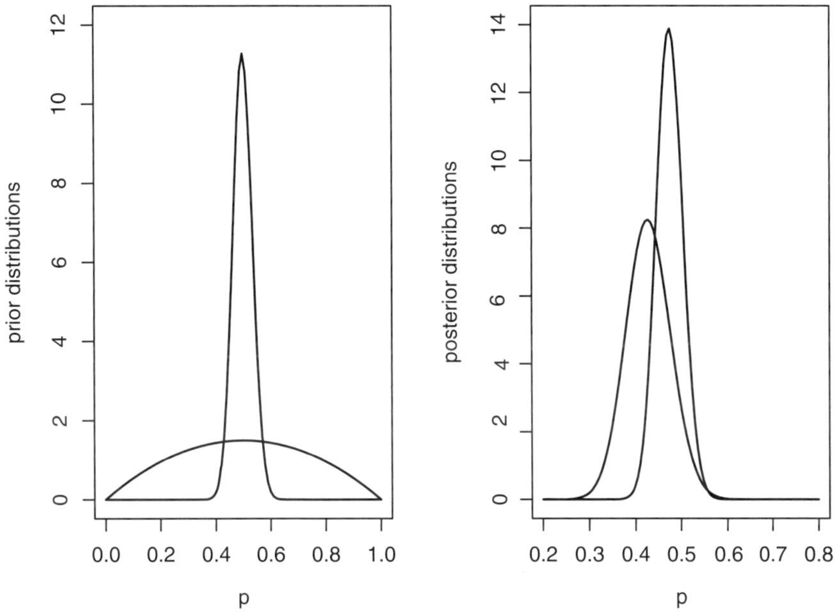

Fig. 2.3. Graphs of prior distributions (left panel) and posterior distributions (right panel) of Example 2.15. The beta $(2,2)$ prior is very flat and results in a skewed posterior that is indistinguishable from the likelihood function. The beta $(100,100)$ prior is very peaked and results in a peaked and symmetric posterior distribution.

(b) Show that if $p = \frac{1}{2}$, equation (2.6) can be simplified to $(n_1 - n_2)^2/N$.

2.2 For the situation of Example 2.8:

(a) Verify the formula for $\hat{\sigma}_0^2$.

(b) Show that

$$\frac{\hat{\sigma}^2}{\hat{\sigma}_0^2} = \frac{\sum_{i=1}^n (y_i - \bar{y})^2}{\sum_{i=1}^n (y_i - \mu_0)^2} = \frac{1}{1 + \frac{(\bar{y}-\mu_0)^2}{\hat{\sigma}^2}}.$$

(For the second equality, use the identity $\sum_{i=1}^n (y_i - \mu_0)^2 = \sum_{i=1}^n (y_i - \bar{y})^2 + n(\bar{y} - \mu_0)^2$.

(c) Verify that, in this last Form, we will reject H_0 if $|\bar{y} - \mu + 0|/\hat{\sigma}$ is large, making this test equivalent to the t–test.

2.3 (a) Verify the MLEs of equation (2.10).

3

Linkage Analysis and Map Construction

3.1 Introduction

Linkage is the tendency for genes to be inherited together because they are located near one another on the same chromosome. Linkage analysis of markers lays a foundation for the construction of a genetic linkage map and the subsequent molecular dissection of quantitative traits using the map. Linkage analysis is based on the cosegregation of adjacent markers and their cotransmission to the next progeny generation. The prerequisite of linkage analysis between any two markers is their known allelic arrangements (i.e., linkage phases) on the homologous chromosome so that parental (or nonrecombinant) vs. nonparental (or recombinant) haplotypes can be readily distinguished. In many domesticated plants and animals, phase-known mapping pedigrees can be established using a segregating population, such as the backcross or F_2, derived from two homologous inbred lines. (Recall the definitions in Section 1.5.) Theories for linkage analysis in such segregating pedigrees have been well-developed.

The linkage of markers can be measured in terms of their recombination fraction or genetic distance. The function of linkage analysis is to detect the relative locations of two or more markers on the same chromosome. Linkage analysis can be performed for a pair of markers (two-point analysis) or three markers simultaneously (three-point analysis). Two- or three-point analyses provide fundamental information for the construction of a genetic linkage map that cover partly or entirely the genome. The map function that converts the recombination fraction to genetic distance can be derived from three-point analysis. Different forms of the map function are available that depend on the assumption about the presence or absence of the interference of crossovers between adjacent marker intervals.

In this chapter, we will describe the basic principle for linkage analysis in a pedigree initiated with two inbred lines. A detailed procedure for the estimation and test of linkage between different markers will be derived theoretically and demonstrated through live examples. We will illustrate the step procedure for deriving the map functions. Analyses of human genetic linkage can be found in the textbooks by Ott (1991) and Lange (1997). Algorithms and software for map construction are introduced at the end of this chapter.

3.2 Experimental Design

The tendency of alleles of different genes on the same chromosome to pass into the same haplotype at meiosis is related to the degree of linkage between these genes. Thus, the development of linkage analysis critically relies upon a segregating pedigree in which both recombinant and nonrecombinant gamete types can be counted. A pedigree comprising a backcross or an F_2 population, initiated with two contrasting inbred lines, has proven a most powerful and efficient tool for linkage analysis.

In practice, two inbred lines that are homologous for two alternative alleles of each gene are crossed as parents P_1 and P_2 to generate an F_1 progeny. Thus, all F_1 individuals are heterozygous at all genes. These heterozygous F_1's can either be backcrossed to each of their parents to generate two backcrosses (B_1 and B_2) or the F_1 individuals can be crossed with each other to produce the F_2 generation. A diagram illustrating this crossing procedure is illustrated in Fig. 3.1.

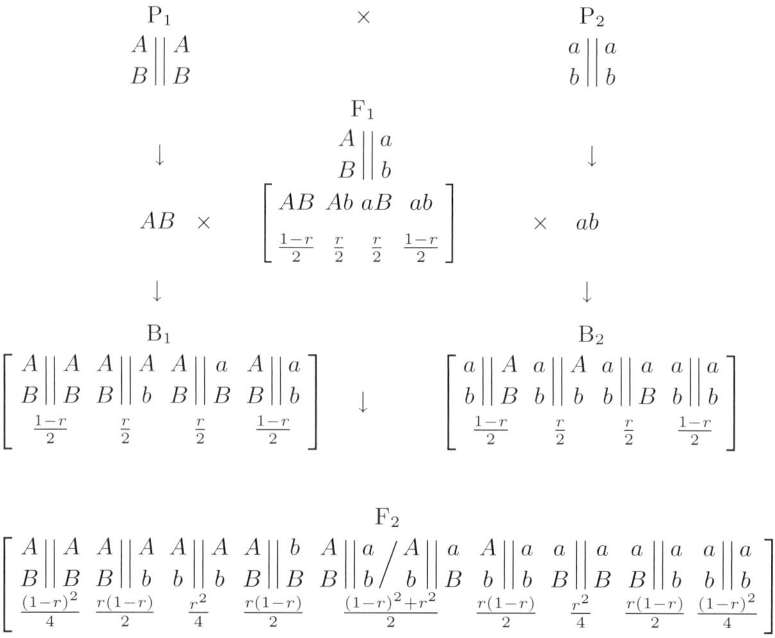

Fig. 3.1. Experimental design used for linkage analysis of markers.

Consider two markers, **A**, with alleles A and a, and **B**, with two alleles B and b. Two inbred line parents, P_1 and P_2, are homozygous for the large and small alleles of these two genes, respectively. Parent P_1 generates gamete or haplotype AB during meiosis, whereas parent P_2 generates gamete ab. These two gametes are combined to form the heterozygous F_1 of genotype $AaBb$. The F_1 will generate four different gametes, two of which (AB and ab) are of nonrecombinant type and the two other (Ab and aB) of recombinant type. The recombination fraction between the two genes is denoted by r. Thus, these two groups of gametes have the frequencies of $(1-r)/2$ (nonrecombinant) and $r/2$ (recombinant). When the F_1 is backcrossed to one of the pure parents, four backcross genotypes will be generated with the same frequencies as those of the F_1 gametes. Intercrossing the F_1 generates the F_2 in which 16 gamete combinations are collapsed into nine genotypes with frequencies combined for the same genotypes.

3.3 Mendelian Segregation

One of the first tasks in a genomic mapping project is to determine whether single markers follow Mendelian segregation ratios in an experimental pedigree. Only after the nature of the single marker ratios is determined can the subsequent linkage analysis be performed using appropriate statistical methods.

Suppose we consider a general case in which a certain mating, initiated with two contrasting inbred lines, is expected to produce k genotypes at a marker in the expected ratio of $\lambda_1 : \ldots : \lambda_k$. The expected relative frequency of any genotype class i is calculated by $\phi_i = \lambda_i/(\sum_{i=1}^{k} \lambda_i)$. The numbers actually observed in the m classes are $n_1, \ldots, n_k$, respectively, where $n = n_1 + \ldots + n_k$, and we wish to compare the observed segregation ratio with the expected value. For a codominant marker, the expected ratio is 1:1 in the backcross and 1:2:1 in the F_2. For a dominant marker, the ratio is 1:1 in the backcross toward the pure recessive and 3:1 in the F_2.

The basic methods for testing marker segregation patterns include the binomial test, the standard normal test, the Pearson chi-squared test, and the likelihood ratio chi-squared test. The first two tests are used in situations involving two classes of genotypes in a pedigree, whereas the latter two can be generalized to situations in which there are more than two classes. Here, more general Pearson chi-squared and likelihood ratio chi-squared tests are described.

3.3.1 Testing Marker Segregation Patterns

The hypothesis for marker segregation patterns can be tested by either the Pearson chi-squared test (2.6) or the likelihood ratio test. In the latter case, the likelihood function, given that different numbers of individuals are observed out of N offspring, is derived from the multinomial distribution and given by

$$L(p_1 \cdots p_k) = \frac{n!}{n_1! \cdots n_k!} \prod_{i=1}^{k} p_i^{n_i}. \tag{3.1}$$

The value of p_i that maximizes the log-likelihood function (and therefore the likelihood function) is $\hat{p}_i = n_i/n$, that is, the actual proportion observed in the sample. The values $\hat{p}_i$ are the maximum likelihood estimates (MLEs) of p_i. To test

$$H_0 : p_1 = p_{10}, \ldots, p_k = p_{k0},$$

where the p_{i0} are specified, we use the likelihood ratio statistic

$$\begin{aligned} -2\log\lambda &= 2(\ln L_1 - \ln L_0) \\ &= 2\left[\sum_{i=1}^{k} n_i \ln\left(\frac{n_i}{n}\right) - \sum_{i=1}^{k} n_i \ln(p_{i0})\right], \end{aligned} \tag{3.2}$$

where L_0 is the likelihood with the hypothesized values substituted for the p_i's and L_1 is the likelihood with the MLEs substituted for the p_i's.

The p-value is then given by the probability that a chi-squared random variable with $k-1$ degrees of freedom will exceed $-2\log\lambda$.

Example 3.1. (**DH Population**). Two inbred lines, semi-dwarf IR64 and tall Azucena, were crossed to generate an F_1 progeny population. By doubling haploid chromosomes of the gametes derived from the heterozygous F_1, a doubled haploid (DH) population of 123 lines was founded (Huang et al. 1997). Such a DH population is equivalent to a backcross population because its marker segregation follows 1:1. With 123 DH lines, Huang et al. genotyped a total of 175 polymorphic markers (including 146 RFLPs, 8 isozymes, 14 RAPDs, and 12 cloned genes) to construct a linkage map representing a good coverage of 12 rice chromosomes.

Let n_1 and n_0 be the number of plants for two different genotypes in the DH population. We now apply the χ^2 test of equation (2.6) and likelihood ratio test of equation (3.2) to test whether the segregation of these testcross markers follows the Mendelian ratio 1:1. Table 3.1 gives the results for six markers on rice chromosome 1. The results from thc likclihood ratio test are consistent with those from the Pearson test. Based on the p-values calculated from the χ^2 distribution with one degree of freedom, we detected that markers RG472, RG246, and U10 segregate 1:1 and that markers K5, RG532, and W1 deviate from the 1:1 ratio.

Example 3.2. (**Intercross $\mathbf{F}_2$**). Cheverud et al. (1996) genotyped 75 microsatellite markers in a population of 535 F_2 progeny derived from two strains, the Large (LG/J) and Small (SM/J). As an example for segregation tests, we choose nine markers located on mouse chromosome 2. Let n_2, n_1, and n_0 be the numbers of mice for three genotypes at each marker in this F_2 population. Both the χ^2 and likelihood ratio tests have consistent results, suggesting that nine markers from the second mouse chromosome segregate in the Mendelian 1:2:1 ratio (Table 3.1) at the .01 significance level. Note that the test statistics calculated in the F_2 are χ^2-distributed with two degrees of freedom because three genotypes present three independent categories.

3.4 Segregation Patterns in a Full-Sib Family

For a marker that is segregating in a full-sib family derived from two outbred parents, we will have many different types of segregation. Up to four marker alleles, besides a

Table 3.1. Pearson and likelihood ratio test statistics for testing Mendelian segregation 1:1 for the doubled haploid population in rice and 1:3:1 for the F_2 population in mice.

				Pearson		Likelihood	
Marker	n_2	n_1	n_0	χ^2	p-value	$-2\log\lambda$	p-value
DH population							
RG472		52	58	0.327	0.592	0.327	0.592
RG246		62	54	0.552	0.408	0.552	0.407
K5		66	41	5.841	0.009	5.895	0.009
U10		52	38	2.178	0.091	2.187	0.090
RG532		69	44	5.531	0.011	5.577	0.010
W1		83	34	20.521	0.000	21.168	<0.001
F_2							
D2Mit362	121	236	120	0.057	0.486	0.057	0.486
D2Mit72	121	262	115	1.502	0.236	1.511	0.235
D2Mit205	132	255	125	0.199	0.453	0.198	0.453
D2Mit38	139	232	148	6.141	0.023	6.122	0.023
D2Mit93	133	244	116	1.223	0.271	1.212	0.273
D2Mit389	128	253	136	0.482	0.393	0.477	0.394
D2Mit17	133	226	144	5.652	0.030	5.617	0.030
D2Mit260	143	250	122	2.150	0.171	2.103	0.175
D2Mit25	121	236	155	7.641	0.011	7.327	0.013

null allele, may be segregating at a single locus. Furthermore, the number of alleles may vary over loci. We assume that each of the marker alleles, symbolized by a, b, c, and d, is codominant with respect to each other but dominant with respect to the null allele, symbolized by o. We assume that all markers undergo precise Mendelian segregation. Depending on how different alleles are combined in the two parents used for the cross, there exists a total of 18 possible cross types for a marker locus (Table 3.2). Based on both parental and offspring marker band patterns, these cross types can be classified into seven groups (see also Maliepaard et al. 1997):

A. Loci that are heterozygous in both parents and segregate in a 1:1:1:1 ratio, involving either four alleles $ab \times cd$, three nonnull alleles $ab \times ac$, three nonnull alleles and a null allele $ab \times co$, or two null alleles and two nonnull alleles $ao \times bo$;

B. Loci that are heterozygous in both parents and segregate in a 1:2:1 ratio, which include three groups:

B_1. One parent has two different dominant alleles and the other has one dominant allele and one null allele, e.g., $ab \times ao$;

B_2. The reciprocal of B_1;

B_3. Both parents have the same genotype of two codominant alleles, $ab \times ab$;

C. Loci that are heterozygous in both parents and segregate in a 3:1 ratio, $ao \times ao$;

D. Loci that are in the testcross configuration between the parents and segregate in a 1:1 ratio, which include two groups:

D_1. Heterozygous in one parent and homozygous in the other, including three alleles $ab \times cc$, two alleles $ab \times aa$, $ab \times oo$ and $bo \times aa$, and one allele (with three null alleles) $ao \times oo$;

D_2. The reciprocals of D_1.

Marker cross type A produces all four possible marker genotypes in the progeny and is regarded as being *fully informative*. The other marker cross types are all *partially informative* because the four possible progeny genotypes are collapsed due to indistinguishable phenotypes. Note that marker cross type D can be viewed as fully informative if only the heterozygous parent is concerned. Marker types B_3 and C each have the same genotypes in both parents and therefore are called *symmetrical* marker cross types. The other marker types have parent-specific marker genotypes and are called *asymmetrical* marker cross types. Marker cross types D_1 and D_2 are called two-way pseudo-test backcrosses (Grattapaglia and Sederoff 1994). Because dominant markers are as informative as codominant markers in such designs, pseudo-test backcrosses are broadly used in full-sib family mapping of outcrossing species, in which it is difficult or even impossible to generate homozygous inbred lines.

For those partially informative markers whose cross types are asymmetrical between the two parents, the reciprocals (e.g., B_1 vs. B_2 and D_1 vs. D_2) supply different information for the characterization of linkage phase when these markers are paired and thus are presented as two distinct groups (see Chapter 4).

For a mapping project, both the parents and progeny are usually genotyped. Based on the segregation pattern of marker band data, we can determine the cross type to which a given marker belongs, as given in Table 3.2. For example, if we observe bands a, b, ac, and bc in the offspring and bands ab and c for the parents, it can be easily inferred that this marker belongs to cross type 3 (A). For the asymmetrical marker cross types (A, B_1, B_2, D_1, and D_2), the marker information of both the parents and their offspring is needed to infer marker cross types. However, for the symmetrical marker cross types (B_3 and C), only the pattern of marker segregation in the offspring is needed for this inference.

Given a marker cross type, one can use the approaches introduced in the previous section to determine whether the marker follows a particular segregation pattern

Table 3.2. Possible marker genotype cross combinations and observed marker band patterns for parents and their offspring.

Cross			Parent			Offspring		
Type			Cross	Observed Band	Remark	Observed Bands	Segregation	No. Phenotypes
A		1	$ab \times cd$	$ab \times cd$	asymmetry	ac, ad, bc, bd	1:1:1:1	4
		2	$ab \times ac$	$ab \times ac$	asymmetry	a, ac, ba, bc	1:1:1:1	4
		3	$ab \times co$	$ab \times c$	asymmetry	ac, a, bc, b	1:1:1:1	4
		4	$ao \times bo$	$a \times b$	asymmetry	ab, a, b, o	1:1:1:1	4
B	B_1	5	$ab \times ao$	$ab \times a$	asymmetry	$ab, 2a, b$	1:2:1	3
	B_2	6	$ao \times ab$	$a \times ab$	asymmetry	$ab, 2a, b$	1:2:1	3
	B_3	7	$ab \times ab$	$ab \times ab$	symmetry	$a, 2ab, b$	1:2:1	3
C		8	$ao \times ao$	$a \times a$	symmetry	$3a, o$	3:1	2
D	D_1	9	$ab \times cc$	$ab \times c$	asymmetry	ac, bc	1:1	2
		10	$ab \times aa$	$ab \times a$	asymmetry	a, ab	1:1	2
		11	$ab \times oo$	$ab \times o$	asymmetry	a, b	1:1	2
		12	$bo \times aa$	$b \times a$	asymmetry	ab, a	1:1	2
		13	$ao \times oo$	$a \times o$	asymmetry	a, o	1:1	2
	D_2	14	$cc \times ab$	$c \times ab$	asymmetry	ac, bc	1:1	2
		15	$aa \times ab$	$a \times ab$	asymmetry	a, ab	1:1	2
		16	$oo \times ab$	$o \times ab$	asymmetry	a, b	1:1	2
		17	$aa \times bo$	$a \times b$	asymmetry	ab, a	1:1	2
		18	$oo \times ao$	$o \times a$	asymmetry	a, o	1:1	2

(Table 3.2). For marker cross type A, for example, there are four possible classes of markers segregating in the 1:1:1:1 ratio, rather than 2 classes (segregating 1:1) in the example given in Table 3.1.

3.5 Two-Point Analysis

Two-point analysis is a statistical approach for estimating and testing the recombination fraction between two different markers. Two-point analysis provides a basis for the derivation of the map function and the construction of genetic linkage maps. Here, we will present statistical methods for linkage analysis separately for the backcross and F_2 populations, because these types of populations need different analytical strategies.

3.5.1 Double Backcross

Using the design described in Fig. 3.1, we make two backcrosses by crossing the F_1 toward each of the two parents. Let us consider the backcross to parent P_2. The expected frequencies and observed numbers of the four genotypes generated in this backcross can be tabulated as follows:

Genotype	AB/ab	ab/ab	Ab/ab	aB/ab
Frequency	$\frac{1}{2}(1-r)$	$\frac{1}{2}(1-r)$	$\frac{1}{2}r$	$\frac{1}{2}r$
Observed number	n_1	n_2	n_3	n_4
	Nonrecombinant		Recombinant	
Frequency	$1-r$		r	
Observed number	$n_{NR}=n_1+n_2$		$n_R=n_3+n_4$	

Since the F_1 has a known linkage phase (that is, allele A is coupled with allele B), the numbers of backcross genotypes from nonrecombinant or recombinant gametes can be calculated. These are denoted by n_{NR} and n_R, respectively. The sum of n_{NR} and n_R is the total number of genotypes, n. The recombination fraction between the two genes can be estimated from the observed numbers of the four different backcross genotypes. The likelihood of r given the marker data $n=(n_{NR}, n_R)$ is

$$L(r|n) = \frac{n!}{n_{NR}!n_R!}(1-r)^{n_{NR}}r^{n_R}. \tag{3.3}$$

The maximum likelihood estimate (MLE) of r can be obtained by differentiating the log-likelihood of equation (3.3) with respect to r, setting the derivative equal to zero and solving the log-likelihood equation. Doing this, we obtain the MLE of r as

$$\hat{r} = \frac{n_R}{n} = \frac{n_3+n_4}{n_1+n_2+n_3+n_4}. \tag{3.4}$$

From Theorem A.3, the variance of $\hat{r}$ can be approximated by (Exercise 3.1)

$$\text{Var}(\hat{r}) = -\frac{1}{\frac{\partial^2}{\partial r^2}\log L(r|n)\big|_{r=\hat{r}}} = \frac{\hat{r}(1-\hat{r})}{n}. \tag{3.5}$$

It is seen that the precision of $\hat{r}$ depends on the size of the sample used and the size of the r value. The precision of the estimation of the recombination fraction is affected by two factors. First, increasing sample sizes always lead to increased estimation precision of r. Second, the recombination fraction can be better estimated for a pair of markers displaying a tight linkage (low r value) than for those displaying a loose linkage (high r value).

Note also that the probability model for the random variable n_R is binomial, with n trials and success probability r, $n_R \sim$ binomial–(n,r). Thus $E(n_R)=nr$ and

$\mathrm{Var}(n_R) = nr(1-r)$, which agree with the likelihood answers. An approximate 95 percent confidence interval for r is

$$\hat{r} - 2\sqrt{\hat{r}(1-\hat{r})/n} \le r \le \hat{r} + 2\sqrt{\hat{r}(1-\hat{r})/n}. \tag{3.6}$$

The MLE of r can also be used to determine the degree of linkage between the two markers. If there is evidence that the two markers are completely linked (that is, $\hat{r} = 0$), then a doubly heterozygous F_1 produces only nonrecombinant gametes. If there is evidence that linkage is absent (free recombination), so $\hat{r} = 0.5$, then the F_1 produces both recombinant and nonrecombinant haplotypes in equal proportions. Generally, the degree of linkage between two given markers can be statistically tested by formulating two alternative hypotheses:

$$\begin{cases} H_0 : r = 0.5 \\ H_1 : r < 0.5 \end{cases}, \tag{3.7}$$

where H_0 corresponds to the reduced model, in which the data are fit with the constraint $r = 0.5$, and H_1 corresponds to the full model, in which the data are fit with no such constraint. The test statistic for testing these two hypotheses is the log-likelihood ratio (LR):

$$\begin{aligned} LR &= -2\log\left[\frac{L(r=0.5|n)}{L(\hat{r}|n)}\right] \\ &= 2\left[n_R \log\left(\frac{n_R}{n}\right) + n_{NR}\log\left(\frac{n_{NR}}{n}\right) - n\log\left(\frac{1}{2}\right)\right], \end{aligned} \tag{3.8}$$

which is asymptotically χ^2-distributed with one degree of freedom.

A similar test statistic used to detect linkage is the LOD score (Ott 1991). The LOD score is a transformation of the likelihood ratio statistic, given by

$$\mathrm{LOD} = \log_{10}\left[\frac{L(\hat{r}|n)}{L(r=0.5|n)}\right] = 0.217\ \mathrm{LR}.$$

The value of LOD = 3 is suggested as the critical threshold to declare the existence of linkage.

Example 3.3. . Revisit Example 3.1. Consider the first six markers on rice chromosome 1 (Huang et al. 1997). As an illustration, we only consider the linkage between two adjacent markers. For each pair of markers considered, all the 123 DH lines are sorted into recombinant (n_R) and nonrecombinant groups (n_{NR}), tabulated in Table 3.3.

Using equations (3.4), (3.5), and (3.6), we estimated the recombination fractions for each pair of markers, their sampling errors and confidence intervals in the backcross. The likelihood ratio tests were performed by calculating the LR and LOD scores. We also conducted Pearson's χ^2 test based on

$$\chi^2 = \frac{(n_R - n/2)^2}{n/2} + \frac{(n_{NR} - n/2)^2}{n/2} = \frac{(n_R - n_{NR})^2}{n},$$

which is χ^2-distributed with one degree of freedom. Although there are four genotypes for two markers, only two categories, recombinant and nonrecombinant, are independent, implying one degree of freedom.

The three tests provide consistent results about the recombination fractions. These marker pairs are significantly linked, as the hypothesis test indicates that the recombination fractions are significantly less than 0.5.

Table 3.3. Linkage analysis of markers in a DH population of rice

Marker Pair	n_R	n_{NR}	$\hat{r}$	SE($\hat{r}$)	Confidence Interval	LR	LOD	χ^2	p-Value
RG472–RG246	90	19	0.17	0.04	(0.10, 0.25)	50.2	10.9	46.2	$< .001$
RG246–K5	90	15	0.14	0.03	(0.07, 0.21)	59.4	12.9	53.6	$< .001$
K5–U10	76	4	0.05	0.02	(0.00, 0.10)	79.1	17.2	64.8	$< .001$
U10–RG532	81	5	0.06	0.03	(0.01, 0.11)	81.1	17.6	67.2	$< .001$
RG532–W1	93	17	0.15	0.03	(0.09, 0.22)	57.8	12.5	52.5	$< .001$

3.5.2 Double Intercross–$\mathbf{F_2}$

When two heterozygous F_1s are crossed, a segregating F_2 population will be produced, in which 16 combinations from four female gametes and four male gametes at any two markers are collapsed into nine distinguishable genotypes. The observed numbers of these nine genotypes can be arrayed, in matrix notation, as

$$\mathbf{n} = \begin{array}{c} \\ BB \\ Bb \\ bb \end{array} \begin{array}{c} \begin{array}{ccc} AA & Aa & aa \end{array} \\ \begin{bmatrix} n_{22} & n_{12} & n_{02} \\ n_{21} & n_{11} & n_{01} \\ n_{20} & n_{10} & n_{00} \end{bmatrix} \end{array}, \tag{3.9}$$

where the first and second subscripts of n denote the number of large alleles at markers **A** and **B**, respectively. The resulting double heterozygote $AaBb$ is the mixture of two possible *diplotypes*, one, $[AB][ab]$, derived from the combination of gametes AB and ab, and the other, $[Ab][aB]$, from the combination of gametes Ab and aB.

Except for the double heterozygote, the frequencies of the eight other genotypes can be calculated as the products of the corresponding gamete frequencies. Note that the genotype frequencies should be two times these products when two gametes that unite to form a genotype are different. The frequency of the double heterozygote is the summation of the frequencies of two diplotypes, $\frac{1}{2}(1-r)^2$ for the diplotype due to the combination of gametes AB and ab and $\frac{1}{2}r^2$ for the diplotype due to the combination of gametes Ab and aB. The genotype frequencies for markers **A** and **B** can be arrayed as

$$\mathbf{F} = \begin{array}{c} \\ BB \\ Bb \\ bb \end{array} \begin{array}{c} \begin{array}{ccc} AA & Aa & aa \end{array} \\ \left[\begin{array}{ccc} \frac{1}{4}(1-r)^2 & \frac{1}{2}r(1-r) & \frac{1}{4}r^2 \\ \frac{1}{2}r(1-r) & \frac{1}{2}[(1-r)^2+r^2] & \frac{1}{2}r(1-r) \\ \frac{1}{4}r^2 & \frac{1}{2}r(1-r) & \frac{1}{4}(1-r)^2 \end{array}\right] \end{array}. \tag{3.10}$$

For each of these genotypes, we can count the number of recombinants involved. Genotypes $AABB$ and $aabb$, each due to the union of the two nonrecombinant gametes, have no recombinant. Genotypes $AABb$, $AaBB$, $aaBb$, and $Aabb$ contain one recombinant, and genotypes $aaBB$ and $AAbb$ contain two recombinants. For genotype $AaBb$, the two different diplotypes contain different numbers of recombinants, 0 for the diplotype due to the union of gametes AB and ab and 2 for the diplotype due to the union of gametes Ab and aB. Based on the diplotype frequencies, the expected number of recombinants for genotype $AaBb$ is calculated as

$$\phi = \frac{0 \times \frac{1}{2}(1-r)^2 + 2 \times \frac{1}{2}r^2}{\frac{1}{2}(1-r)^2 + \frac{1}{2}r^2} = \frac{2r^2}{(1-r)^2+r^2}. \tag{3.11}$$

The expected numbers of recombinants in the nine genotypes can be arrayed as

$$\mathbf{R} = \begin{array}{c} \\ BB \\ Bb \\ bb \end{array} \begin{array}{c} \begin{array}{ccc} AA & Aa & aa \end{array} \\ \left[\begin{array}{ccc} 0 & 1 & 2 \\ 1 & \phi & 1 \\ 2 & 1 & 0 \end{array}\right] \end{array}. \tag{3.12}$$

Using the matrices **n** and **F** from matrices (3.9) and (3.10), the likelihood function of r given the marker data is

$$\begin{aligned} L(r|n) = {} & \frac{n!}{n_{22}!...n_{00}!} \left[\frac{1}{4}(1-r)^2\right]^{n_{22}+n_{00}} \left[\frac{1}{2}r(1-r)\right]^{n_{12}+n_{21}+n_{01}+n_{10}} \\ & \times \left[\frac{1}{4}r^2\right]^{n_{02}+n_{20}} \left[\frac{1}{2}(1-r)^2+\frac{1}{2}r^2\right]^{n_{11}}. \end{aligned} \tag{3.13}$$

The MLE of the recombination fraction can be obtained by differentiating $\log L(r|n)$ with respect to r, setting the derivative equal to zero, and solving the resulting cubic function (Exercise 3.5).

Alternatively, the estimation of r can be obtained by implementing the EM algorithm (Lander and Green 1987). If we split the $AaBb$ cell into two diplotypes, $[AB][ab]$ and $[Ab][aB]$, we introduce missing data $z \sim$ binomial $(n_{11}, \phi/2)$, resulting in the complete-data likelihood

$$L(r|n,z) = \frac{n!}{n_{22}!...n_{00}!}\left[\frac{1}{4}(1-r)^2\right]^{n_{22}+n_{00}}\left[\frac{1}{2}r(1-r)\right]^{n_{12}+n_{21}+n_{01}+n_{10}}$$

$$\times\left[\frac{1}{4}r^2\right]^{n_{02}+n_{20}}\left[\frac{1}{2}(1-r)^2\right]^{n_{11}-z}\left[\frac{1}{2}r^2\right]^{z}. \tag{3.14}$$

The EM algorithm proceeds as follows. In the expected log complete-data likelihood, we replace z with $n_{11}\phi/2$. Based on the number of recombinants contained in each genotype (matrix $\mathbf{R}$) and the observations of different genotypes (matrix $\mathbf{n}$), we have the EM sequence converging to the MLE of r,

$$\hat{r} = \frac{1}{2n}[n_{12}+n_{21}+n_{01}+n_{10}+2(n_{02}+n_{20})+\phi n_{11}]. \tag{3.15}$$

Equations (3.11) and (3.15) represent two subsequent steps in the EM algorithm. In the E step, the expected number of recombinants, ϕ, is calculated using equation (3.11). In the M step, the estimated ϕ is used to update the estimate of r using equation (3.15). This procedure is repeated until the estimate converges at a stable value.

The variance of $\hat{r}$ can be approximated directly from the observed data likelihood as

$$\mathrm{Var}(\hat{r}) \approx -\frac{1}{\frac{\partial^2}{\partial r^2}\log L(r|n)\big|_{r=\hat{r}}} = \frac{\hat{r}(1-\hat{r})(1-2\hat{r}+2\hat{r}^2)}{2(1-3\hat{r}+3\hat{r}^2)n}. \tag{3.16}$$

As in the backcross, the degree of linkage between the two markers under consideration can be tested by formulating two alternative hypotheses. The LR test statistic or LOD score can be calculated using the likelihoods under the full model $L(r|n)$ of equation (3.13) and the reduced model $L(r=0.5)$ expressed as

$$L(r=0.5) = \frac{n!}{n_{22}!...n_{00}!}\left[\frac{1}{16}\right]^{n_{00}+n_{02}+n_{20}+n_{22}}$$

$$\times\left[\frac{2}{16}\right]^{n_{12}+n_{21}+n_{01}+n_{10}}\left[\frac{4}{16}\right]^{n_{11}}. \tag{3.17}$$

The LR value is chi-square distributed with one degree of freedom because there is only one parameter r that makes the full and reduced models different.

Example 3.4. Revisit Example 3.2. We use the F_2 data provided by Cheverud et al. (1996) to estimate the linkage for seven markers on mouse chromosome 2. For each pair of markers, all the F_2 mice are sorted into nine different genotypes (see matrix (3.9)). According to matrix (3.12), the nine genotypes are further divided into four groups that contain two recombinants ($n_{2R} = n_{20}+n_{02}$), one recombinant ($n_{1R} = n_{21}+n_{12}+n_{10}+n_{01}$), no recombinant ($n_{0R} = n_{22}+n_{00}$) and an uncertain number of recombinants (n_{11}). The observations of each of these four groups are tabulated in Table 3.4.

Table 3.4. Linkage analysis of markers in an F_2 population of mice.

Marker Pair	n_{2R}	n_{1R}	n_{0R}	n_{11}	$\hat{r}$	SE($\hat{r}$)	LR	LOD	χ^2	p-value
D2mit362–D2mit72	9	120	151	169	0.17	0.014	252.2	54.8	253.4	$< .001$
D2mit72–D2mit205	9	97	178	196	0.13	0.012	363.2	78.9	386.8	$< .001$
D2mit205–D2mit38	0	69	226	202	0.07	0.008	599.0	130.1	656.3	$< .001$
D2mit38–D2mit93	1	26	239	218	0.03	0.006	786.5	170.8	819.4	$< .001$
D2mit93–D2mit389	3	88	195	194	0.10	0.010	451.4	98.0	472.7	$< .001$
D2mit389–D2mit17	1	108	206	173	0.12	0.011	431.7	93.7	480.9	$< .001$
D2mit17–D2mit260	8	103	200	175	0.13	0.012	389.5	84.6	455.0	$< .001$
D2mit260–D2mit25	12	151	171	160	0.20	0.015	241.2	52.4	279.4	$< .001$

Using n_{2R}, n_{1R}, n_{0R}, and n_{11} and equations (3.15) and (3.16), the MLEs of the recombination fractions, along with their sampling errors, were estimated with the EM algorithm. The estimated recombination fractions were tested on the basis of the likelihood ratio (LR and LOD) and χ^2 test approaches. As in the backcross, only two categories of gametes that form these nine genotypes, recombinant and nonrecombinant, are independent, although there are nine different genotypes in the F_2, suggesting that there is only one degree of freedom. The three approaches provided consistent results for linkage tests (Table 3.4).

To demonstrate the calculation procedure by the EM algorithm, we detail the iterative steps using the first pair of markers in Table 3.4. From equations (3.11) and (3.15), we have:

Iteration	E Step	M Step
0	–	$r^{(0)} = 0.1$
1	$\phi^{(1)} = 0.01220$	$r^{(1)}$=0.158265
2	$\phi^{(2)} = 0.03415$	$r^{(2)}$=0.166527
3	$\phi^{(3)} = 0.03839$	$r^{(3)}$=0.168123
3	$\phi^{(3)} = 0.03924$	$r^{(3)}$=0.168445
4	$\phi^{(4)} = 0.03942$	$r^{(4)}$=0.168511
5	$\phi^{(5)} = 0.03945$	$r^{(5)}$=0.168524
6	$\phi^{(6)} = 0.03946$	$r^{(6)}$=0.168527

Given an initial value for $r = 0.1$, calculate the expected number of recombinants for the double heterozygote, $\phi^{(1)}$, as the first step using equation (3.11). The calculated $\phi^{(1)}$ is used to calculate $r^{(1)}$ of the same step with equation (3.15). Such iterations that contain both the E and M steps are repeated until the difference of the estimated r values between the two successive iterations is less than a small value, for example 10^{-5}. In this example, $r^{(5)}$ is estimated as 0.168524 in iteration 5, which is just less than $r^{(6)}$ estimated in iteration 6 by $0.000003 < 10^{-5}$. We therefore stop the iteration. The estimate of $r = 0.1685$ in iteration 6 is used as the MLE of r.

3.6 Three-Point Analysis

For a linkage analysis between two markers, we only need to estimate one parameter, the recombination fraction. This reflects the relative frequencies of recombinant and nonrecombinant gametes with respect to the two markers. However, when three markers are simultaneously considered in a *three-point analysis*, we will need to estimate three recombination fractions for three possible pairs of markers. Compared with a two-point analysis, a three-point analysis has two advantages: (1) it may increase the precision of the estimates of the recombination fractions when markers are not fully informative (Thompson 1984; Wu et al. 2002b); and (2) it provides a way of determining the optimal order of different markers. In this section, we discuss a general approach for three-point analysis.

Consider three markers, **A**, **B**, and **C**, without a particular order for a triply heterozygous F_1, from which a triple backcross or F_2 is generated. Let us first consider a backcross $ABC/abc \times abc/abc$. A total of eight groups of marker genotypes in the backcross progeny can be classified into four groups based on the number of recombinants between marker pair **A** and **B** and between marker pair **B** and **C**. These four groups are genotypes AbC/abc and aBc/abc (one recombinant from each pair), Abc/abc and aBc/abc (one recombinant only from the first pair), ABc/abc and abC/abc (one recombinant only from the second pair), and ABC/abc and abc/abc (no recombinant for each pair). Assume that n_{ij} is the number of genotypes containing i recombinants between markers **A** and **B** and j recombinants between markers **B** and **C** and that g_{ij} is the corresponding joint recombination fraction. Both n_{ij} and g_{ij} can be expressed as

	Pair **B** and **C**		
Pair **A** and **B**	Recombinant	Nonrecombinant	Total
Recombinant	n_{11}	n_{10}	$n_{1\cdot}$
Nonrecombinant	n_{01}	n_{00}	$n_{0\cdot}$
Total	$n_{\cdot 1}$	$n_{\cdot 0}$	n
Recombinant	g_{11}	g_{10}	$r_{\mathbf{AB}}$
Nonrecombinant	g_{01}	g_{00}	$1 - r_{\mathbf{AB}}$
Total	$r_{\mathbf{BC}}$	$1 - r_{\mathbf{BC}}$	

The recombination fraction between markers **A** and **B**, $r_{\mathbf{AB}}$, reflects the frequencies of the recombinant genotypes of these two markers, regardless of whether or not there is a recombinant between markers **B** and **C**. Similarly, the recombination fraction between markers **B** and **C**, $r_{\mathbf{BC}}$, reflects the frequencies of the recombinant genotypes of these two markers, regardless of the types of genotypes between markers **A** and **B**. However, the recombination fraction between marker **A** and **C**, $r_{\mathbf{AC}}$, reflects the frequencies of the recombinant genotypes of these two markers, regardless of such a recombinant event occurring between the first pair of markers or between the second pair. Based on these relationships, the recombination fractions are expressed in terms of the marginals of g_{ij}.

The likelihood function is

$$L(\mathbf{n}|g_{11}, g_{10}, g_{01}, g_{00}) = \frac{n!}{n_{11}!n_{10}!n_{01}!n_{00}!} g_{11}^{n_{11}} g_{10}^{n_{10}} g_{01}^{n_{01}} g_{00}^{n_{00}},$$

and it should be clear that the MLEs of the joint recombination probabilities g_{ij} are given by $\hat{g}_{ij} = n_{ij}/n$, and the MLEs of $r_{\mathbf{AB}}, r_{\mathbf{BC}}$, and $r_{\mathbf{AC}}$ can be estimated simultaneously as

$$\begin{aligned} \hat{r}_{\mathbf{AB}} &= \hat{g}_{10} + \hat{g}_{11} = \frac{n_{10} + n_{11}}{n}, \\ \hat{r}_{\mathbf{BC}} &= \hat{g}_{01} + \hat{g}_{11} = \frac{n_{01} + n_{11}}{n}, \\ \hat{r}_{\mathbf{AC}} &= \hat{g}_{01} + \hat{g}_{10} = \frac{n_{01} + n_{10}}{n}. \end{aligned} \tag{3.18}$$

Solving for g_{11}, g_{10}, g_{01}, and g_{00} results in

$$\begin{aligned} g_{11} &= \tfrac{1}{2}(r_{\mathbf{AB}} + r_{\mathbf{BC}} - r_{\mathbf{AC}}), \\ g_{10} &= \tfrac{1}{2}(r_{\mathbf{AB}} - r_{\mathbf{BC}} + r_{\mathbf{AC}}), \\ g_{01} &= \tfrac{1}{2}(-r_{\mathbf{AB}} + r_{\mathbf{BC}} + r_{\mathbf{AC}}), \\ g_{00} &= 1 - g_{11} - g_{10} - g_{01} = 1 - \tfrac{1}{2}(r_{\mathbf{AB}} + r_{\mathbf{BC}} + r_{\mathbf{AC}}), \end{aligned} \tag{3.19}$$

and to ensure the joint recombination probabilities $g_{ij} \geq 0$, the following inequality restrictions should hold:

$$\begin{aligned} r_{\mathbf{AC}} &\leq r_{\mathbf{AB}} + r_{\mathbf{BC}}, \\ r_{\mathbf{BC}} &\leq r_{\mathbf{AB}} + r_{\mathbf{AC}}, \\ r_{\mathbf{AB}} &\leq r_{\mathbf{BC}} + r_{\mathbf{AC}}. \end{aligned}$$

This means that the recombination fraction between the markers at two endpoints is equal to or less than the sum of the recombination fractions between any two adjacent markers.

Theorem 3.5. *For three ordered markers, the recombination fraction between the two markers, each at an end, is always greater than or equal to that between two adjacent markers. Thus, for a given marker order* **A**-**B**-**C**, $r_{\mathbf{AC}} \geq r_{\mathbf{AB}}$ *and* $r_{\mathbf{AC}} \geq r_{\mathbf{BC}}$.

Proof. Consider the marker order **A**-**B**-**C**. According to the definition of the recombination fraction ≤ 0.5, the number of nonrecombinants between markers **A** and **B** is always greater than or equal to the number of recombinants conditional on the same event occurring between markers **B** and **C**; that is, $g_{01} \geq g_{11}$ and $g_{00} \geq g_{10}$. Adding g_{10} to both sides of the first equality yields

$$g_{01} + g_{10} > g_{11} + g_{10},$$

which leads to $r_{\mathbf{AC}} \geq r_{\mathbf{AB}}$ based on equation (3.18). Similarly, we can prove $r_{\mathbf{AC}} \geq r_{\mathbf{BC}}$.

A three-point analysis in an F_2 population becomes more complicated. The strategy proposed above for the backcross design is actually based on the segregation of the F_1's gamete genotypes because the recurrent parent makes no contribution to the segregation of backcross genotypes. However, in the F_2 progeny, both parents contribute to genotype segregation, and thus three-point analysis of the F_2 should be based on the segregation of zygote genotypes. In Chapter 4, we will develop a general statistical model that covers the F_2 case for estimating the recombination fractions in three-point analyses.

3.7 Multilocus Likelihood and Locus Ordering

For a given data set containing multiple markers, marker order is not known *a priori*. An optimal marker order, which is important to linkage analysis, can be determined by comparing multilocus likelihoods for all possible orders. Again, consider a triple backcross $ABC/abc \times abc/abc$ with no information about marker order. No matter how these three markers are ordered, this backcross includes eight genotypes, which are classified into four groups in terms of the number of recombinants between different marker pairs (Section 3.6). Let $r_{\mathbf{AB}}$, $r_{\mathbf{AC}}$, and $r_{\mathbf{BC}}$ be the recombination fractions between marker pair **A** and **B**, marker pair **A** and **C**, and marker pair **B** and **C**, respectively. These four groups of backcross genotypes are tabulated in Table 3.5, along with their observed numbers and expected frequencies, under each of the three possible orders. Note that the derivation of the expected frequency of a three-marker gamete is based on the assumption that the recombination events between different marker intervals are independent. Considering the first group of gametes, for example, we have, under this assumption,

$$\begin{aligned} &\text{Prob(no recombination in } ABC \text{ or } abc) \\ &= \text{Prob(no recombination in } AB \text{ or } ab) \times \text{Prob(no recombination in } BC \text{ or } bc) \\ &= (1 - r_{\mathbf{AB}})(1 - r_{\mathbf{BC}}). \end{aligned}$$

Table 3.5. The expected frequencies of four groups of gametes in the backcross under three possible gene orders.

Gamete Type	Observation	Expected Frequency under Order **A-B-C**	**A-C-B**	**B-A-C**
ABC or *abc*	n_{00}	$(1-r_{\mathbf{AB}})(1-r_{\mathbf{BC}})$	$(1-r_{\mathbf{AC}})(1-r_{\mathbf{BC}})$	$(1-r_{\mathbf{AB}})(1-r_{\mathbf{AC}})$
ABc or *abC*	n_{01}	$(1-r_{\mathbf{AB}})r_{\mathbf{BC}}$	$r_{\mathbf{AC}}r_{\mathbf{BC}}$	$(1-r_{\mathbf{AB}})r_{\mathbf{AC}}$
Abc or *aBC*	n_{10}	$r_{\mathbf{AB}}(1-r_{\mathbf{BC}})$	$r_{\mathbf{AC}}(1-r_{\mathbf{BC}})$	$r_{\mathbf{AB}}r_{\mathbf{AC}}$
AbC or *aBc*	n_{11}	$r_{\mathbf{AB}}r_{\mathbf{BC}}$	$(1-r_{\mathbf{AC}})r_{\mathbf{BC}}$	$r_{\mathbf{AB}}(1-r_{\mathbf{AC}})$

The MLEs of the three recombination fractions for each order can be obtained by maximizing the likelihood function under that order. Note that, in the backcross design, the expression of the MLEs does not depend on marker order, which is expressed as

$$\begin{aligned}\hat{r}_{\mathbf{AB}} &= \frac{n_{10}+n_{11}}{n} = \hat{g}_{10}+\hat{g}_{11},\\ \hat{r}_{\mathbf{BC}} &= \frac{n_{01}+n_{11}}{n} = \hat{g}_{01}+\hat{g}_{11},\\ \hat{r}_{\mathbf{AC}} &= \frac{n_{01}+n_{10}}{n} = \hat{g}_{01}+\hat{g}_{10}.\end{aligned}$$

When we calculate the likelihood value of the observations, it can be seen in Table 3.5 that it will differ depending on the order of the markers. For example, for a particular order **A-B-C**, we have

$$\begin{aligned}L_{\mathbf{ABC}} &\propto (1-r_{\mathbf{AB}})^{n_{00}+n_{10}}(1-r_{\mathbf{BC}})^{n_{00}+n_{10}}(r_{\mathbf{AB}})^{n_{10}+n_{11}}(r_{\mathbf{BC}})^{n_{01}+n_{11}}\\ &= \left[(1-r_{\mathbf{AB}})^{n_{00}+n_{10}}(r_{\mathbf{AB}})^{n_{10}+n_{11}}\right]\left[(1-r_{\mathbf{BC}})^{n_{00}+n_{10}}(r_{\mathbf{BC}})^{n_{01}+n_{11}}\right].\end{aligned} \tag{3.20}$$

The likelihood is actually a product of binomials and can be solved for the MLEs $\hat{r}_{\mathbf{AB}}, \hat{r}_{\mathbf{BC}}$. The sampling variances of the estimates of the recombination fractions ($r_{\mathbf{AB}}$ and $r_{\mathbf{BC}}$) are

$$\begin{aligned}\mathrm{Var}(\hat{r}_{\mathbf{AB}}) &\approx \frac{\hat{r}_{\mathbf{AB}}(1-\hat{r}_{\mathbf{AB}})}{n},\\ \mathrm{Var}(\hat{r}_{\mathbf{BC}}) &\approx \frac{\hat{r}_{\mathbf{BC}}(1-\hat{r}_{\mathbf{BC}})}{n}.\end{aligned} \tag{3.21}$$

Similarly, the likelihood values can be calculated for the two other marker orders (see Table 3.5), denoted by $L_{\mathbf{ACB}}$ and $L_{\mathbf{CAB}}$ (Exercise 3.6).

The marker order that corresponds to the maximum likelihood value can be regarded as the optimal order supported by the data. Thus, by comparing the three likelihood values, $L_{\mathbf{ABC}}$, $L_{\mathbf{ACB}}$, and $L_{\mathbf{CAB}}$, we can determine the most likely marker order. A similar procedure can be used to determine an optimal marker order for an F_2 progeny, but, unlike the backcross, the MLEs of the recombination fractions will differ among the orders. We will consider this issue in Chapter 4.

The calculation of the likelihood (3.20) is based on the assumption of independent crossovers in different marker intervals. Yet, as shown by Speed et al. (1992), the ordering of the loci that maximizes the likelihood under the assumption of no interference is a virtually consistent estimate of the true order even when interference actually exists. This property expands the utility of the procedure above for gene ordering if an adequately large sample size is used.

Example 3.6. (**Three-point Analysis**). Revisit Example 3.1 for a rice mapping population of $n = 123$ plants. Consider three markers, RG472, RG246, and K5, on rice chromosome 1. The heterozygous F_1 $AaBbCc$ derived from genotypes $AABBCC$ and $aabbcc$ generates eight different types of haploid gametes. Doubled haploids (DH) were then observed for each gamete type as follows ($n = 100$ after deleting the observations missing in the three markers):

DH	ABC/ABC	ABc/ABc	AbC/AbC	Abc/Abc	aBC/aBC	aBc/aBc	abC/abC	abc/abc
Obs.	38	2	2	5	11	1	10	31

These eight gamete types are sorted into four groups (see Table 3.5) based on the distribution of recombinants between markers. Observations in each group are $n_{00} = 38 + 31 = 69$, $n_{01} = 2 + 10 = 12$, $n_{10} = 5 + 11 = 16$, and $n_{11} = 2 + 1 = 3$. Using equation (3.18), the MLEs of three recombination fractions between each pair of these three markers are estimated as

$$\hat{r}_{\mathbf{AB}} = \frac{16 + 3}{100} = 0.19,$$

$$\hat{r}_{\mathbf{AC}} = \frac{12 + 3}{100} = 0.15,$$

$$\hat{r}_{\mathbf{BC}} = \frac{16 + 12}{100} = 0.28,$$

with the sampling variances, $\text{Var}(\hat{r}_{\mathbf{AB}})$, $\text{Var}(\hat{r}_{\mathbf{AC}})$, and $\text{Var}(\hat{r}_{\mathbf{BC}})$, approximated by $0.19(1 - 0.19)/100 = 0.0015$, $0.15(1 - 0.15)/100 = 0.0013$, and $0.28(1 - 0.28)/100 = 0.0020$, respectively.

Taking the log of the likelihood values under each marker order, calculated with $\hat{r}_{\mathbf{AB}}$, $\hat{r}_{\mathbf{AC}}$, and $\hat{r}_{\mathbf{BC}}$ by equation (3.20), we have

$$
\begin{aligned}
&\log L_{\mathbf{ABC}} \\
&= [(69+12)\log(1-0.19) + (16+3)\log(0.19)] \\
&\quad +[(69+16)\log(1-0.15) + (12+3)\log(0.15)] = -90.89, \\
&\log L_{\mathbf{ACB}} \\
&= [(69+3)\log(1-0.28) + (12+16)\log(0.28)] \\
&\quad +[(69+16)\log(1-0.15) + (12+3)\log(0.15)] = -101.57, \\
&\log L_{\mathbf{BAC}} \\
&= [(69+12)\log(1-0.19) + (16+3)\log(0.19)] \\
&\quad +[(69+3)\log(1-0.28) + (12+16)\log(0.28)] = -107.92.
\end{aligned}
$$

Because $\log L_{\mathbf{ABC}}$ is the largest among the three values, we conclude that these three markers have an order **A-B-C**. Although an optimal marker order has been determined, another unsolved important issue is how much more likely these three markers are to have order **A-B-C** than the two other orders, **A-C-B** and **B-A-C**. We will address this issue in Section 3.9.

3.8 Estimation with Many Loci

In principle, the problems of locus ordering and interloci distance estimation can be tackled simultaneously by comparing the likelihoods maximized over interloci distances for all possible locus orders. However, the number of possible orders, as well as the computer time and memory required for each multilocus likelihood calculation, increases rapidly with the number of loci. Even after taking into account the equivalence of two orders that are the reverse of each other, there are still $m!/2$ possible orders for m loci. Evaluation of the likelihoods of all possible orders rapidly becomes impractical as the number of loci increases ($2!/2 = 1, \dots, 10!/2 = 1,814,400$). It is therefore necessary to generate a small number of approximate orders before proceeding to a formal likelihood analysis of these orders.

There are two main approaches to the generation of approximate orders. One approach is to start with a small number of markers whose order can be established by a likelihood analysis and then proceed to place the remaining markers, one at a time, into one of the intervals between the markers already in the map. At each stage, the effect on the likelihood of placing the additional marker in each of the possible intervals is evaluated, and the placement that produces the highest likelihood is chosen. If one starts with two markers, for example, then the first additional marker can be placed in one of three intervals, the next into one of four, and so on, so that the entire procedure will require only $3 + 4 + ... + m = (m-1)(m+3)/2$ likelihood evaluations.

The second approach for generating approximate orders is to analyze all pairs of loci using two-point linkage analysis and then subject the $m(m-1)/2$ recombination fraction estimates (or maximum LOD scores) to some method of seriation. The problem of converting a square matrix of dissimilarities (or distance) between objects into a set of coordinates that specify the relative positions of the objects is well known in multivariate exploratory analysis. The general class of methods for tackling this problem is known as multidimensional scaling. Any one of the several methods of multidimensional scaling can be applied to an m-by-m matrix of recombination fraction estimates to produce a two-dimensional representation of the m points. Except for a very short chromosomal segment, the m points will fall on a horseshoe-shaped curve rather than on a straight line because the recombination fraction between the two farthest markers is at most 1/2 but the sum of the recombination fractions between adjacent loci, arranged in any order, is likely to exceed 1/2. Several variations of this approach have been proposed. Once approximate orders have been generated, these can be subjected to formal multipoint likelihood analysis.

3.9 Mixture Likelihoods and Order Probabilities

A natural question is that of estimating the gene order probabilities. Here, we derive the full likelihood, with which we can then jointly estimate the recombination fractions and order probabilities in a three-point analysis.

The likelihoods for each of the three orders are expressed as

$$\begin{aligned}
L_{\mathbf{ABC}} &\propto (1-r_{\mathbf{AB}})^{n_{00}+n_{10}}(1-r_{\mathbf{BC}})^{n_{00}+n_{10}}(r_{\mathbf{AB}})^{n_{10}+n_{11}}(r_{\mathbf{BC}})^{n_{01}+n_{11}},\\
L_{\mathbf{ACB}} &\propto (1-r_{\mathbf{AC}})^{n_{00}+n_{11}}(1-r_{\mathbf{BC}})^{n_{00}+n_{10}}(r_{\mathbf{AC}})^{n_{01}+n_{10}}(r_{\mathbf{BC}})^{n_{01}+n_{11}},\\
L_{\mathbf{BAC}} &\propto (1-r_{\mathbf{AB}})^{n_{00}+n_{10}}(1-r_{\mathbf{AC}})^{n_{00}+n_{11}}(r_{\mathbf{AB}})^{n_{10}+n_{11}}(r_{\mathbf{AC}})^{n_{01}+n_{10}};
\end{aligned}$$

however, none of these likelihood functions is the likelihood function for the full model for the data. For that model, we obtain an observation y from one of the three orders and the order is given to us with a certain probability, the unknown probability that it is the true order. Thus, if we denote these probabilities by p, with the appropriate subscript, the likelihood function for the full model is

$$L_F = p_{\mathbf{ABC}}L_{\mathbf{ABC}} + p_{\mathbf{ACB}}L_{\mathbf{ACB}} + p_{\mathbf{BAC}}L_{\mathbf{BAC}}, \tag{3.22}$$

a mixture model.

We know that we will get a gene order with a certain probability, and if we knew that order we would know the correct piece of the likelihood to use. This suggests that we can introduce missing data $z = (z_1, z_2, z_3)$ to tell us the gene order. For example, we specify that z can have exactly one 1 and two 0's, with the 1 denoting the gene order. The joint complete-data likelihood of (n, z) is

$$L_C = [p_{\mathbf{ABC}}L_{\mathbf{ABC}}]^{z_1}[p_{\mathbf{ACB}}L_{\mathbf{ACB}}]^{z_2}[p_{\mathbf{BAC}}L_{\mathbf{BAC}}]^{z_3}, \tag{3.23}$$

and the resulting missing-data density is

$$k(z,n) = \frac{[p_{\mathbf{ABC}}L_{\mathbf{ABC}}]^{z_1}[p_{\mathbf{ACB}}L_{\mathbf{ACB}}]^{z_2}[p_{\mathbf{BAC}}L_{\mathbf{BAC}}]^{z_3}}{p_{\mathbf{ABC}}L_{\mathbf{ABC}} + p_{\mathbf{ACB}}L_{\mathbf{ACB}} + p_{\mathbf{BAC}}L_{\mathbf{BAC}}},$$

which we should recognize as a multinomial distribution with one observation to be put in one of three cells, with cell probabilities given by

$$\mathrm{E}(z_1) = P_{\mathbf{ABC}} = \frac{p_{\mathbf{ABC}} L_{\mathbf{ABC}}}{p_{\mathbf{ABC}} L_{\mathbf{ABC}} + p_{\mathbf{ACB}} L_{\mathbf{ACB}} + p_{\mathbf{BAC}} L_{\mathbf{BAC}}}, \tag{3.24}$$

and we similarly define $P_{\mathbf{ACB}}$ and $P_{\mathbf{BAC}} = 1 - P_{\mathbf{ABC}} - P_{\mathbf{ACB}}$.

Using equations (3.23) and (3.24), we construct an EM algorithm as follows. The expected complete-data log likelihood is

$$\begin{aligned} E(L_C) &= P_{\mathbf{ABC}} \log(p_{\mathbf{ABC}} L_{\mathbf{ABC}}) \\ &\quad + P_{\mathbf{ACB}} \log(p_{\mathbf{ACB}} L_{\mathbf{ACB}}) + P_{\mathbf{BAC}} \log(p_{\mathbf{BAC}} L_{\mathbf{BAC}}). \end{aligned} \tag{3.25}$$

Differentiating $E(L_C)$ with respect to the recombination fractions yields the MLEs of r. With those, we can estimate the $P_{\mathbf{ABC}}$ using

$$\hat{P}_{\mathbf{ABC}} = \frac{p_{\mathbf{ABC}} L_{\mathbf{ABC}}}{p_{\mathbf{ABC}} L_{\mathbf{ABC}} + p_{\mathbf{ACB}} L_{\mathbf{ACB}} + p_{\mathbf{BAC}} L_{\mathbf{BAC}}},$$

and similarly for $P_{\mathbf{ACB}}$ and $P_{\mathbf{BAC}} = 1 - P_{\mathbf{ABC}} - P_{\mathbf{ACB}}$. (Notice the similarity to the mixture model of Section 2.2.)

Example 3.7. Revisit Example 3.6 for the three-point analysis of markers RG472, RG246, and K5 in a rice HD population. Based on the observations of each gamete type, we have $n_{00} = 69$, $n_{01} = 12$, $n_{10} = 16$, and $n_{11} = 3$, which provide the MLEs of the recombination fractions

$$\hat{r}_{\mathbf{AB}} = 0.19, \quad \hat{r}_{\mathbf{BC}} = 0.15, \quad \hat{r}_{\mathbf{AC}} = 0.28.$$

To estimate P, the values of p must be specified. Since we do not have information, we cannot do much but set $p_{ABC} = p_{ACB} = p_{BAC} = 0.333$. Then the estimate of P_{ABC} is

$$\hat{P}_{\mathbf{ABC}} = \frac{\frac{1}{3} L_{\mathbf{ABC}}}{\frac{1}{3} L_{\mathbf{ABC}} + \frac{1}{3} L_{\mathbf{ACB}} + \frac{1}{3} L_{\mathbf{BAC}}} = 0.999977,$$

with $\hat{P}_{\mathbf{ACB}}$ and $\hat{P}_{\mathbf{BAC}}$ being near 0. These results suggest that there is a very high probability (0.999977) for the data to support marker order **A-B-C**.

3.10 Map Functions

The map function is a mathematical function that converts the recombination fraction (r) between two loci to the genetic distance separating them (d). The recombination fraction is not an additive distance measure. Consider three markers **A**, **B**, and **C**. If the recombination fraction between markers **A** and **B** and that between markers **B** and **C** are each assumed to be equal to $r = 0.30$, then the recombination fraction between markers **A** and **C** cannot be $2r$ since that value would exceed 50 percent. One therefore needs to transform the recombination fraction, r, into the additive map distance, d.

Definition 3.8. [Map Distance] The *map distance* between the two loci is defined as the expected number of crossovers occurring between them on a single chromatid during meiosis.

The two nonalleles, each from a locus, will be derived from the same parental chromosomes if no crossover or an even number of crossovers occurs between the two loci, and from the different parental chromosomes if an odd number of crossovers occurs between the two loci. Therefore, we can formulate a theoretical model to express the recombination fraction between two loci in terms of their map distance or length by using the number of crossover events.

3.10.1 Mather's Formula

Mather (1938) derived a formula connecting the recombination fraction between two loci **A** and **B** to the random number of chiasmata (that is, crossovers) X occurring on the interval [**A**, **B**] of the chromatid bundle. According to his derivation, the recombination fraction between two loci r is half the probability of chiasmata occurring in all four strands of tetrads between the loci. Mathematically, this can be expressed as

$$r = \frac{1}{2}\text{Prob}(X > 0) = \frac{1}{2}[1 - \text{Prob}(X = 0)], \tag{3.26}$$

where $\text{Prob}(X = 0)$ is the probability of no chiasma between two loci. The genetic map distance d separating **A** and **B** is defined as $\frac{1}{2}\text{E}(X)$, the expected number of chiasmata on [**A**, **B**] for the tetrad as a whole, because each crossover involves two chromatids.

Mather's formula (3.26) can be proven by first noting that a gamete is recombinant between two loci **A** and **B** if and only if an odd number of crossovers occurs on the gamete between the loci. Denote the probability that the gamete is recombinant after x chiasmata between **A** and **B** by r_x. It is clear that $r_0 = 0$. For $x > 0$, we have the recurrence

$$r_x = \frac{1}{2}r_{x-1} + \frac{1}{2}(1 - r_{x-1}) = \frac{1}{2}$$

because a gamete is recombinant after x crossovers if it is recombinant after $x - 1$ crossovers and does not participate in crossover x or if it is nonrecombinant after $x-1$ crossovers and does participate in crossover x, and these probabilities are equally likely. Thus, it follows that $r_x = \frac{1}{2}$ for all $x > 0$. This thus suggests that the probability of obtaining a recombinant strand is 1/2 as long as there is at least one chiasma between the two loci. This proves Mather's formula.

3.10.2 The Morgan Map Function

The Morgan map function is the simplest map function, which assumes that (1) there is at most one crossover occurring on the interval of two loci, and (2) the probability of a crossover on an interval is proportional to the map length of the interval (Morgan 1928). Under these assumptions, the probability of a chiasma occurring in a distance of d map units is equal to the expected number of crossovers per gamete in this distance and therefore to $2d$ (see the definition of d above), which gives

$$r = \tfrac{1}{2}[1 - \text{Prob}(X = 0)] = \tfrac{1}{2}[1 - (1 - 2d)] = d.$$

This function holds only when $0 \leq d \leq 1/2$ since for $d > 1/2$ it results in recombination fractions of greater than 1/2. It may therefore be used as an approximation for short distances but is not applicable for long segments of chromosomes.

3.10.3 The Haldane Map Function

The Haldane map function, the second simplest map function, assumes that crossovers occur at random and independently of each other (Haldane 1919). With this assumption, the occurrence of crossovers between two loci on a chromosome can be viewed as a Poisson process (i.e., they are equally probable at any point between the loci), so that the number of crossovers between the loci can be modelled by a Poisson distribution. Since map distance d is defined as the average number of crossovers per chromatid in a given interval, the average number of crossovers for the tetrad as a whole is $2d$. The assumption of a Poisson process implies that the probability of no chiasma in the interval, $\text{Prob}(X = 0)$, is e^{-2d}. Using Mather's formula, this gives the Haldane map function

$$r = \tfrac{1}{2}\left[1 - \text{Prob}(X = 0)\right] = \tfrac{1}{2}\left(1 - e^{-2d}\right), \tag{3.27}$$

whose inverse is

$$d = -\tfrac{1}{2}\ln(1 - 2r). \tag{3.28}$$

The Haldane map function can be derived in another way. Because the genetic distance (d) between two loci is defined as the average number of crossovers, the probability of the number of crossovers can be expressed using the Poisson distribution in terms of d as follows:

Crossover	0	1	2	3	$\cdots$	x	$\cdots$
Probability	e^{-d}	$\frac{d}{1!}e^{-d}$	$\frac{d^2}{2!}e^{-d}$	$\frac{d^3}{3!}e^{-d}$	$\cdots$	$\frac{d^x}{x!}e^{-d}$	$\cdots$

Because the value of the recombination fraction r for a genetic distance of d is the sum of the probabilities of all odd numbers of crossovers, we have

$$r = e^{-d}\left(\frac{d}{1!} + \frac{d^3}{3!} + \frac{d^5}{5!} + \cdots\right)$$
$$= \tfrac{1}{2}\left(1 - e^{-2d}\right).$$

The additivity of the Haldane function can be established by assuming that three loci are in the order **A-B-C**. A gamete is a recombinant with respect to **A** and **C** if and only if it is a recombinant with respect to **A** and **B** but not **B** and **C** or if it is a recombinant with respect to **B** and **C** but not **A** and **B**. Therefore, with the assumption of independence, three possible recombination fractions among these three loci have the following relationship:

$$r_{\mathbf{AC}} = r_{\mathbf{AB}}(1 - r_{\mathbf{BC}}) + r_{\mathbf{BC}}(1 - r_{\mathbf{AB}}) = r_{\mathbf{AB}} + r_{\mathbf{BC}} - 2r_{\mathbf{AB}}r_{\mathbf{BC}}, \tag{3.29}$$

or

$$1 - 2r_{\mathbf{AC}} = (1 - 2r_{\mathbf{AB}})(1 - 2r_{\mathbf{BC}}).$$

Given $r_{\mathbf{AB}} = \frac{1}{2}(1 - e^{-2d_{\mathbf{AB}}})$ and $r_{\mathbf{BC}} = \frac{1}{2}(1 - e^{-2d_{\mathbf{BC}}})$, where the d's are the map distances between the corresponding loci, we have

$$\begin{aligned}
&r_{\mathbf{AC}} \\
&= r_{\mathbf{AB}} + r_{\mathbf{BC}} - 2r_{\mathbf{AB}}r_{\mathbf{BC}} \\
&= \tfrac{1}{2}(1 - e^{-2d_{\mathbf{AB}}}) + \tfrac{1}{2}(1 - e^{-2d_{\mathbf{BC}}}) - 2 \cdot \tfrac{1}{2}(1 - e^{d_{\mathbf{AB}}})\tfrac{1}{2}(1 - e^{d_{\mathbf{BC}}}) \\
&= \tfrac{1}{2}(1 - e^{-2d_{\mathbf{AB}}} + 1 - e^{-2d_{\mathbf{BC}}} - 1 + e^{-2d_{\mathbf{AB}}} + e^{-2d_{\mathbf{BC}}} - e^{-2d_{\mathbf{AB}}}e^{-2d_{\mathbf{BC}}}) \\
&= \tfrac{1}{2}[1 - e^{-2(d_{\mathbf{AB}} + d_{\mathbf{BC}})}] \\
&= \tfrac{1}{2}(1 - e^{-2d_{\mathbf{AC}}}),
\end{aligned}$$

which leads to $d_{\mathbf{AC}} = d_{\mathbf{AB}} + d_{\mathbf{BC}}$.

In practice, the Haldane map function may not be accurate at small distances. Empirical observations show that the probability of having two crossovers occur in close proximity to each other is often less than that predicted by the Haldane map function. However, in his 1919 paper, Haldane also introduced a differential equation method that generalized the construction of various map functions. One of the applications of this generalization was the derivation of Kosambi's (1944) map function, which is both simple and justifiable in practice.

3.10.4 The Kosambi Map Function

At very short distances, interference appears to be complete, so that assuming that locus **B** is between loci **A** and **C**, the recombination between **A** and **B** implies non-recombination between **B** and **C**, and vice versa, and thus either $r_{\mathbf{AB}}$ or $r_{\mathbf{BC}}$ is zero. Recombination fractions therefore become approximately additive at short distances, satisfying

$$r_{\mathbf{AC}} = r_{\mathbf{AB}} + r_{\mathbf{BC}}, \tag{3.30}$$

whereas at long distances equation (3.29) is more accurate. When the markers are located at moderate distances, the relationship between the recombination fractions is expressed as

$$r_{\mathbf{AC}} = r_{\mathbf{AB}} + r_{\mathbf{BC}} - r_{\mathbf{AB}}r_{\mathbf{BC}}. \tag{3.31}$$

In sum, a general model describing the relationship can be written as

$$r_{\mathbf{AC}} = r_{\mathbf{AB}}(1 - r_{\mathbf{BC}}) + r_{\mathbf{BC}}(1 - r_{\mathbf{AB}}) = r_{\mathbf{AB}} + r_{\mathbf{BC}} - 2cr_{\mathbf{AB}}r_{\mathbf{BC}},$$

where, based on equation (3.19),

$$\begin{aligned} c &= \frac{\frac{1}{2}(r_{\mathbf{AB}} + r_{\mathbf{BC}} - r_{\mathbf{AC}})}{r_{\mathbf{AB}}r_{\mathbf{BC}}} \\ &= \frac{g_{11}}{r_{\mathbf{AB}}r_{\mathbf{BC}}}. \end{aligned} \tag{3.32}$$

Equation (3.32) measures the deviation of observed recombinations in different intervals between markers from the recombinations that are assumed to occur independently (Muller 1916). This deviation (c) from independence is defined as the coefficient of coincidence, while $I = 1-c$ is called interference. The absence of interference ($I = 0$) and positive ($I > 0$) and negative interference ($I < 0$) correspond to $c = 1$, $c < 1$, and $c > 1$, respectively (Ott 1991). Coefficients c and I are a property of intervals $[\mathbf{A}, \mathbf{B}]$ and $[\mathbf{B}, \mathbf{C}]$.

Next, we want to find a function $r = f(d)$ that can reflect the relationship of the recombination fractions, as described by equations (3.29)–(3.31), at different genetic distances. Assume that f satisfies the relationship

$$f(d + h) = f(d) + f(h) - 2cf(d)f(h).$$

Recalling Haldane's (1919) differential equation, we have

$$\frac{f(d+h) - f(d)}{h} = \frac{f(h)}{h} - 2cf(d)\frac{f(h)}{h}. \tag{3.33}$$

If we require $r = f(d) = d$ at short distances, then as h tends to 0, $f(h)/h$ tends to 1, and we have the derivative based on equation (3.33),

$$f'(d) = 1 - 2c_0 f(d) = 1 - 2c_0 r,$$

where c_0 is known as a marginal coincidence, distinguished from c because it is the limit as one of the two intervals approaches 0. When c_0 is a nonzero constant, this differential equation yields the solution

$$r = \int \frac{1}{1 - 2c_0 r} dr = -\frac{1}{2c_0} \ln(1 - 2c_0 r),$$

which is the Haldane map function (3.27) when $c_0 = 1$. However, interference suggests that smaller values of c_0 are appropriate for smaller values of r. This would lead to an infinite number of map functions, but Kosambi (1944) noted that if c_0 were allowed to be an appropriate function of r, then a single mapping function could be derived. The simplest function of r that increases in the interval $0 < r < 1$ and takes the value 0 at $r = 0$ and the value 1 when $r = 1/2$ is $c_o = 2r$. Then the differential equation becomes

$$f'(d) = 1 - 2c_0 r = 1 - 4r^4.$$

Integration then yields the function

$$d = \frac{1}{4}\ln\frac{1+2r}{1-2r} \tag{3.34}$$

with inverse

$$r = \frac{1}{2}\frac{e^{2d} - e^{-2d}}{e^{2d} + e^{-2d}}. \tag{3.35}$$

This is known as the Kosambi map function, which has been widely used in linkage mapping. From equation (3.33), we have Kosambi's addition formula for the recombination fractions of the loci **A-B-C**:

$$\begin{aligned} r_{\mathbf{AC}} &= \frac{1}{2}\tanh 2d_{\mathbf{AC}} \\ &= \frac{1}{2}\tanh\,(2d_{\mathbf{AB}} + 2d_{\mathbf{BC}}) \\ &= \frac{\frac{1}{2}\tanh 2d_{\mathbf{AB}} + \frac{1}{2}\tanh 2d_{\mathbf{BC}}}{1 + \tanh 2d_{\mathbf{AB}}\tanh 2d_{\mathbf{BC}}} \\ &= \frac{r_{\mathbf{AB}} + r_{\mathbf{BC}}}{1 + 4r_{\mathbf{AB}}r_{\mathbf{BC}}}. \end{aligned} \tag{3.36}$$

This is similar to the velocity addition rule in the special theory of relativity.

Many modifications of these classic map functions have been made for different situations, mostly by considering interference. Carter and Falconer's (1951) function considered relatively strong interference. A variable level of interference was modeled by Felsenstein (1979). All of these functions can be derived from Haldane's (1919) differential equation by assuming different marginal coincidences c (Liberman and Karlin 1984).

Example 3.8. Revisit Example 3.6 for a three-point analysis in rice. The recombination fractions for three markers, RG472 (**A**), RG246 (**B**), and K5 (**C**), were estimated in Example 3.6. The best order of the three markers **A-B-C** was determined in Example 3.7. Here, we use the Haldane and Kosambi map functions to estimate the genetic distances for these three markers as follows:

Marker	Haldane		Kosambi	
Pair	r	d (cM)	r	d (cM)
A-B	0.190	23.902	0.190	20.003
B-C	0.150	17.834	0.150	15.476
A-C	0.280	41.049	0.280	31.642
(**A-C**)	(0.283)	(41.736)	(0.305)	(35.479)

The recombination fraction between markers **A** and **C** at the two ends can be estimated *directly* on the basis of the procedure described in three-point analysis. This recombination fraction can also be estimated *indirectly* using equations (3.29) and (3.36) based on the estimates of the other recombination fractions, $r_{\mathbf{AB}}$ and $r_{\mathbf{BC}}$. We then calculate the genetic distance between markers **A** and **C** using these estimates from the indirect approach (shown in parentheses). The estimates of the genetic distance between **A** and **C** are consistent between the direct and indirect approaches for the Haldane map function, whereas these are different for the Kosambi function. This is due to different assumptions used for the derivation of these two map functions. The Haldane map function assumes no interference between two adjacent marker intervals, whereas this assumption is not necessary for the Kosambi map function.

At the end of this section, we summarize the Haldane and Kosambi map functions as follows:

Function	$r(d)$	$d(r)$
Haldane	$r = \frac{1}{2}(1 - e^{-2d})$	$d = -\frac{1}{2}\ln(1 - 2r)$
Kosambi	$r = \frac{1}{2}\frac{e^{2d} - e^{-2d}}{e^{2d} + e^{-2d}}$	$d = \frac{1}{4}\ln\frac{1+2r}{1-2r}$

The comparison between these two functions is made by plotting the genetic distance against the recombination fraction and vice versa (Fig. 3.2). For two highly linked markers (corresponding to a small r value), these two functions obtain similar genetic distances. However, the divergence in the genetic distance estimated by the two functions increases with increasing r values. When the r is close to 0.5, the two functions tend to converge.

3.11 Exercises

3.1 Verify that the MLE of r given in equation (3.4) is obtained from differentiation of the likelihood (3.3).

3.2 Referring to two-point analysis, assume an F_2 population in which two markers are genotyped. The two markers form nine distinguishable genotypes, each with observations as follows:

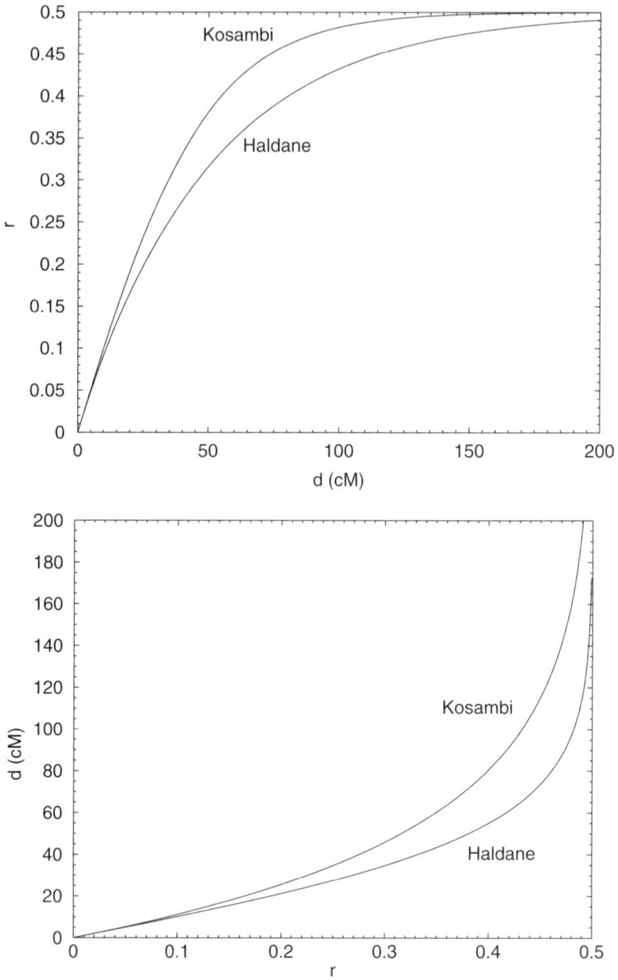

Fig. 3.2. The genetic distance (d) as a function of the recombination fraction (r) and vice versa estimated by the Haldane and Kosambi functions.

$$\mathbf{n} = \begin{array}{c} \\ BB \\ Bb \\ bb \end{array} \begin{array}{c} \begin{array}{ccc} AA & Aa & aa \end{array} \\ \left[\begin{array}{ccc} 107 & 20 & 3 \\ 24 & 175 & 18 \\ 5 & 41 & 93 \end{array} \right] \end{array}$$

(a) Using the EM algorithm, estimate the recombination fraction between these two markers.

(b) Test whether these two markers are significantly linked using χ^2 and likelihood ratio test approaches.

(c) Estimate the sampling error of the estimated recombination fraction and the confidence interval of the recombination fraction.

3.3 Referring to marker order, three different markers were genotyped for a backcross population toward parent *aabbcc*. The genotypes at the three markers were observed as follows:

DH	ABC/abc	ABc/abc	AbC/abc	Abc/abc	aBC/abc	aBc/abc	abC/abc	abc/abc
Obs.	60	1	2	1	14	0	9	15

(a) Estimate the recombination fractions between each pair of markers.
(b) Test the significance of these estimated recombination fractions.
(c) Determine the best order of these three markers.
(d) Calculate the genetic distances between each pair of marker using the Haldane and Kosambi map functions.
(e) How can you estimate the recombination fraction and genetic distance between markers at the two ends? Compare the results from different approaches.

3.4 Referring to two-point analysis, in a second F_2 population, one observes nine genotypes for two markers as follows:

$$\mathbf{n} = \begin{array}{c} \\ BB \\ Bb \\ bb \end{array} \begin{array}{c} \begin{array}{ccc} AA & Aa & aa \end{array} \\ \left[\begin{array}{ccc} 0 & 8 & 113 \\ 18 & 202 & 27 \\ 18 & 202 & 27 \end{array}\right] \end{array}$$

Using the EM algorithm, estimate the recombination fraction. If your estimate is greater than 0.5, which violates the definition of this parameter, how can you explain it?

3.5 Find the MLE from equation (3.13) by (a) solving the cubic equation and (b) using the EM algorithm.

3.6 Referring to the likelihood of equation (3.20):
(a) Show that it can be written as a product of binomial likelihoods and find the MLEs $\hat{r}_{\mathbf{AB}}$ and $\hat{r}_{\mathbf{BC}}$. Show that equation (3.21) is an approximation of their variances.
(b) Find the MLEs for the other two orders (see Table 3.5). Which are the overall MLEs?

3.12 Notes: Algorithms and Software for Map Construction

Genetic linkage maps have been used as a powerful tool to study the structure and organization of the genome for a species and detect and identify loci that are responsible for quantitative traits. The construction of linkage maps includes two steps: (1) grouping markers into linkage groups and (2) ordering the markers within each linkage group. Markers can be placed into linkage groups based on their linkage relationships. The degree of linkage between any two markers measured in terms of the recombination value can be estimated and tested by the approaches for two- or three-point analyses. The markers can be grouped by a cluster analysis on the basis of a matrix consisting of all possible pairwise marker recombinations. Specific cutoff criteria are used to determine the number of linkage groups.

After marker groups are specified, the order of the markers within a group is obtained by minimizing the differences between recombination fractions from the pairwise data and calculated fractions in the map. However, when the number of markers increases, the number of possible marker orders increases rapidly, which presents a considerable computational challenge. In fact, the ordering of markers can be regarded as a special case of the travelling salesman problem (TSP) (Wilson 1988; Olson and Boehnke 1990; Falk 1992; Liu 1998; Gaspin and Schiex 1997); Mester et al. 2003), a classical nondeterministic polynomial-complete problem in mathematics and computer science. The problem of marker ordering can be handled by performing exhaustive searches. It is extremely time-consuming to exhaustively search all possible orders when the number of markers is large, say >30. For this reason, an algorithm attempted to obtain an approximate optimal solution has been a practical approach for large-scale linkage analysis.

Several approximation algorithms available for map construction have been introduced in Tan and Fu (2006). They include seriation (Buetow and Chakravarti 1987), simulated annealing (Thompson 1984; Weeks and Lange 1987), branch and bound (Lathrop et al. 1985), the Lander-Green algorithm (Lander and Green 1987,) and stepwise-likelihood (Lathrop et al. 1984). Software packages that implemented these algorithms are LINKAGE (Lathrop et al. 1984), MAPMAKER/EXP (Lander et al. 1987), LINKAGE MAP (Eppig and Eicher 1983), JoinMap (Stam, P. 1993), LINKAGE-1 (Suiter et al. 1983), and GMendel (Echt et al. 1992). In general, the performance of all these algorithms and software may be affected by experimental errors, sample size, and interference of recombination and double crossovers.

It is worthwhile to mention two recent approaches for map construction. The first is Mester et al.'s (2003) genetic and evolutionary algorithm (GEA) that searches for an optimal solution adaptively by mimicking the evolutionary process of a population including mutation, recombination, and selection. The second was proposed by Tan and Fu (2006), who used the principle of neighbor mapping (Ellis 1997) to construct a linkage map by starting with a small map and adding markers into it one at a time. Compared with other algorithms, Tan and Fu's sequential algorithm displays important advantages in computational speed and the accuracy of detecting a true marker order.

There are specific statistical principles behind each of the map construction algorithms and software packages mentioned above. To help the reader understand, we provide some explanation of the statistical principles underlying Lander and Green's algorithm which has been widely used for map construction for experimental organisms. Readers are also referred to Sham (1998), who reviewed the application of Lander and Green's algorithm with a comparison with Elston and Stewart's (1971) algorithm.

The Lander-Green algorithm is based on a hidden Markov formulation of the pattern of inheritance at several ordered markers. The linkage information of a haplotype can be summarized as a string of f alleles derived from the parent's paternal haplotype and m alleles derived from the parent's maternal haplotype. At a single marker, the linkage information of each nonfounding member can be summarized by just two binary variables, one for the allele in the paternal haplotype and the other for the allele in the maternal haplotype. A pedigree with N nonfounders contains $2N$ potentially informative gametes, so that the pattern of inheritance at a single marker can be described by a vector of $2N$ elements. Each element of this inheritance vector, $\mathbf{v}$, describes the parental status of the allele at the marker in one gamete: it is arbitrarily defined as 0 if the allele is paternal and 1 if the allele is maternal. There are therefore 2^{2N} possible inheritance vectors, each describing a different pattern of allele transmission at the locus. For each possible inheritance vector at the locus, the probability of the phenotypic data relevant to the locus can be decomposed as a sum of products of the conditional probabilities of phenotypes ($\mathbf{x}$) given genotype ($\mathbf{g}$) and the conditional probabilities of genotype ($\mathbf{g}$) given the inheritance vector ($\mathbf{v}$),

$$P(\mathbf{x}|\mathbf{v}) = \sum_{G} P(\mathbf{x}|\mathbf{g})P(\mathbf{g}|\mathbf{v}),$$

where the summation is taken over all combinations of possible genotypes ($\mathbf{G}$) at the locus. This probability is therefore a function of the penetrance and population parameters of the locus. The 2^{2N} such probabilities, one for each possible inheritance vector, can be arranged as the diagonal elements of a square $2^{2N} \times 2^{2N}$ matrix, denoted as $\mathbf{Q}$. Since all 2^{2N} possible inheritance vectors are equally likely prior to the consideration of the phenotypic data, the likelihood of the data at a single locus is proportional to the sum of these 2^{2N} probabilities. This can be written in matrix form as

$$P(\mathbf{x}) \propto \mathbf{1}^T\mathbf{Q}\mathbf{1},$$

where $\mathbf{x}$ represents the phenotypic data at the locus and $\mathbf{1}$ is a column vector with all 2^{2N} elements equal to 1. When k ordered markers are considered jointly in a multipoint analysis, the joint likelihood can be factored as

$$P(\mathbf{x}_1, \mathbf{x}_2, ..., \mathbf{x}_k) = P(\mathbf{x}_1)P(\mathbf{x}_1|\mathbf{x}_2)P(\mathbf{x}_3|\mathbf{x}_1, \mathbf{x}_2)...P(\mathbf{x}_k|\mathbf{x}_1, \mathbf{x}_2, ..., \mathbf{x}_{k-1}),$$

where $\mathbf{x}_1, \mathbf{x}_2, ..., \mathbf{x}_k$ represent phenotypic data at the k ordered markers. The necessity of conditioning on all preceding markers can be avoided by recognizing that, at each locus, the probability of the phenotypic data is a function of its own inheritance vector, which is conditionally independent of the inheritance vectors of all preceding markers given the inheritance vector of the immediately preceding locus. In other words, the k inheritance vectors of the k ordered markers constitute a hidden Markov chain. The conditional probability of an inheritance vector $\mathbf{v}_{i+1}$ at markers $i+1$, given an inheritance vector $\mathbf{v}_i$ at markers i, is $r_i^j(1-r_i)^{2N-j}$, where r_i is the recombination fraction between markers i and $i+1$, and j is the number of changes in the elements of the inheritance vector from $\mathbf{v}_i$ to $\mathbf{v}_{i+1}$. There are $[2^{2N}]^2$ possible combinations of $\mathbf{v}_i$ and $\mathbf{v}_{i+1}$, each associated with a characteristic conditional probability, so that there are $[2^{2N}]^2$ conditional probabilities that can be arranged in a square transition matrix $\mathbf{T}_i$. The joint likelihood of the multilocus phenotype data is then given by

$$P(\mathbf{x}_1, \mathbf{x}_2, ..., \mathbf{x}_k) \propto \mathbf{1}^T\mathbf{Q}_1\mathbf{T}_1\mathbf{Q}_2\mathbf{T}_2...\mathbf{T}_{k-1}\mathbf{Q}_k\mathbf{1}.$$

This is the basic form of the Lander-Green algorithm, which has various refinements added to reduce the number of necessary arithmetic operations.

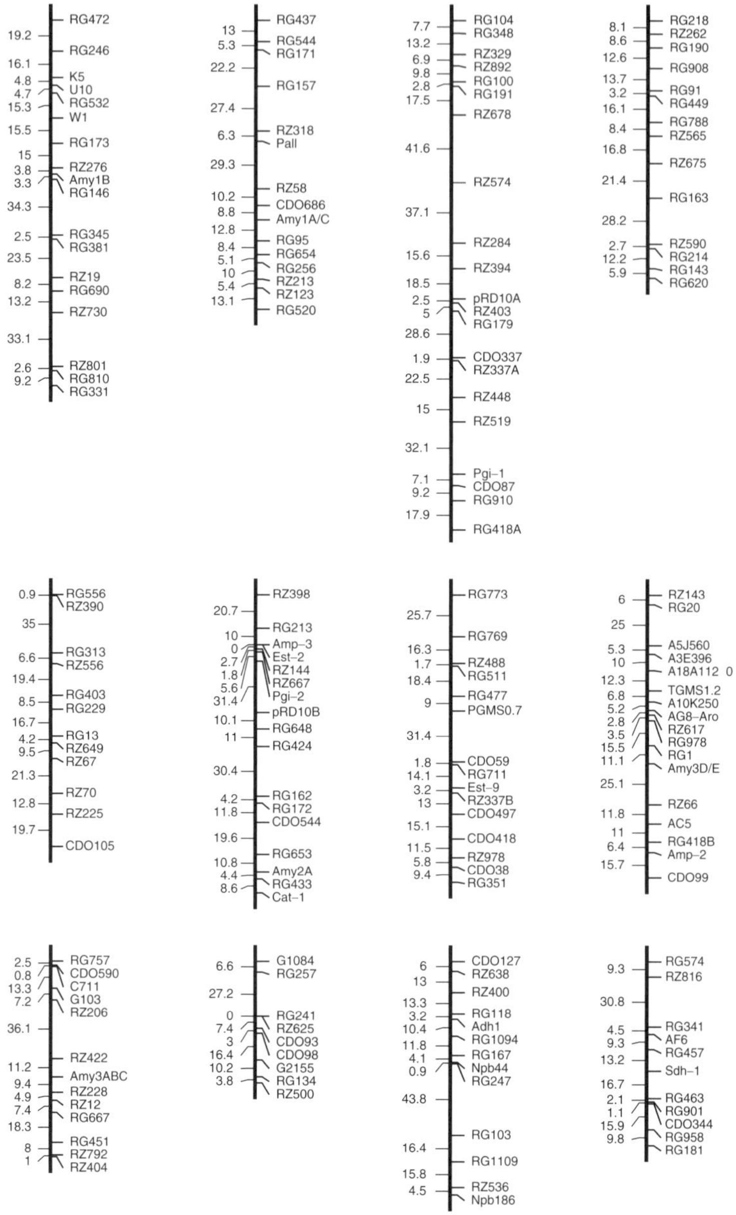

Fig. 3.3. Genetic linkage maps constructed from 135 RFLP and 40 isozyme and RAPD markers for 123 DH plants derived from tall Azucena and short IR64 parents.

Example 3.9. (**Genetic Linkage Map**). Revisit Example 3.1. One hundred twenty-three DH plants derived from two inbred lines, semi-dwarf IR64 and tall Azucena, were genotyped for a total of 175 polymorphic markers (Huang et al. 1997). Two-point analysis was performed to estimate pairwise recombination fractions for these 175 markers. Based on a cluster analysis of the 175 $\times$ 175 matrix for recombination fractions, these markers are sorted into 12 different groups, each representing a rice chromosome (Fig. 3.3).

The Lander-Green algorithm was used to determine the best order for markers clustered in each group (Fig. 3.3). The genetic distances between all adjacent markers were estimated and the corresponding genetic distances were calculated using a map function. In this example, the Kosambi map function was used. In Fig. 3.3 are given the genetic distances and marker names at the left and right sides of chromosomes.

4

A General Model for Linkage Analysis in Controlled Crosses

4.1 Introduction

Statistical methods for linkage analysis in a backcross or F_2 population derived from two inbred lines were described in the previous chapter. An advantage of linkage analysis using these inbred line crosses is that the parental linkage phase between different genetic loci is known and therefore the patterns of marker segregation can be determined and the linkage measured in terms of the recombination fraction tested. However, this inbred line-based analysis is not appropriate for outcrossing species in which it is not possible to generate homozygous lines through successive inbreeding.

Outcrossing populations have two significant characteristics that make their linkage analyses qualitatively different from those in inbred line crosses. The first characteristic is that the number of alleles and the inheritance mode of markers generally vary from locus to locus. Some markers may have more alleles than others, some markers are codominant, whereas others are dominant; and some markers are heterozygous in one parent but fixed in the other parent, whereas the opposite can be true for other markers. The second characteristic is the uncertainty about linkage phases between different loci. Because the estimation of the linkage between different markers relies upon information about linkage phases, it is essential to determine a correct linkage phase prior to the linkage estimation.

A traditional strategy for estimating linkage phases is to account for all the possible linkage phases for given marker pairs and choose the most likely one based on the minimum recombination fraction and maximum likelihood value (Ritter et al. 1990; Ritter and Salamini 1996; Maliepaard et al. 1997; Ridout et al. 1998). However, this strategy is not always statistically effective because the minimum estimate of the recombination fraction may be obtained from an incorrect linkage phase. Wu et al. (2002b) derived a general algorithm for simultaneously estimating linkage and parental linkage phases over all linked molecular markers of any kind in a full-sib family derived from two outbred parents. Lu et al. (2004) proposed a unifying model for characterizing the linkage, parental linkage phase, and gene order for any type of marker. In this chapter, a general framework for the simultaneous estimation of the linkage and linkage phases is presented that can be viewed as a generalization of linkage analysis in inbred line crosses. Much of our presentation is derived from Lu et al.'s (2004) paper.

4.2 Fully Informative Markers: A Diplotype Model

In a full-sib family derived from two parents, P and Q, of an outcrossing species, up to four marker alleles, besides a null allele, may be segregating at a single locus. Furthermore, the number of alleles may vary over loci. We assume that each of the marker alleles, symbolized by a, b, c, and d, is codominant with respect to each other but dominant with respect to the null allele, symbolized by o. We assume that all markers undergo Mendelian segregation without distortion. Depending on how different alleles are combined in the two parents used for the cross, there exist a total of 18 possible cross types for a marker locus (Table 3.2). In Section 3.4, we have shown how these cross types are classified into seven groups based on both parental and offspring marker band patterns. Segregation analysis allows us to determine a likely cross type from raw data.

In this section, we will present a general model for fully informative markers in which there are four phenotypically distinguishable genotypes in the full-sib family. The models for linkage analysis based on partially informative markers, in which some different genotypes are phenotypically identical, will be discussed in the subsequent sections.

4.2.1 Two-Point Analysis

Recall the definition of linkage phase (Definition 1.1). Linkage phase describes the configuration of alleles at a pair of heterozygous loci on homologous chromosomes in a single parent. The linkage phase between any two linked markers can be determined if we know what alternative allele one of the homologous chromosomes carries for each marker in a parent. Thus, the question of determining linkage phase becomes a question of labelling parental chromosomes using the alleles at given markers. The association of the marker alleles and homologous chromosomes can be anchored by calculating the probabilities of the genotypes of a marker conditional on the state at linked markers in the full-sib family.

Consider two fully informative markers, **A** and **B**, in a full-sib family. For the first marker **A**, the parental chromosomes can be arbitrarily labelled by its alleles. Assume that the parental chromosomes for marker **A** are labelled as $A_1| \quad |A_2$ (or $1| \quad |2$ for simplicity) for parent P and $A_3| \quad |A_4$ (or $3| \quad |4$ for simplicity) for parent Q, where $| \quad |$ stands for two homologous chromosomes on the left and right, respectively. The cross of parents P and Q leads to four different progeny genotypes at this marker, A_1A_3, A_1A_4, A_2A_3, and A_2A_4 or 13, 14, 23, and 24. The linkage phase between the alleles of markers **A** and **B** can be determined by assigning the alternative alleles of marker **B** to a different homologous chromosome given the defined label of marker **A**. For each parent, there are two possible linkage phases. Thus, when the two parents are crossed, four phase combinations are possible, one of which can be schematically expressed as

$$\mathbf{\Phi} = \begin{array}{c} \\ \mathbf{A} \\ \mathbf{B} \end{array} \begin{array}{ccc} \mathrm{P} & \times & \mathrm{Q} \\ 1 \| 2 & \times & 3 \| 4 \\ 1 \| 2 & & 3 \| 4 \end{array} \quad \begin{array}{c} \mathrm{P} \times \mathrm{Q} \\ \text{or } [11][22] \times [33][44]. \end{array} \tag{4.1}$$

For a particular parent, the combination of phased chromosomes is called a *parental diplotype*, which is symbolized by $[\cdot\cdot][\cdot\cdot]$, as shown by display (4.1). Let r be the recombination fraction between the two markers. Assuming that the diplotypes for the two parents are known, as shown above, the cosegregation pattern of the two markers can be expressed in matrix notation as

$$\mathbf{H} = \begin{array}{c} \\ 13 \\ 14 \\ 23 \\ 24 \end{array} \begin{array}{c} \begin{array}{cccc} 13 & 14 & 23 & 24 \end{array} \\ \begin{bmatrix} \frac{(1-r)^2}{4} & \frac{r(1-r)}{4} & \frac{r(1-r)}{4} & \frac{r^2}{4} \\ \frac{r(1-r)}{4} & \frac{(1-r)^2}{4} & \frac{r^2}{4} & \frac{r(1-r)}{4} \\ \frac{r(1-r)}{4} & \frac{r^2}{4} & \frac{(1-r)^2}{4} & \frac{r(1-r)}{4} \\ \frac{r^2}{4} & \frac{r(1-r)}{4} & \frac{r(1-r)}{4} & \frac{(1-r)^2}{4} \end{bmatrix} \end{array}, \tag{4.2}$$

where each cell represents a two-marker genotype in the full-sib =progeny. The columns correspond to marker **A**, whereas the rows correspond to marker **B**.

The expected number of recombination events (i.e., the number of r) occurring between the two markers can also be expressed in matrix notation, as

$$\mathbf{D} = \begin{array}{c} \\ 13 \\ 14 \\ 23 \\ 24 \end{array} \begin{array}{c} \begin{array}{cccc} 13 & 14 & 23 & 24 \end{array} \\ \begin{bmatrix} 0 & 1 & 1 & 2 \\ 1 & 0 & 2 & 1 \\ 1 & 2 & 0 & 1 \\ 2 & 1 & 1 & 0 \end{bmatrix} \end{array}. \tag{4.3}$$

Let

$$\mathbf{n} = (n_{j_1 j_2})_{4\times 4} = \begin{array}{c} \\ 13 \\ 14 \\ 23 \\ 24 \end{array} \begin{array}{c} \begin{array}{cccc} 13 & 14 & 23 & 24 \end{array} \\ \begin{bmatrix} n_{11} & n_{12} & n_{13} & n_{14} \\ n_{21} & n_{22} & n_{23} & n_{24} \\ n_{31} & n_{32} & n_{33} & n_{34} \\ n_{41} & n_{42} & n_{43} & n_{44} \end{bmatrix} \end{array} \tag{4.4}$$

be the matrix for the observations of progeny, where $j_1, j_2 = 1$ for 13, 2 for 14, 3 for 23, or 4 for 34 denote the marker phenotypes at **A** and **B**, respectively. Note that a marker has a particular "phenotype" determined by its genotype. With $n_{j_1 j_2}$ following a multinomial distribution, the likelihood function of marker genotypes under the parental diplotype combination shown in display (4.1) is expressed as

$$L(r|\mathbf{n}) \propto r^{(2n_2+n_3+n_4)}(1-r)^{(2n_1+n_3+n_4)}, \tag{4.5}$$

where

$$
\begin{aligned}
n_1 &= n_{11} + n_{22} + n_{33} + n_{44}, \\
n_2 &= n_{14} + n_{23} + n_{32} + n_{41}, \\
n_3 &= n_{12} + n_{21} + n_{34} + n_{43}, \\
n_4 &= n_{13} + n_{31} + n_{24} + n_{42}.
\end{aligned} \tag{4.6}
$$

The MLEs of the recombination fraction r with their large-sample variances are thus

$$
\begin{aligned}
\hat{r} &= \frac{2n_2 + n_3 + n_4}{2n}, \\
\text{Var}\ (\hat{r}) &= \frac{\hat{r}(1-\hat{r})}{n}.
\end{aligned}
$$

The hypothesis about the existence of the linkage can be formulated as

$$
H_0 : r = 0.5 \quad \text{vs.} \quad H_1 < 0.5, \tag{4.7}
$$

where H_0 corresponds to the $r = 0.5$; i.e., no significant linkage exists.

The test statistics for testing the hypotheses (4.7) are calculated as the log-likelihood ratio (LR) of the full model over the reduced model:

$$
\text{LR} = -2\log\left[\frac{L(\hat{r}|\mathbf{n})}{L(r=0.5|\mathbf{n})}\right].
$$

The test statistic LR can be viewed as being asymptotically χ^2-distributed with one degree of freedom.

Example 4.1. Assume a full-sib family derived from two outbred parents with known diplotypes [11][22] and [33][44], respectively. Two fully informative markers **A** and **B** are genotyped for each full-sib, with observations of a total of 16 two-marker genotypes as follows:

$$
\mathbf{n} = \begin{array}{c} \\ 13 \\ 14 \\ 23 \\ 24 \end{array}
\begin{array}{c} \begin{array}{cccc} 13 & 14 & 23 & 24 \end{array} \\
\left[\begin{array}{cccc}
n_{11} = 24 & n_{12} = 8 & n_{13} = 7 & n_{14} = 1 \\
n_{21} = 11 & n_{22} = 36 & n_{23} = 4 & n_{24} = 8 \\
n_{31} = 7 & n_{32} = 0 & n_{33} = 35 & n_{34} = 2 \\
n_{41} = 2 & n_{42} = 6 & n_{43} = 12 & n_{44} = 37
\end{array}\right] \end{array}.
$$

Based on the expected frequency for each cell in terms of the recombination fraction r between the two markers, we can formulate a likelihood function as described by equation (4.5). It is not difficult to derive the maximum likelihood estimator of r as

$$
\begin{aligned}
\hat{r} &= \frac{2n_2 + n_3 + n_4}{2n} \\
&= \frac{2 \times 7 + 33 + 28}{200 \times 2} = 0.1875,
\end{aligned} \tag{4.8}
$$

where $n_2 = 1 + 4 + 0 + 2 = 7$, $n_3 = 8 + 11 + 2 + 12 = 33$, and $n_4 = 7 + 7 + 8 + 6 = 28$ are defined by equation (4.6).

The test statistic

$$\text{LR} = -2[\log(L(\hat{r}|\mathbf{n})) - \log(L(r = 0.5|\mathbf{n}))] = -2(-193.0 + 277.3) = 168.6,$$

which is greater than $\chi^2_{0.05}(1) = 3.84$. Therefore, the linkage between the two markers is significant.

It can be seen from the above that the estimate of r for two fully informative markers can be obtained with only one step and does not need the iterative EM algorithm at all. The implementation of the EM algorithm below, with equations (4.12) and (4.13), aims to derive a general framework for estimating the linkage between any types of markers, including partially informative ones.

4.2.2 A More General Formulation

Let G_{j_1} and G_{j_2} $(j_1, j_2 = 1, 2, 3, 4)$ denote the four progeny genotypes in the order given in the matrices **H** and **D** above for markers **A** and **B**, respectively. Assuming that n offspring in the full-sib family are independent, we rewrite the likelihood of the marker data **n**, given by equation (4.5), under the parental diplotype combination (display (4.1)) as

$$\begin{aligned} L(r|\mathbf{n}) &= \prod_{i=1}^{n} L_i(r|\mathbf{n}) \\ &= \prod_{i=1}^{n} \sum_{j_1=1}^{4} \sum_{j_2=1}^{4} x_{ij_1} P(G_{j_1} G_{j_2}) x_{ij_2}, \end{aligned} \tag{4.9}$$

where x_{ij_k} is the indicator variable describing the j_kth genotype of marker $\mathbf{M}^k$ for offspring i, which is one if the marker genotype observed is compatible with G_{j_k} and zero otherwise, and $P(G_{j_1} G_{j_2})$ is the joint probability of the j_1th genotype of marker **A** and the j_2th genotype of marker **B** (as in the matrix (4.2)). Equation (4.9) can be written in matrix form as

$$L(r|\mathbf{n}) = \prod_{i=1}^{n} \mathbf{m}_{ij_1}^{\mathrm{T}} \mathbf{H} \mathbf{m}_{ij_2}, \tag{4.10}$$

where $\mathbf{m}_{ij_k}$ is the four-dimensional vector of the indicator variable x_{ij_k} for marker $\mathbf{M}^k$.

The likelihood function of observable marker phenotypes (**n**) given by equation (4.10) is constructed on the basis of the recombination fraction matrix (**H**) derived from distinguishable genotypes for fully informative markers. We define an incidence matrix **I** that relates the marker genotypes **H** to marker phenotypes **P**. Thus, we have

$$\begin{aligned} L(r|\mathbf{n}) &= \prod_{i=1}^{n} \mathbf{m}_{ij_1}^{\mathrm{T}} (\mathbf{I}_{b_1}^{\mathrm{T}} \mathbf{H} \mathbf{I}_{b_2}) \mathbf{m}_{ij_2} \\ &= \prod_{i=1}^{N} \mathbf{m}_{ij_1}^{\mathrm{T}} \mathbf{P} \mathbf{m}_{ij_2}, \end{aligned} \tag{4.11}$$

where b_k is the number of distinguishable genotypes (phenotypes) in the offspring at marker $\mathbf{M}^k$, which is 4 for fully informative markers, $\mathbf{P} = \mathbf{I}_{b_1}^{\mathrm{T}} \mathbf{H} \mathbf{I}_{b_2}$ is a $(b_1 \times b_2)$ matrix of the joint phenotype probability of the two markers, and $\mathbf{I}_{b_k}$ is a $(4 \times b_k)$ incidence matrix that is designed to specify the segregation pattern of a marker type under a given parental diplotype combination. For fully informative markers, we have

$$\mathbf{I}_4 = \begin{bmatrix} 1 & 0 & 0 & 0 \\ 0 & 1 & 0 & 0 \\ 0 & 0 & 1 & 0 \\ 0 & 0 & 0 & 1 \end{bmatrix}$$

when the parental diplotype combination (4.1) is considered. As seen earlier, similar incidence matrices can also be designed for other marker cross types (Table 3.2) and other diplotype combinations.

Wu et al. (2002b) presented a general method for estimating the recombination fraction between any marker types by maximizing the log-likelihood function of equation (4.11). This method was implemented with the EM algorithm, with the procedure given as follows.

E Step: At step τ, using the matrix $\mathbf{H}$ based on the current estimate $r^{(\tau)}$, calculate the expected number of recombination events between markers $\mathbf{A}$ and $\mathbf{B}$ for offspring i under a parental diplotype combination,

$$D_{ij_1j_2}^{(\tau+1)} = \frac{\mathbf{m}_{ij_1}^{\mathrm{T}} [\mathbf{I}_{b_1}^{\mathrm{T}} (\mathbf{H} \circ \mathbf{D}) \mathbf{I}_{b_2}] \mathbf{m}_{ij_2}}{\mathbf{m}_{ij_1}^{\mathrm{T}} \mathbf{P} \mathbf{m}_{ij_2}}, \tag{4.12}$$

where $\circ$ denotes an elementwise product of two matrices.

M Step: Calculate $r^{(\tau+1)}$ under the given parental diplotype combination using the equation

$$r^{(\tau+1)} = \frac{1}{2n} \sum_{i=1}^{n} \sum_{j_1=1}^{b_1} \sum_{j_2=1}^{b_2} D_{ij_1j_2}^{(\tau+1)}. \tag{4.13}$$

These iterations are repeated between equations (4.12) and (4.13) until r converges to a stable value. This stable value represents the MLE of the recombination fraction between markers $\mathbf{A}$ and $\mathbf{B}$ under the given parental diplotype combination.

For any marker pair, we will have multiple parental diplotype combinations under each of which the recombination fraction is estimated and the plug-in likelihood value calculated. A most likely diplotype combination can be obtained on the basis of the minimum recombination fraction and maximum likelihood.

4.2.3 Three-Point Analysis

Statistical algorithms for estimating the recombination fraction based on two-point analysis may not be powerful, especially in the case where partially informative markers are involved. Ridout et al. (1998) demonstrated an example in which three-point analysis can detect more linkage relationships between three loci than two-point analysis.

Consider three markers in the order **A-B-C**. Relative to marker **A**, marker **B** has two possibilities to assign its alleles to two homologous chromosomes. Similarly, there are also two such allelic configurations for marker **C** when marker **B** is fixed. Thus, for one parent, there are $2 \times 2 = 4$ possible diplotypes. We assume that the diplotypes of two parents, P and Q, are known, as shown below:

$$\begin{array}{cc} & \text{P} \times \text{Q} \\ \Phi = & \begin{array}{c} \mathbf{A}\,1 \\ \mathbf{B}\,1 \\ \mathbf{C}\,1 \end{array}\left|\right|\begin{array}{c} 2 \quad\;\; 3 \\ 2 \times 3 \\ 2 \quad\;\; 3 \end{array}\left|\right|\begin{array}{c} 4 \\ 4 \\ 4 \end{array} \end{array} \text{ or } \begin{array}{c} \text{P} \times \text{Q} \\ [111][222] \times [333][444]. \end{array} \tag{4.14}$$

Let $r_{\mathbf{AB}}$, $r_{\mathbf{BC}}$, and $r_{\mathbf{AC}}$ be the recombination fractions between markers **A** and **B**, between markers **B** and **C**, and between markers **A** and **C**, respectively. These recombination fractions are associated with the probabilities with which a crossover occurs between markers **A** and **B** and between markers **B** and **C**. The event when a crossover or no crossover occurs in each interval is denoted by $\mathcal{G}_{11}$ and $\mathcal{G}_{00}$, respectively, whereas the event when a crossover occurs only in the first interval or in the second interval is denoted by $\mathcal{G}_{10}$ and $\mathcal{G}_{01}$, respectively. The probabilities of these events are denoted by g_{00}, g_{01}, g_{10}, and g_{11}, respectively, whose sum equals 1. According to the definition of the recombination fraction as the probability of a crossover between a pair of loci, it is clear that

$$\begin{aligned} r_{\mathbf{AB}} &= g_{10} + g_{11} \\ r_{\mathbf{BC}} &= g_{01} + g_{11} \\ r_{\mathbf{AC}} &= g_{01} + g_{10} \end{aligned} \tag{4.15}$$

and

$$\begin{aligned} g_{11} &= \tfrac{1}{2}(r_{\mathbf{AB}} + r_{\mathbf{BC}} - r_{\mathbf{AC}}), \\ g_{10} &= \tfrac{1}{2}(r_{\mathbf{AB}} + r_{\mathbf{AC}} - r_{\mathbf{BC}}), \\ g_{01} &= \tfrac{1}{2}(r_{\mathbf{BC}} + r_{\mathbf{AC}} - r_{\mathbf{AB}}), \\ g_{00} &= 1 - \tfrac{1}{2}(r_{\mathbf{AB}} + r_{\mathbf{AC}} + r_{\mathbf{BC}}). \end{aligned} \tag{4.16}$$

The cross between the two parents shown by display 4.14 yield a total of $4 \times 4 \times 4 = 64$ genotypes for three fully informative markers. The frequencies of each genotype can be expressed in terms of g_{00}, g_{01}, g_{10}, and g_{11} (see Table 4.1) and therefore $r_{\mathbf{AB}}$, $r_{\mathbf{BC}}$, and $r_{\mathbf{AC}}$ as shown by equation 4.16. For each genotype, the number of crossovers that occur between each pair of adjacent markers is tabulated in Table 4.2, from which the closed forms for estimating g's can be derived. In the next example, we provide a procedure for estimating g's, and therefore the recombination fractions with equation 4.15.

Table 4.1. Joint probability matrix (**H**) among three markers in terms of the number of crossovers between **A** and **B** as well as between **B** and **C** under a particular parental diplotype combination as shown by display (4.14).

Marker	Marker	Marker **C**			
A	**B**	13	14	23	24
13	13	g_{00}^2	$g_{00}g_{01}$	$g_{01}g_{00}$	g_{01}^2
13	14	$g_{00}g_{11}$	$g_{00}g_{10}$	$g_{01}g_{11}$	$g_{01}g_{10}$
13	23	$g_{11}g_{00}$	$g_{11}g_{01}$	$g_{10}g_{00}$	$g_{10}g_{01}$
13	24	g_{11}^2	$g_{11}g_{10}$	$g_{10}g_{11}$	g_{10}^2
14	13	$g_{00}g_{10}$	$g_{00}g_{11}$	$g_{01}g_{10}$	$g_{01}g_{11}$
14	14	$g_{00}g_{01}$	g_{00}^2	g_{01}^2	$g_{01}g_{00}$
14	23	$g_{11}g_{10}$	g_{11}^2	g_{10}^2	$g_{10}g_{11}$
14	24	$g_{11}g_{01}$	$g_{11}g_{00}$	$g_{10}g_{01}$	$g_{10}g_{00}$
23	13	$g_{10}g_{00}$	$g_{10}g_{01}$	$g_{11}g_{00}$	$g_{11}g_{01}$
23	14	$g_{10}g_{11}$	g_{10}^2	g_{11}^2	$g_{11}g_{10}$
23	23	$g_{01}g_{00}$	g_{01}^2	g_{00}^2	$g_{00}g_{01}$
23	24	$g_{01}g_{11}$	$g_{01}g_{10}$	$g_{00}g_{11}$	$g_{00}g_{10}$
24	13	g_{10}^2	$g_{10}g_{11}$	$g_{11}g_{10}$	g_{11}^2
24	14	$g_{10}g_{01}$	$g_{10}g_{00}$	$g_{11}g_{01}$	$g_{11}g_{00}$
24	23	$g_{01}g_{10}$	$g_{01}g_{11}$	$g_{00}g_{10}$	$g_{00}g_{11}$
24	24	g_{01}^2	$g_{01}g_{00}$	$g_{00}g_{01}$	g_{00}^2

Example 4.2. Three fully informative markers, **A**, **B**, and **C** are genotyped for a full-sib family derived from two outbred parents with known diplotypes [111][222] and [333][444], respectively. Table 4.3 tabulates observations of a total of 64 three-marker genotypes.

Based on the expected frequency for each cell in terms of the crossover probabilities (g's) between the three markers as shown in Table 4.1, we can formulate a likelihood function to be described by equation (4.21). We can derive the maximum likelihood estimator of g's as

$$\hat{g}_{00} = \frac{2(n_{111} + n_{222} + n_{333} + n_{444}) + n_1}{2n} = \frac{2 \times 225 + 147}{2 \times 400} = 0.7462, \tag{4.17}$$

where $n_1 = n_{121} + n_{131} + n_{211} + n_{221} + n_{311} + n_{331} + n_{112} + n_{122} + n_{212} + n_{242} + n_{422} + n_{442} + n_{113} + n_{133} + n_{313} + n_{343} + n_{433} + n_{443} + n_{224} + n_{244} + n_{334} + n_{344} + n_{424} + n_{434}$;

Table 4.2. Matrices ($\mathbf{G}_{00}$, $\mathbf{G}_{01}$, $\mathbf{G}_{10}$, and $\mathbf{G}_{11}$) for interval-specific crossover events among three markers **A**, **B**, and **C** under a particular parental diplotype combination as shown by display (4.14).

		Marker **C**															
Marker	Marker	$\mathbf{G}_{00}$				$\mathbf{G}_{01}$				$\mathbf{G}_{10}$				$\mathbf{G}_{11}$			
A	**B**	13	14	23	24	13	14	23	24	13	14	23	24	13	14	23	24
13	13	2	1	1	0	0	1	1	2	0	0	0	0	0	0	0	0
13	14	1	1	0	0	0	0	1	1	0	1	0	1	1	0	1	0
13	23	1	0	1	0	0	1	0	1	0	0	1	1	1	1	0	0
13	24	0	0	0	0	0	0	0	0	0	1	1	2	2	1	1	0
14	13	1	1	0	0	0	0	1	1	1	0	1	0	0	1	0	1
14	14	1	2	0	1	1	0	2	1	0	0	0	0	0	0	0	0
14	23	0	0	0	0	0	0	0	0	1	0	2	1	1	2	0	1
14	24	0	1	0	1	1	0	1	0	0	0	1	1	1	1	0	0
23	13	1	0	1	0	0	1	0	1	1	1	0	0	0	0	1	1
23	14	0	0	0	0	0	0	0	0	1	2	0	1	1	0	2	1
23	23	1	0	2	1	1	2	0	1	0	0	0	0	0	0	0	0
23	24	0	0	1	1	1	1	0	0	0	1	0	1	1	0	1	0
24	13	0	0	0	0	0	0	0	0	2	1	1	0	0	1	1	2
24	14	0	1	0	1	1	0	1	0	1	1	0	0	0	0	1	1
24	23	0	0	1	1	1	1	0	0	1	0	1	0	0	1	0	1
24	24	0	1	1	2	2	1	1	0	0	0	0	0	0	0	0	0

$$\hat{g}_{01} = \frac{2(n_{441} + n_{332} + n_{223} + n_{114}) + n_2}{2n} = \frac{2 \times 5 + 68}{2 \times 400} = 0.0975, \tag{4.18}$$

where $n_2 = n_{221} + n_{241} + n_{331} + n_{341} + n_{421} + n_{431} + n_{112} + n_{132} + n_{312} + n_{342} + n_{432} + n_{442} + n_{113} + n_{123} + n_{213} + n_{243} + n_{423} + n_{443} + n_{124} + n_{134} + n_{214} + n_{224} + n_{314} + n_{334}$;

$$\hat{g}_{10} = \frac{2(n_{441} + n_{332} + n_{223} + n_{114}) + n_3}{2n} = \frac{2 \times 5 + 101}{2 \times 400} = 0.1388, \tag{4.19}$$

where $n_3 = n_{211} + n_{231} + n_{311} + n_{321} + n_{421} + n_{431} + n_{122} + n_{142} + n_{312} + n_{342} + n_{412} + n_{422} + n_{133} + n_{143} + n_{213} + n_{243} + n_{413} + n_{433} + n_{124} + n_{134} + n_{234} + n_{244} + n_{324} + n_{344}$;

$$\hat{g}_{11} = \frac{2(n_{441} + n_{332} + n_{223} + n_{114}) + n_4}{2n} = \frac{2 + 14}{2 \times 400} = 0.0175, \tag{4.20}$$

Table 4.3. Observations for three-marker genotypes in a full-sib family derived from display (4.14).

Marker	Marker	Marker **C**			
A	**B**	13	14	23	24
13	13	$n_{111} = 49$	$n_{112} = 3$	$n_{113} = 7$	$n_{114} = 1$
13	14	$n_{121} = 2$	$n_{122} = 6$	$n_{123} = 0$	$n_{124} = 1$
13	23	$n_{131} = 1$	$n_{132} = 1$	$n_{133} = 15$	$n_{134} = 1$
13	24	$n_{141} = 0$	$n_{142} = 1$	$n_{143} = 0$	$n_{144} = 3$
14	13	$n_{211} = 17$	$n_{212} = 2$	$n_{213} = 3$	$n_{214} = 0$
14	14	$n_{221} = 6$	$n_{222} = 64$	$n_{223} = 4$	$n_{224} = 9$
14	23	$n_{231} = 0$	$n_{232} = 0$	$n_{233} = 0$	$n_{234} = 0$
14	24	$n_{241} = 0$	$n_{242} = 0$	$n_{243} = 2$	$n_{244} = 11$
23	13	$n_{311} = 13$	$n_{312} = 1$	$n_{313} = 0$	$n_{314} = 1$
23	14	$n_{321} = 1$	$n_{322} = 1$	$n_{323} = 0$	$n_{324} = 0$
23	23	$n_{331} = 11$	$n_{332} = 0$	$n_{333} = 53$	$n_{334} = 5$
23	24	$n_{341} = 0$	$n_{342} = 2$	$n_{343} = 1$	$n_{344} = 11$
24	13	$n_{411} = 1$	$n_{412} = 0$	$n_{413} = 1$	$n_{414} = 0$
24	14	$n_{421} = 2$	$n_{422} = 4$	$n_{423} = 0$	$n_{424} = 1$
24	23	$n_{431} = 1$	$n_{432} = 0$	$n_{433} = 8$	$n_{434} = 2$
24	24	$n_{441} = 0$	$n_{442} = 6$	$n_{443} = 6$	$n_{444} = 59$

where $n_4 = n_{121} + n_{131} + n_{231} + n_{241} + n_{321} + n_{341} + n_{132} + n_{142} + n_{212} + n_{242} + n_{412} + n_{432} + n_{123} + n_{143} + n_{313} + n_{343} + n_{413} + n_{423} + n_{214} + n_{234} + n_{314} + n_{324} + n_{424} + n_{434}$. The recombination fractions are then estimated as

$$\hat{r}_{\mathbf{AB}} = 0.1563, \quad \hat{r}_{\mathbf{BC}} = 0.1150, \quad \hat{r}_{\mathbf{AC}} = 0.2363.$$

Since the estimates of g's for three fully informative markers can be obtained with only one step, the iterative EM algorithm is not needed. With equations (4.22) and (4.23) below, we derive a general EM framework for estimating the linkage between any types of markers, including partially informative ones.

4.2.4 A More General Formulation

As in a two-point analysis, the likelihood of the crossover probability, $\mathbf{g} = (g_{00}, g_{01}, g_{10}, g_{11})$, given the three-marker data under the parental diplotype combination of equation (4.14) is expressed as

$$L(\mathbf{g}|(\mathbf{n}) = \prod_{i=1}^{n} (\mathbf{m}_{ij_1}^{\mathrm{T}} \otimes \mathbf{m}_{ij_2}^{\mathrm{T}})(\mathbf{I}_{p_1}^{\mathrm{T}} \otimes \mathbf{I}_{p_2}^{\mathrm{T}})(\mathbf{H}\mathbf{I}_{p_3})\mathbf{m}_{ij_3}, \tag{4.21}$$

where $j_1, j_2, j_3 = 1$ for 13, 2 for 14, 3 for 23, and 4 for 34 stand for genotypes at markers **A**, **B**, and **C**, respectively, $\otimes$ denotes the Kronecker product, and the vectors and matrices are as defined for the two-point analysis. Table 4.1 tabulates the joint three-marker genotype probability matrix, $\mathbf{H} = (H_{j_1 j_2 j_3})_{16\times4}$, while Table 4.2 lists the matrices for the numbers of crossovers ($\mathbf{G}_{00}$, $\mathbf{G}_{01}$, $\mathbf{G}_{10}$, and $\mathbf{G}_{11}$) on two intervals **A-B** and **B-C** under the parental diplotype combination of display (4.14).

The EM algorithm is used to obtain the MLEs of the crossover probabilities (and therefore the recombination fractions) between the three markers. The general equations formulating the iteration of the $(\tau+1)$th EM step are given as follows

E Step: Calculate the expected number of interval-specific crossovers associated with $\mathbf{G}_{00} = (\mathcal{G}^{00}_{j_1 j_2 j_3})_{16\times4}$, $\mathbf{G}_{01} = (\mathcal{G}^{01}_{j_1 j_2 j_3})_{16\times4}$, $\mathbf{G}_{10} = (\mathcal{G}^{10}_{j_1 j_2 j_3})_{16\times4}$, and $\mathbf{G}_{11} = (\mathcal{G}^{11}_{j_1 j_2 j_3})_{16\times4}$, respectively, for offspring i under a given parental diplotype combination:

$$\mathcal{G}^{00(\tau+1)}_{ij_1j_2j_3} = \frac{[\mathbf{m}^{\mathrm{T}}_{ij_1} \otimes \mathbf{m}^{\mathrm{T}}_{ij_2}][\mathbf{I}^{\mathrm{T}}_{b_1} \otimes \mathbf{I}^{\mathrm{T}}_{b_2}][(\mathbf{G}^{(\tau)}_{00} \circ \mathbf{H}^{(\tau)})\mathbf{I}_{b_3}]\mathbf{m}_{ij_3}}{[\mathbf{m}^{\mathrm{T}}_{ij_1} \otimes \mathbf{m}^{\mathrm{T}}_{ij_2}][\mathbf{I}^{\mathrm{T}}_{b_1} \otimes \mathbf{I}^{\mathrm{T}}_{b_2}](\mathbf{H}^{(\tau)}\mathbf{I}_{b_3})\mathbf{m}_{ij_3}},$$

$$\mathcal{G}^{01(\tau+1)}_{ij_1j_2j_3} = \frac{[\mathbf{m}^{\mathrm{T}}_{ij_1} \otimes \mathbf{m}^{\mathrm{T}}_{ij_2}][\mathbf{I}^{\mathrm{T}}_{b_1} \otimes \mathbf{I}^{\mathrm{T}}_{b_2}][(\mathbf{G}^{(\tau)}_{01} \circ \mathbf{H}^{(\tau)})\mathbf{I}_{b_3}]\mathbf{m}_{ij_3}}{[\mathbf{m}^{\mathrm{T}}_{ij_1} \otimes \mathbf{m}^{\mathrm{T}}_{ij_2}][\mathbf{I}^{\mathrm{T}}_{b_1} \otimes \mathbf{I}^{\mathrm{T}}_{b_2}](\mathbf{H}^{(\tau)}\mathbf{I}_{b_3})\mathbf{m}_{ij_3}},$$

(4.22)

$$\mathcal{G}^{10(\tau+1)}_{ij_1j_2j_3} = \frac{[\mathbf{m}^{\mathrm{T}}_{ij_1} \otimes \mathbf{m}^{\mathrm{T}}_{ij_2}][\mathbf{I}^{\mathrm{T}}_{b_1} \otimes \mathbf{I}^{\mathrm{T}}_{b_2}][(\mathbf{G}^{(\tau)}_{10} \circ \mathbf{H}^{(\tau)})\mathbf{I}_{b_3}]\mathbf{m}_{ij_3}}{[\mathbf{m}^{\mathrm{T}}_{ij_1} \otimes \mathbf{m}^{\mathrm{T}}_{ij_2}][\mathbf{I}^{\mathrm{T}}_{b_1} \otimes \mathbf{I}^{\mathrm{T}}_{b_2}](\mathbf{H}^{(\tau)}\mathbf{I}_{b_3})\mathbf{m}_{ij_3}},$$

$$\mathcal{G}^{11(\tau+1)}_{ij_1j_2j_3} = \frac{[\mathbf{m}^{\mathrm{T}}_{ij_1} \otimes \mathbf{m}^{\mathrm{T}}_{ij_2}][\mathbf{I}^{\mathrm{T}}_{b_1} \otimes \mathbf{I}^{\mathrm{T}}_{b_2}][(\mathbf{G}_{11} \circ \mathbf{H}^{(\tau)})\mathbf{I}_{b_3}]\mathbf{m}_{ij_3}}{[\mathbf{m}^{\mathrm{T}}_{ij_1} \otimes \mathbf{m}^{\mathrm{T}}_{ij_2}][\mathbf{I}^{\mathrm{T}}_{b_1} \otimes \mathbf{I}^{\mathrm{T}}_{b_2}](\mathbf{H}^{(\tau)}\mathbf{I}_{b_3})\mathbf{m}_{ij_3}}.$$

M Step: Calculate the probabilities of interval-specific crossovers g_{00}, g_{01}, g_{10}, and g_{11} using

$$g^{(\tau+1)}_{00} = \frac{1}{2n}\sum_{i=1}^{n}\sum_{j_1=1}^{b_1}\sum_{j_2=1}^{b_2}\sum_{j_3=1}^{b_3}\mathcal{G}^{00(\tau+1)}_{ij_1j_2j_3},$$

(4.23)

$$g^{(\tau+1)}_{01} = \frac{1}{2n}\sum_{i=1}^{n}\sum_{j_1=1}^{b_1}\sum_{j_2=1}^{b_2}\sum_{j_3=1}^{b_3}\mathcal{G}^{01(\tau+1)}_{ij_1j_2j_3},$$

$$g^{(\tau+1)}_{10} = \frac{1}{2n}\sum_{i=1}^{n}\sum_{j_1=1}^{b_1}\sum_{j_2=1}^{b_2}\sum_{j_3=1}^{b_3}\mathcal{G}^{10(\tau+1)}_{ij_1j_2j_3},$$

$$g^{(\tau+1)}_{11} = \frac{1}{2n}\sum_{i=1}^{n}\sum_{j_1=1}^{b_1}\sum_{j_2=1}^{b_2}\sum_{j_3=1}^{b_3}\mathcal{G}^{11(\tau+1)}_{ij_1j_2j_3}.$$

The E and M steps are repeated among equations (4.22) and (4.23) until these probabilities converge to stable values. The MLEs of the g's can be transformed to give the MLEs of recombination fractions $r_{\mathbf{AB}}$, $r_{\mathbf{BC}}$, and $r_{\mathbf{AC}}$ because the MLEs are invariant under parameter transformation. Because all possible recombination fractions among the three markers are estimated, the three-point analysis provides important information about marker ordering. For example, if either $r_{\mathbf{AB}}$ or $r_{\mathbf{BC}}$ estimated under assumed marker order **A-B-C** is greater than $r_{\mathbf{AC}}$, then the assumed order is likely to be wrong based on Theorem 3.1. When every three adjacent markers from a linkage group are subject to the three-point analysis and the marker ordering has been confirmed, we can obtain two different estimates of the recombination fraction for the same marker interval (except for the intervals at the two ends of a linkage group). The best way to combine these estimates is to take a weighted mean, with the weights being the reciprocals of the variances of the two separate estimates (Ridout et al. 1998).

The most likely parental diplotype combination is determined among three markers by choosing the maximum of the likelihood values under all possible diplotype combinations. Only under the most likelydiplotype combination are the MLEs of the recombination fractions optimal.

4.3 Fully Informative Markers: A Genotype Model

In Section 4.2, we presented a general procedure for linkage analysis of fully informative markers in a full-sib family derived from two phased parents. Since the two crossed parents are assumed to have known diplotypes, the patterns of marker cosegregation in their offspring can be predicted. In practice, parental diplotypes cannot be directly observed. Rather, only genotypes can be observed, so that statistical models need to be developed for estimating the linkage based on observable genotypes. One approach for linkage analysis of genotypic data is to calculate the likelihood values under all possible parental diplotypes and estimate the recombination fractions under a most likely diplotype combination. In this section, a joint model that incorporates the probabilities of parental diplotypes is introduced.

4.3.1 Parental Diplotypes

Consider two ordered markers **A-B**. Each parent, P or Q, has two different diplotypes between the two markers. Thus, when the two parents are crossed, there are four diplotype combinations ($\boldsymbol{\Phi}$), expressed as

$$
\begin{array}{rcl}
 & & \mathrm{P} \quad \times \quad \mathrm{Q} \\
\mathbf{\Phi}_{11} & = & \begin{array}{c} \mathbf{A} \\ \mathbf{B} \end{array} \; \begin{array}{c|c|c} 1 & & 2 \\ 1 & & 2 \end{array} \times \begin{array}{c|c|c} 3 & & 4 \\ 3 & & 4 \end{array} \\[2ex]
\mathbf{\Phi}_{12} & = & \begin{array}{c} \mathbf{A} \\ \mathbf{B} \end{array} \; \begin{array}{c|c|c} 1 & & 2 \\ 1 & & 2 \end{array} \times \begin{array}{c|c|c} 3 & & 4 \\ 4 & & 3 \end{array} \\[2ex]
\mathbf{\Phi}_{21} & = & \begin{array}{c} \mathbf{A} \\ \mathbf{B} \end{array} \; \begin{array}{c|c|c} 1 & & 2 \\ 2 & & 1 \end{array} \times \begin{array}{c|c|c} 3 & & 4 \\ 3 & & 4 \end{array} \\[2ex]
\mathbf{\Phi}_{22} & = & \begin{array}{c} \mathbf{A} \\ \mathbf{B} \end{array} \; \begin{array}{c|c|c} 1 & & 2 \\ 2 & & 1 \end{array} \times \begin{array}{c|c|c} 3 & & 4 \\ 4 & & 3 \end{array}
\end{array}
\Bigg\} , \tag{4.24}
$$

where the first (ξ) and second subscripts (ζ) of $\mathbf{\Phi}$ denote two possible diplotypes of parent P and Q, respectively. Each of these diplotype combinations generates a different pattern of marker co-segregation, expressed in matrix notation with the recombination fraction r, as

$$
\begin{array}{l}
\mathbf{H}_{11} = \mathbf{A} \begin{array}{c} \\ 13 \\ 14 \\ 23 \\ 24 \end{array} \!\!\! \begin{array}{c} \begin{array}{cccc} \mathbf{B}\;13 & 14 & 23 & 24 \end{array} \\ \left[\begin{array}{cccc}
\frac{(1-r)^2}{4} & \frac{r(1-r)}{4} & \frac{r(1-r)}{4} & \frac{r^2}{4} \\
\frac{r(1-r)}{4} & \frac{(1-r)^2}{4} & \frac{r^2}{4} & \frac{r(1-r)}{4} \\
\frac{r(1-r)}{4} & \frac{r^2}{4} & \frac{(1-r)^2}{4} & \frac{r(1-r)}{4} \\
\frac{r^2}{4} & \frac{r(1-r)}{4} & \frac{r(1-r)}{4} & \frac{(1-r)^2}{4}
\end{array} \right] \end{array} , \\
\mathbf{H}_{12} = \mathbf{A} \begin{array}{c} 13 \\ 14 \\ 23 \\ 24 \end{array} \left[\begin{array}{cccc}
\frac{r(1-r)}{4} & \frac{(1-r)^2}{4} & \frac{r^2}{4} & \frac{r(1-r)}{4} \\
\frac{(1-r)^2}{4} & \frac{r(1-r)}{4} & \frac{r(1-r)}{4} & \frac{r^2}{4} \\
\frac{r^2}{4} & \frac{r(1-r)}{4} & \frac{r(1-r)}{4} & \frac{(1-r)^2}{4} \\
\frac{r(1-r)}{4} & \frac{r^2}{4} & \frac{(1-r)^2}{4} & \frac{r(1-r)}{4}
\end{array} \right] , \\
\mathbf{H}_{21} = \mathbf{A} \begin{array}{c} 13 \\ 14 \\ 23 \\ 24 \end{array} \left[\begin{array}{cccc}
\frac{r(1-r)}{4} & \frac{r^2}{4} & \frac{r^2}{4} & \frac{r(1-r)}{4} \\
\frac{r^2}{4} & \frac{r(1-r)}{4} & \frac{r(1-r)}{4} & \frac{r^2}{4} \\
\frac{(1-r)^2}{4} & \frac{r(1-r)}{4} & \frac{r(1-r)}{4} & \frac{(1-r)^2}{4} \\
\frac{r(1-r)}{4} & \frac{(1-r)^2}{4} & \frac{(1-r)^2}{4} & \frac{r(1-r)}{4}
\end{array} \right] ,
\end{array}
\tag{4.25}
$$

$$\mathbf{H}_{22} = \mathbf{A} \begin{array}{c} \\ 13 \\ 14 \\ 23 \\ 24 \end{array} \overset{\begin{array}{cccc} \mathbf{B}\ 13 & 14 & 23 & 24 \end{array}}{\begin{bmatrix} \frac{r^2}{4} & \frac{r(1-r)}{4} & \frac{r(1-r)}{4} & \frac{(1-r)^2}{4} \\ \frac{r(1-r)}{4} & \frac{r^2}{4} & \frac{(1-r)^2}{4} & \frac{r(1-r)}{4} \\ \frac{r(1-r)}{4} & \frac{(1-r)^2}{4} & \frac{r^2}{4} & \frac{r(1-r)}{4} \\ \frac{(1-r)^2}{4} & \frac{r(1-r)}{4} & \frac{r(1-r)}{4} & \frac{r^2}{4} \end{bmatrix}}.$$

The expected number of recombination events (i.e., the number of r) occurring between the two markers under different diplotype combinations can also be expressed in matrix notation, as

$$\mathbf{D}_{11} = \mathbf{A} \begin{array}{c} 13 \\ 14 \\ 23 \\ 24 \end{array} \overset{\begin{array}{cccc} \mathbf{B}\ 13 & 14 & 23 & 24 \end{array}}{\begin{bmatrix} 0 & 1 & 1 & 2 \\ 1 & 0 & 2 & 1 \\ 1 & 2 & 0 & 1 \\ 2 & 1 & 1 & 0 \end{bmatrix}},$$

$$\mathbf{D}_{12} = \mathbf{A} \begin{array}{c} 13 \\ 14 \\ 23 \\ 24 \end{array} \begin{bmatrix} 1 & 0 & 2 & 1 \\ 0 & 1 & 1 & 2 \\ 2 & 1 & 1 & 0 \\ 1 & 2 & 0 & 1 \end{bmatrix},$$

$$\mathbf{D}_{21} = \mathbf{A} \begin{array}{c} 13 \\ 14 \\ 23 \\ 24 \end{array} \begin{bmatrix} 1 & 2 & 0 & 1 \\ 2 & 1 & 1 & 0 \\ 0 & 1 & 1 & 2 \\ 1 & 0 & 2 & 1 \end{bmatrix},$$

$$\mathbf{D}_{22} = \mathbf{A} \begin{array}{c} 13 \\ 14 \\ 23 \\ 24 \end{array} \begin{bmatrix} 2 & 1 & 1 & 0 \\ 1 & 2 & 0 & 1 \\ 1 & 0 & 2 & 1 \\ 0 & 1 & 1 & 2 \end{bmatrix}.$$

The two markers are genotyped for the full-sib family with observations $\mathbf{n} = (n_{j_1j_2})_{4\times4}$ described by matrix (4.4). Thus, the likelihoods for the four diplotype combinations are expressed as

$$\begin{aligned} L_{11}(r|\mathbf{n}) &\propto r^{(2n_2+n_3+n_4)}(1-r)^{(2n_1+n_3+n_4)}, \\ L_{12}(r|\mathbf{n}) &\propto r^{(2n_4+n_1+n_2)}(1-r)^{(2n_3+n_1+n_2)}, \\ L_{21}(r|\mathbf{n}) &\propto r^{(2n_3+n_1+n_2)}(1-r)^{(2n_4+n_1+n_2)}, \\ L_{22}(r|\mathbf{n}) &\propto r^{(2n_1+n_3+n_4)}(1-r)^{(2n_2+n_3+n_4)} \end{aligned} \quad (4.26)$$

with n_1, n_2, n_3, and n_4 defined by equation (4.5).

It can be seen that the MLE of r under diplotype combination $\mathbf{\Phi}_{11}$ is equal to one minus the MLE of r under $\mathbf{\Phi}_{22}$, and the same relation holds between $\mathbf{\Phi}_{12}$ and $\mathbf{\Phi}_{21}$. Although there are identical plug-in likelihood values between $\mathbf{\Phi}_{11}$ and $\mathbf{\Phi}_{22}$ as well as between $\mathbf{\Phi}_{12}$ and $\mathbf{\Phi}_{21}$, one can still choose an appropriate MLE of r within each of these two pairs because one of a pair of the r MLEs is greater than 0.5. Traditional approaches for estimating the linkage and parental diplotypes are to estimate the recombination fractions and likelihood values under each of the four combinations and choose one legitimate estimate of r with a higher likelihood.

Example 4.3. Assume a full-sib family derived from two outbred parents. The two-marker genotypes for these two parents are 12/12 and 34/34, but we do not know about their diplotypes. Two fully informative markers **A** and **B** are genotyped for each full-sib, with observations of a total of 16 two-marker genotypes as follows:

$$\mathbf{n} = \begin{array}{c} \\ 13 \\ 14 \\ 23 \\ 24 \end{array} \begin{array}{c} \begin{array}{cccc} 13 & 14 & 23 & 24 \end{array} \\ \begin{bmatrix} n_{11}=3 & n_{12}=18 & n_{13}=2 & n_{14}=4 \\ n_{21}=21 & n_{22}=5 & n_{23}=4 & n_{24}=0 \\ n_{31}=1 & n_{32}=2 & n_{33}=6 & n_{34}=13 \\ n_{41}=4 & n_{42}=1 & n_{43}=13 & n_{44}=3 \end{bmatrix} \end{array}.$$

We have $n_1 = n_{11} + n_{22} + n_{33} + n_{44} = 17$, $n_2 = n_{14} + n_{23} + n_{32} + n_{41} = 14$, $n_3 = n_{12} + n_{21} + n_{34} + n_{43} = 65$, and $n_4 = n_{13} + n_{31} + n_{24} + n_{42} = 4$. For the MLEs of r, we have

$$\begin{aligned} \hat{r} &= 0.4850 \text{ under } \mathbf{\Phi}_{11},\ L_{11} = -138.5394, \\ \hat{r} &= 0.1950 \text{ under } \mathbf{\Phi}_{12},\ L_{12} = \ -98.6785, \\ \hat{r} &= 0.8050 \text{ under } \mathbf{\Phi}_{21},\ L_{21} = \ -98.6785, \\ \hat{r} &= 0.5150 \text{ under } \mathbf{\Phi}_{22},\ L_{22} = -138.5394. \end{aligned}$$

The correct parental diplotypes are $\mathbf{\Phi}_{12}$; i.e. [12][12] for parent P and [34][43] for parent Q.

as in equation (4.11), we write the likelihood in matrix notation with the consideration of a parental diplotype combination $\mathbf{\Phi}_{\xi\zeta}$; i.e.,

$$\begin{aligned} L_{\xi\zeta}(r|\mathbf{n}) &= \prod_{i=1}^{n} \mathbf{m}_{ij_1}^{\mathrm{T}} (\mathbf{I}_{p_1}^{\mathrm{T}} \mathbf{H}_{\xi\zeta} \mathbf{I}_{p_2}) \mathbf{m}_{ij_2} \\ &= \prod_{i=1}^{n} \mathbf{m}_{ij_1}^{\mathrm{T}} \mathbf{P}_{\xi\zeta} \mathbf{m}_{ij_2}. \end{aligned} \quad (4.27)$$

The EM algorithm has been developed to estimate the crossover probabilities and recombination fractions based on equations (4.22) and (4.23).

4.4 Joint modeling of the Linkage, Parental Diplotype, and Gene Order

For any two fully informative markers, their segregation pattern in a full-sib family is determined by the diplotypes of the two outbred parents. Thus, the linkage between these two markers resulting from their cosegregating pattern can be correctly estimated when a true parental diplotype combination is determined. The models presented in Section 4.3 allow for the simultaneous estimation of the linkage and parental diplotypes. However, when the number of markers studied is three or greater, a joint model of the linkage and parental diplotype is not sufficient because different gene orders may also blur our inference and estimation of the linkage. In this section, a model that incorporates the linkage, parental diplotype, and gene order will be introduced.

Consider three markers in a linkage group that have three possible orders **A-B-C**, **A-C-B**, and **B-A-C**. Let o_1, o_2, and o_3 be the corresponding probabilities of occurrence of these orders in the parental genome. Without loss of generality, for a given order, the allelic arrangement of the first marker between the two homologous chromosomes can be fixed for a parent. Thus, the change of the allelic arrangements at the two other markers will lead to $2 \times 2 = 4$ parental diplotypes. The three-marker genotype (12/12/12) of parent P may have four possible diplotypes, [111][222], [112][221], [121][212], and [122][211]. Relative to the fixed allelic arrangement [1··][2··] of the first marker, the probabilities of allelic arrangements [·1·][·2·] and [·2·][·1·] for the second marker are denoted as p_1 and $1 - p_1$ and those of allelic arrangements [··1][··2] and [··2][··1] for the third marker are denoted as p_2 and $1 - p_2$, respectively. Assuming that allelic arrangements are independent between the second and third markers, the probabilities of these four three-marker diplotypes can be described by p_1p_2, $p_1(1-p_2)$, $(1-p_1)p_2$, and $(1-p_1)(1-p_2)$, respectively. The four diplotypes of parent Q can also be constructed, whose probabilities are defined as q_1q_2, $q_1(1-q_2)$, $(1-q_1)q_2$, and $(1-q_1)(1-q_2)$, respectively. Thus, there are $4 \times 4 = 16$ possible diplotype combinations (whose probabilities are the product of the corresponding diplotype probabilities) when parents P and Q are crossed.

Let $r_{\mathbf{AB}}$ denote the recombination fraction between markers **A** and **B**, with $r_{\mathbf{BC}}$ and $r_{\mathbf{AC}}$ defined similarly. Let G_{00}, G_{01}, G_{10}, and G_{11} denote no crossover between markers **A** and **B** and between markers **B** and **C**, only one crossover in the second interval, only one crossover in the first interval, and one in each interval, respectively. As previously shown, the probabilities of these events, denoted by g_{00}, g_{01}, g_{10}, and g_{11}, respectively, can be used to define the three recombination fractions.

Table 4.1 describes a (16×4)–matrix for three-marker genotype frequencies under a particular parental diplotype combination [111][222] × [333][444] and gene order **A-B-C**. For any order $(\mathcal{O}_k)$, there are 16 diplotype combinations and therefore 16 such matrices, denoted by $\mathbf{H}_{x_1^k x_2^k y_1^k y_2^k}$, where $x_1^k = 1$ for [11·][22·] or 2 for [12·][21·] denotes the two alternative allelic arrangements of marker **B** in parent P, $x_2^k = 1$ for [1·1][2·2] or 2 for [1·2][2·1] denotes the two alternative allelic arrangements of marker **C** in parent P, and $y_1^k, y_2^k = 1$ or 2 have similar means for parent Q. According to Ridout et al. (1998) and Wu et al. (2002b), elements in $\mathbf{H}_{x_1^k x_2^k y_1^k y_2^k}$ are expressed in terms of g_{00}, g_{01}, g_{10}, and g_{11} (Table 4.1). Similarly, there are 16 (16×4)–matrices for the expected number of crossovers that have occurred for G_{00}, G_{01}, G_{10}, and G_{11} for a given marker order, denoted by $\mathbf{G}^{00}_{x_1^k x_2^k y_1^k y_2^k}$, $\mathbf{G}^{01}_{x_1^k x_2^k y_1^k y_2^k}$, $\mathbf{G}^{10}_{x_1^k x_2^k y_1^k y_2^k}$, and $\mathbf{G}^{11}_{x_1^k x_2^k y_1^k y_2^k}$, respectively (see Table 4.2).

The joint genotype frequencies of the three markers can be viewed as a mixture of 16 diplotype combinations and three orders, weighted by their probabilities of occurring, and expressed as

$$\mathbf{H} = \sum_{k=1}^{3} o_k \sum_{x_1^k=1}^{2} \sum_{x_2^k=1}^{2} \sum_{y_1^k=1}^{2} \sum_{y_2^k=1}^{2} p_1^{2-x_1^k}(1-p_1)^{x_1^k-1} p_2^{2-x_2^k}(1-p_2)^{x_2^k-1}$$

$$q_1^{2-y_1^k}(1-q_1)^{y_1^k-1} q_2^{2-y_2^k}(1-q_2)^{y_2^k-1} \mathbf{H}_{x_1^k x_2^k y_1^k y_2^k}. \tag{4.28}$$

Similarly, the expected number of recombination events contained within a progeny genotype is the mixture of the different diplotype and order combinations, expressed as

$$\mathbf{G}_{00} = \sum_{k=1}^{3} o_k \sum_{x_1^k=1}^{2} \sum_{x_2^k=1}^{2} \sum_{y_1^k=1}^{2} \sum_{y_2^k=1}^{2} p_1^{2-x_1^k}(1-p_1)^{x_1^k-1} p_2^{2-x_2^k}(1-p_2)^{x_2^k-1}$$
$$q_1^{2-y_1^k}(1-q_1)^{y_1^k-1} q_2^{2-y_2^k}(1-q_2)^{y_2^k-1} \mathbf{G}^{00}_{x_1^k x_2^k y_1^k y_2^k},$$

$$\mathbf{G}_{01} = \sum_{k=1}^{3} o_k \sum_{x_1^k=1}^{2} \sum_{x_2^k=1}^{2} \sum_{y_1^k=1}^{2} \sum_{y_2^k=1}^{2} p_1^{2-x_1^k}(1-p_1)^{x_1^k-1} p_2^{2-x_2^k}(1-p_2)^{x_2^k-1}$$
$$q_1^{2-y_1^k}(1-q_1)^{y_1^k-1} q_2^{2-y_2^k}(1-q_2)^{y_2^k-1} \mathbf{G}^{01}_{x_1^k x_2^k y_1^k y_2^k},$$

(4.29)

$$\mathbf{G}_{10} = \sum_{k=1}^{3} o_k \sum_{x_1^k=1}^{2} \sum_{x_2^k=1}^{2} \sum_{y_1^k=1}^{2} \sum_{y_2^k=1}^{2} p_1^{2-x_1^k}(1-p_1)^{x_1^k-1} p_2^{2-x_2^k}(1-p_2)^{x_2^k-1}$$
$$q_1^{2-y_1^k}(1-q_1)^{y_1^k-1} q_2^{2-y_2^k}(1-q_2)^{y_2^k-1} \mathbf{G}^{10}_{x_1^k x_2^k y_1^k y_2^k},$$

$$\mathbf{G}_{11} = \sum_{k=1}^{3} o_k \sum_{x_1^k=1}^{2} \sum_{x_2^k=1}^{2} \sum_{y_1^k=1}^{2} \sum_{y_2^k=1}^{2} p_1^{2-x_1^k}(1-p_1)^{x_1^k-1} p_2^{2-x_2^k}(1-p_2)^{x_2^k-1}$$
$$q_1^{2-y_1^k}(1-q_1)^{y_1^k-1} q_2^{2-y_2^k}(1-q_2)^{y_2^k-1} \mathbf{G}^{11}_{x_1^k x_2^k y_1^k y_2^k}.$$

Also define the matrices

$$\mathbf{P}_1 = \sum_{k=1}^{3} o_k \sum_{x_2^k=1}^{2} \sum_{y_1^k=1}^{2} \sum_{y_2^k=1}^{2} p_1 p_2^{2-x_2^k}(1-p_2)^{x_2^k-1}$$
$$q_1^{2-y_1^k}(1-q_1)^{y_1^k-1} q_2^{2-y_2^k}(1-q_2)^{y_2^k-1} \mathbf{H}_{x_1^k x_2^k y_1^k y_2^k},$$

(4.30)

$$\mathbf{P}_2 = \sum_{k=1}^{3} o_k \sum_{x_1^k=1}^{2} \sum_{y_1^k=1}^{2} \sum_{y_2^k=1}^{2} p_1^{2-x_2^k}(1-p_1)^{x_2^k-1} p_2$$
$$q_1^{2-y_1^k}(1-q_1)^{y_1^k-1} q_2^{2-y_2^k}(1-q_2)^{y_2^k-1} \mathbf{H}_{x_1^k x_2^k y_1^k y_2^k}$$

$$
\begin{aligned}
\mathbf{Q}_1 = & \sum_{k=1}^3 o_k \sum_{x_1^k=1}^2 \sum_{x_2^k=1}^2 \sum_{y_2^k=1}^2 p_1^{2-x_1^k}(1-p_1)^{x_1^k-1} p_2^{2-x_2^k}(1-p_2)^{x_2^k-1} \\
& q_1 q_2^{2-y_2^k}(1-q_2)^{y_2^k-1}\mathbf{H}_{x_1^k x_2^k y_1^k y_2^k}, \\
\mathbf{Q}_2 = & \sum_{k=1}^3 o_k \sum_{x_1^k=1}^2 \sum_{x_2^k=1}^2 \sum_{y_1^k=1}^2 p_1^{2-x_1^k}(1-p_1)^{x_1^k-1} p_2^{2-x_2^k}(1-p_2)^{x_2^k-1} \\
& q_1^{2-y_1^k}(1-q_1)^{y_1^k-1} q_2 \mathbf{H}_{x_1^k x_2^k y_1^k y_2^k}.
\end{aligned} \tag{4.31}
$$

The probabilities of occurrence of the three marker orders are the mixture of all diplotype combinations, expressed in matrix notation as

$$
\begin{aligned}
\mathbf{O}_1 = & \, o_1 \sum_{x_1^1=1}^2 \sum_{x_2^1=1}^2 \sum_{y_1^1=1}^2 \sum_{y_2^1=1}^2 p_1^{2-x_1^1}(1-p_1)^{x_1^1-1} p_2^{2-x_2^1}(1-p_2)^{x_2^1-1} \\
& q_1^{2-y_1^1}(1-q_1)^{y_1^1-1} q_2^{2-y_2^1}(1-q_2)^{y_2^1-1}\mathbf{H}_{x_1^1 x_2^1 y_1^1 y_2^1}, \\
\mathbf{O}_2 = & \, o_2 \sum_{x_1^2=1}^2 \sum_{x_2^2=1}^2 \sum_{y_1^2=1}^2 \sum_{y_2^2=1}^2 p_1^{2-x_1^2}(1-p_1)^{x_1^2-1} p_2^{2-x_2^2}(1-p_2)^{x_2^2-1} \\
& q_1^{2-y_1^2}(1-q_1)^{y_1^2-1} q_2^{2-y_2^2}(1-q_2)^{y_2^2-1}\mathbf{H}_{x_1^2 x_2^2 y_1^2 y_2^2}, \\
\mathbf{O}_3 = & \, o_3 \sum_{x_1^3=1}^2 \sum_{x_2^3=1}^2 \sum_{y_1^3=1}^2 \sum_{y_2^3=1}^2 p_1^{2-x_1^3}(1-p_1)^{x_1^3-1} p_2^{2-x_2^3}(1-p_2)^{x_2^3-1} \\
& q_1^{2-y_1^3}(1-q_1)^{y_1^3-1} q_2^{2-y_2^3}(1-q_2)^{y_2^3-1}\mathbf{H}_{x_1^3 x_2^3 y_1^3 y_2^3}.
\end{aligned} \tag{4.32}
$$

We implement the EM algorithm to estimate the MLEs of the recombination fractions among the three markers. The general equations formulating the iteration of the $(\tau+1)$th EM step are given as follows:

E Step: As step τ, calculate the expected number of recombination events for individual i associated with $\mathbf{G}_{00} = (\mathcal{G}^{00}_{j_1j_2j_3})_{16\times 4}$, $\mathbf{G}_{01} = (\mathcal{G}^{01}_{j_1j_2j_3})_{16\times 4}$, $\mathbf{G}_{10} = (\mathcal{G}^{10}_{j_1j_2j_3})_{16\times 4}$, and $\mathbf{G}_{11} = (\mathcal{G}^{11}_{j_1j_2j_3})_{16\times 4}$ for the $(j_1j_2j_3)$th progeny genotype (where j_1, j_2, and j_3 denote the progeny genotypes of the three individual markers, respectively) using formulas that have the same forms as equations (4.22), whose $\mathbf{H}$ and $\mathbf{G}$'s are now defined by equations (4.28) and (4.29), respectively. Note that matrices $\mathbf{H}$ and $\mathbf{G}$'s of equations (4.28) and (4.29) are also determined by p_1, p_2, q_1, q_2, and k $(k = 1, 2, 3)$.

M Step: Calculate $g_{00}^{(\tau+1)}, g_{01}^{(\tau+1)}, g_{10}^{(\tau+1)}$, and $g_{11}^{(\tau+1)}$ using the equations that have the same forms as equations (4.23). But here we also need to update $p_1^{(\tau+1)}, p_2^{(\tau+1)}$, $q_1^{(\tau+1)}$, $q_2^{(\tau+1)}$, and $o_k^{(\tau+1)}$ using

$$
p_1^{(\tau+1)} = \frac{1}{n}\sum_{j_1=1}^4 \sum_{j_2=1}^4 \sum_{j_3=1}^4 \frac{p_{1j_1j_2j_3}^{(\tau)}}{h_{j_1j_2j_3}^{(\tau)}} n_{j_1j_2j_3},
$$

$$p_2^{(\tau+1)} = \frac{1}{n}\sum_{j_1=1}^{4}\sum_{j_2=1}^{4}\sum_{j_3=1}^{4}\frac{p_{2j_1j_2j_3}^{(\tau)}}{h_{j_1j_2j_3}^{(\tau)}}n_{j_1j_2j_3},$$

$$q_1^{(\tau+1)} = \frac{1}{n}\sum_{j_1=1}^{4}\sum_{j_2=1}^{4}\sum_{j_3=1}^{4}\frac{q_{1j_1j_2j_3}^{(\tau)}}{h_{j_1j_2j_3}^{(\tau)}}n_{j_1j_2j_3},$$

$$q_2^{(\tau+1)} = \frac{1}{n}\sum_{j_1=1}^{4}\sum_{j_2=1}^{4}\sum_{j_3=1}^{4}\frac{q_{2j_1j_2j_3}^{(\tau)}}{h_{j_1j_2j_3}^{(\tau)}}n_{j_1j_2j_3},$$

$$o_k^{(\tau+1)} = \frac{1}{n}\sum_{j_1=1}^{4}\sum_{j_2=1}^{4}\sum_{j_3=1}^{4}\frac{o_{kj_1j_2j_3}^{(\tau)}}{h_{j_1j_2j_3}^{(\tau)}}n_{j_1j_2j_3},$$

where $n_{j_1j_2j_3}$ denotes the number of progeny with a particular three-marker genotype, $h_{j_1j_2j_3}$, $p_{1j_1j_2j_3}$, $p_{2j_1j_2j_3}$, $q_{1j_1j_2j_3}$, and $q_{2j_1j_2j_3}$ are the $(j_1j_2j_3)$th element of matrices $\mathbf{H}$, $\mathbf{P}_1$, $\mathbf{P}_2$, $\mathbf{Q}_1$, and $\mathbf{Q}_2$, respectively.

The E and M steps are repeated until g_{00}, g_{01}, g_{10}, and g_{11} converge to values with satisfactory precision. From the MLEs of the g's, the MLEs of recombination fractions $r_{\mathbf{AB}}$, $r_{\mathbf{AC}}$, and $r_{\mathbf{BC}}$ can be obtained according to the invariance property of the MLEs.

Example 4.4. (**Jointly modeling the Linkage, Parental Diplotype and Gene Order**). Yin et al. (2004) reported a high-density linkage map constructed with microsatellite and AFLP markers for a backcross, (T × D) × D, between two poplar species, *Populus trichocarpa* (T) and *P. deltoides* (D). Unlike general inbred lines, the poplar parents used for the crosses are heterozygous. Thus, this backcross is analogous to a full-sib family with one parent being the F_1 parent and the other being the D parent. We choose three testcross markers, CA/CGA-580RD (**A**), AM11-1060 (**B**), and A7-690 (**C**), that are heterozygous (12) in the F_1 but homozygous (22) in the D parent. We do not know exactly the diplotype at these three markers for the F_1 parent and their order. The model described in Section 4.4 will be used to simultaneously estimate the linkage, parental diplotype, and gene order. The observations of these three markers are given below:

		Marker **C**	
Marker **A**	Marker **B**	12	22
12	12	33	1
12	22	4	7
22	12	3	4
22	22	0	21

Let p_1, $1-p_1$, and p_2 and $1-p_2$ be the probabilities of four possible diplotypes of parent D, [111][222], [121][212], and [111][222] and [112][221], respectively, and o_1,

o_2, and o_3 be the probabilities of three possible gene orders for these three markers, **A**, **B**, and **C**, respectively. Using g_{00}, g_{01}, g_{10}, and g_{11} to denote the probabilities of the number of interval-specific crossovers, we derive one (4×2)–**H** matrix based on equation (4.28), two (4×2)–**P** matrices based on equations (4.30), and three (4×2)–**O** matrices based on equation (4.32). Meanwhile, four (4×2)–**G** matrices based on equation (4.29) are derived.

With the EM algorithm described above, the MLEs of the unknown parameters are obtained as $\hat{g}_{00} = 0.5621$, $\hat{g}_{01} = 0.1924$, $\hat{g}_{10} = 0.0925$, $\hat{g}_{11} = 0.1507$ and $\hat{p}_1 = 0.9$, $\hat{p}_2 = 0.8$, $\hat{o}_1 = 0$, $\hat{o}_2 = 0.9989$, and $\hat{o}_3 = 0.0011$. The diplotype of parent D for the three markers in order **A-C-B** is constructed by

$$\begin{array}{lr||l} \text{CA/CGA-580RD } (\mathbf{A}) & 1 & 2 \\ \text{A7-690 } (\mathbf{C}) & 1 & 2 \\ \text{AM11-1060 } (\mathbf{B}) & 2 & 1 \end{array}$$

The estimated g values are further used to estimate the MLEs of recombination fractions; i.e., $\hat{r}_{\mathbf{AB}} = 0.35$, $\hat{r}_{\mathbf{BC}} = 0.28$, and $\hat{r}_{\mathbf{AC}} = 0.24$.

4.5 Partially Informative Markers

In Section 4.2.1, we defined a (4×4)–**H** matrix for joint *genotype* frequencies between two fully informative markers. For a fully informative marker, there is 1:1 correspondence between the genotype and phenotype. Many practically useful markers are partially informative, as shown in Table 3.1, for which there is no such 1:1 correspondence. The general models described above can be modified to estimate the linkage and parental diplotypes for partially informative markers.

4.5.1 Joint modeling of the Linkage and Parental Diplotype

Unlike in a fully informative marker, four possible modes of genotype formation for marker cross type 12×12 will yield three "phenotypes" because genotype formations 12 and 21 are phenotypically identical. For this reason, the (4×4)–**H** matrix for the frequencies of genotype formation at two markers will be collapsed into one with lower dimensions for the frequencies of marker phenotypes. Wu et al. (2002b) designed specific incidence matrices (**I**) relating the genotype frequencies to the phenotype frequencies for different types of markers. (The *incidence matrix* (**I**) here is defined as the matrix that relates the marker genotype to marker phenotype for a type of partially informative marker.) Here, we use the notation $\mathbf{H}' = \mathbf{I}_{b_1}^{\mathrm{T}} \mathbf{H} \mathbf{I}_{b_2}$ for a $(b_1 \times b_2)$ matrix of the phenotype frequencies between two partially informative markers, where b_1 and b_2 are the number of distinguishable phenotypes for markers **A** and **B**, respectively. Correspondingly, we have

$$
\begin{aligned}
(\mathbf{DH})' &= \mathbf{I}_{b_1}^{\mathrm{T}}(\mathbf{D} \circ \mathbf{H})\mathbf{I}_{b_2}, \\
\mathbf{P}' &= \mathbf{I}_{b_1}^{\mathrm{T}}\mathbf{P}\mathbf{I}_{b_2}, \\
\mathbf{Q}' &= \mathbf{I}_{b_1}^{\mathrm{T}}\mathbf{Q}\mathbf{I}_{b_2},
\end{aligned}
$$

where we recall that the notation $\circ$ means componentwise products between the two matrices.

The EM algorithm can then be developed to estimate the recombination fraction between any two partial informative markers.

E Step: At step τ, based on the matrix $(\mathbf{DH})'$ derived from the current estimate $r^{(\tau)}$, calculate the expected number of recombination events between the two markers for a given progeny genotype,

$$
D_{j_1j_2}^{(\tau+1)} = \frac{(dh)_{j_1j_2}^{\prime(\tau)}}{h_{j_1j_2}^{\prime(\tau)}} n_{j_1j_2}, \tag{4.33}
$$

where $(dh)'_{j_1j_2}$ and $h'_{j_1j_2}$ are the (j_1j_2)th element of matrices $(\mathbf{DH})'$ and $\mathbf{H}'$, respectively.

M Step: Calculate $r^{(\tau+1)}$, $p^{(\tau+1)}$, and $q^{(\tau+1)}$ using

$$
\begin{aligned}
r^{(\tau+1)} &= \frac{1}{2n}\sum_{j_1=1}^{b_1}\sum_{j_2=1}^{b_2} D_{j_1j_2}^{(\tau+1)}, \\
p^{(\tau+1)} &= \frac{1}{n}\sum_{j_1=1}^{b_1}\sum_{j_2=1}^{b_2} \frac{p_{j_1j_2}^{\prime(\tau)}}{h_{j_1j_2}^{\prime(\tau)}} n_{j_1j_2}, \\
q^{(\tau+1)} &= \frac{1}{n}\sum_{j_1=1}^{b_1}\sum_{j_2=1}^{b_2} \frac{q_{j_1j_2}^{\prime(\tau)}}{h_{j_1j_2}^{\prime(\tau)}} n_{j_1j_2},
\end{aligned} \tag{4.34}
$$

where $p'_{j_1j_2}$ and $q'_{j_1j_2}$ are the (j_1j_2)th element of matrices $\mathbf{P}'$ and $\mathbf{Q}'$, respectively.

In each step, matrices $(\mathbf{DH})'$ and $\mathbf{H}'$ are updated by newly estimated r, p, and q. The E and M steps between equations (4.33) and (4.34) are repeated until the estimate converges to a stable value.

Example 4.5. (**Jointly modeling the Linkage and Parental Diplotype**). We use an example for three dominant markers to demonstrate our unifying model for simultaneous estimation of the linkage and parental diplotype. A cross was made between two triple heterozygotes with genotype $ao/bo/co$ for markers **A**, **B**, and **C**. Because these three markers are dominant, the cross generates 8 (rather than 27) distinguishable genotypes, with observations

		Marker **C**	
Marker **A**	Marker **B**	$c\_$	oo
$a\_$	$b\_$	28	4
$a\_$	oo	12	3
oo	$b\_$	1	8
oo	oo	2	2

We first use two-point analysis to estimate the recombination fractions and parental diplotypes between all possible pairs of the three markers. For a dominant marker cross type $a|o| \times a|o|$ and $a|o| \times o|a|$, the incidence matrix is defined as

$$\mathbf{I}_2 = \begin{bmatrix} 1\,1\,1\,0 \\ 0\,0\,0\,1 \end{bmatrix} \text{ or } \begin{bmatrix} 1\,1\,0\,1 \\ 0\,0\,1\,0 \end{bmatrix},$$

respectively. With the corresponding incidence matrices implemented to collapse the $\mathbf{H}$ matrix, we use the EM algorithm to estimate the recombination fraction between markers **A** and **B** as $r_{\mathbf{AB}} = 0.376$, whose estimated parental diplotypes are $[ao][ob] \times [ab][oo]$ or $[ab][oo] \times [ao][ob]$. Similarly, the two other recombination fractions and the corresponding parental diplotypes are estimated as $r_{\mathbf{BC}} = 0.386$, $[bo][oc] \times [bc][oo]$ or $[bc][oo] \times [bo][oc]$ and $r_{\mathbf{AC}} = 0.184$, $[ac][oo] \times [ac][oo]$. From the two-point analysis, one of the two parents have dominant alleles from markers **A** and **B**, that are repulsed with the dominant alleles from marker **C**.

4.5.2 Joint modeling of the Linkage, Parental Diplotype, and Gene Order

Consider three partially informative markers with the number of distinguishable phenotypes denoted by b_1, b_2, and b_3, respectively. Define $\mathbf{H}' = (\mathbf{I}_{b_1}^{\mathrm{T}} \otimes \mathbf{I}_{b_2}^{\mathrm{T}})\mathbf{H}\mathbf{I}_{b_3}$ as a $(b_1 b_2 \times b_3)$ matrix of genotype frequencies for three partially informative markers. Similarly, we define $(\mathbf{HD}_{00})' = (\mathbf{I}_{b_1}^{\mathrm{T}} \otimes \mathbf{I}_{b_2}^{\mathrm{T}})(\mathbf{D}_{00} \circ \mathbf{H})\mathbf{I}_{b_3}$, $(\mathbf{HD}_{01})' = (\mathbf{I}_{b_1}^{\mathrm{T}} \otimes \mathbf{I}_{b_2}^{\mathrm{T}})(\mathbf{D}_{01} \circ \mathbf{H})\mathbf{I}_{b_3}$, $(\mathbf{HD}_{10})' = (\mathbf{I}_{b_1}^{\mathrm{T}} \otimes \mathbf{I}_{b_2}^{\mathrm{T}})(\mathbf{D}_{10} \circ \mathbf{H})\mathbf{I}_{b_3}$, $(\mathbf{HD}_{11})' = (\mathbf{I}_{b_1}^{\mathrm{T}} \otimes \mathbf{I}_{b_2}^{\mathrm{T}})(\mathbf{D}_{11} \circ \mathbf{H})\mathbf{I}_{b_3}$, $\mathbf{P}_1' = (\mathbf{I}_{b_1}^{\mathrm{T}} \otimes \mathbf{I}_{b_2}^{\mathrm{T}})\mathbf{P}_1\mathbf{I}_{b_3}$, $\mathbf{P}_2' = (\mathbf{I}_{b_1}^{\mathrm{T}} \otimes \mathbf{I}_{b_2}^{\mathrm{T}})\mathbf{P}_2\mathbf{I}_{b_3}$, $\mathbf{Q}_1' = (\mathbf{I}_{b_1}^{\mathrm{T}} \otimes \mathbf{I}_{b_2}^{\mathrm{T}})\mathbf{Q}_1\mathbf{I}_{b_3}$, $\mathbf{Q}_2' = (\mathbf{I}_{b_1}^{\mathrm{T}} \otimes \mathbf{I}_{b_2}^{\mathrm{T}})\mathbf{Q}_2\mathbf{I}_{b_3}$, $\mathbf{O}_1' = (\mathbf{I}_{b_1}^{\mathrm{T}} \otimes \mathbf{I}_{b_2}^{\mathrm{T}})\mathbf{O}_1\mathbf{I}_{b_3}$, $\mathbf{O}_2' = (\mathbf{I}_{b_1}^{\mathrm{T}} \otimes \mathbf{I}_{b_2}^{\mathrm{T}})\mathbf{O}_2\mathbf{I}_{b_3}$, and $\mathbf{O}_3' = (\mathbf{I}_{b_1}^{\mathrm{T}} \otimes \mathbf{I}_{b_2}^{\mathrm{T}})\mathbf{O}_3\mathbf{I}_{b_3}$.

Using the procedure described in Section 4.3, we implement the EM algorithm to estimate the MLEs of the recombination fractions among the three partially informative markers as well as their parental diplotype combinations and gene orders.

Example 4.6. Revisit Example 4.5. We use a three-point analysis that combines parental diplotypes and gene orders to estimate the linkage between these three dominant markers. The estimated gene order is **B-A-C**. The estimated parental diplotypes are

$$\begin{array}{l} \mathbf{B}\ \mathrm{b} \\ \mathbf{A}\ \mathrm{a} \\ \mathbf{C}\ \mathrm{c} \end{array} \left\|\begin{array}{l} \mathrm{o}\\ \mathrm{o}\\ \mathrm{o}\end{array}\right. \times \begin{array}{r} \mathrm{b}\\ \mathrm{o}\\ \mathrm{o}\end{array} \left\|\begin{array}{l}\mathrm{o}\\ \mathrm{a}\\ \mathrm{c}\end{array}\right. \quad \text{and} \quad \begin{array}{l} \mathbf{B}\ \mathrm{b} \\ \mathbf{A}\ \mathrm{o} \\ \mathbf{C}\ \mathrm{o} \end{array} \left\|\begin{array}{l} \mathrm{o}\\ \mathrm{a}\\ \mathrm{c}\end{array}\right. \times \begin{array}{r} \mathrm{b}\\ \mathrm{a}\\ \mathrm{c}\end{array} \left\|\begin{array}{l}\mathrm{o}\\ \mathrm{o}\\ \mathrm{o}\end{array}\right. .$$

The MLEs of the recombination fractions are $\hat{r}_{\mathbf{BA}} = 0.309$, $\hat{r}_{\mathbf{AC}} = 0.184$, and $\hat{r}_{\mathbf{BC}} = 0.379$.

4.6 Exercises

4.1 Fully informative markers
Suppose there is a full-sib family derived from two outbred parents with diplotypes [11][22] and [33][44], respectively, for two fully informative markers **A** and **B**. The observations for each of 16 two-marker genotypes are given as

	13	14	23	24
13	$n_{11} = 70$	$n_{12} = 13$	$n_{13} = 9$	$n_{14} = 2$
14	$n_{21} = 9$	$n_{22} = 82$	$n_{23} = 2$	$n_{24} = 13$
23	$n_{31} = 14$	$n_{32} = 4$	$n_{33} = 75$	$n_{34} = 15$
24	$n_{41} = 4$	$n_{42} = 13$	$n_{43} = 9$	$n_{44} = 66$

with $n_{j_1j_2}$ denoting the number of progeny in a cell.

(a) Write down the **H** matrix for the genotype frequencies in terms of the recombination fraction r between two markers.
(b) Derive and estimate the MLE of r.
(b) Test whether the linkage is significant.

4.2 One F$_2$ codominant marker
One of the two markers (say **B**) is summed to the same allele system between the two parents above. Such markers are called F$_2$ codominant markers. The diplotypes of two parents are [11][22] and [31][42]. In this cross, the genotype formations with the same phenotype will be pooled together, with observations as follows:

	11	12	22
13	$n_{11} = 70$	$n_{12} + n_{13} = 22$	$n_{14} = 2$
14	$n_{21} = 9$	$n_{22} + n_{23} = 84$	$n_{24} = 13$
23	$n_{31} = 14$	$n_{32} + n_{33} = 79$	$n_{34} = 15$
24	$n_{41} = 4$	$n_{42} + n_{43} = 22$	$n_{44} = 66$

(a) Write down the **H** matrix for the genotype frequencies in terms of the recombination fraction r between two markers.
(b) Show how the EM algorithm can be used to estimate r.
(c) Estimate r using the program for the EM algorithm.
(d) Test if these two markers are linked.

Hints: In this situation, the **H** and **D** matrices, respectively, are collapsed as

$$\mathbf{H}_{11} = \begin{array}{c} \\ 13 \\ 14 \\ 23 \\ 24 \end{array} \begin{array}{c} \begin{array}{ccc} 11 & 12 & 22 \end{array} \\ \begin{bmatrix} \frac{(1-r)^2}{4} & \frac{r(1-r)}{4} + \frac{r(1-r)}{4} & \frac{r^2}{4} \\ \frac{r(1-r)}{4} & \frac{(1-r)^2}{4} + \frac{r^2}{4} & \frac{r(1-r)}{4} \\ \frac{r(1-r)}{4} & \frac{r^2}{4} + \frac{(1-r)^2}{4} & \frac{r(1-r)}{4} \\ \frac{r^2}{4} & \frac{r(1-r)}{4} + \frac{r(1-r)}{4} & \frac{(1-r)^2}{4} \end{bmatrix} \end{array},$$

$$\mathbf{D} = \begin{array}{c} \\ 13 \\ 14 \\ 23 \\ 24 \end{array} \begin{array}{c} \begin{array}{ccc} 11 & 12 & 22 \end{array} \\ \begin{bmatrix} 0 & 1 & 2 \\ 1 & 2\phi & 1 \\ 1 & 2\phi & 1 \\ 2 & 1 & 0 \end{bmatrix} \end{array},$$

where $\phi = \frac{r^2}{(1-r)^2+r^2}$. From here, you should figure out the MLE of r and indicate the step for the EM algorithm.

4.3 Two F_2 codominant markers
Do the same things as Problem 4.2 if both markers are F_2 codominant markers.

4.6.1 One dominant marker

If one of the two markers above (say **B**) is dominant, the observations will be further collapsed as follows:

	10	00
13	$n_{11} + n_{12} + n_{13} = 92$	$n_{14} = 2$
14	$n_{21} + n_{22} + n_{23} = 93$	$n_{24} = 13$
23	$n_{31} + n_{32} + n_{33} = 93$	$n_{34} = 15$
24	$n_{41} + n_{42} + n_{43} = 26$	$n_{44} = 66$

Do the same problems as Exercise 4.2.

4.6.2 One dominant marker and one F_2 codominant marker

If one of the above two markers (**B**) is dominant and the other (**A**) is F_2–codominant, the observations will be further collapsed as follows:

	10	00
11	$n_{11} + n_{12} + n_{13} = 92$	$n_{14} = 2$
12	$n_{21} + n_{22} + n_{23} + n_{31} + n_{32} + n_{33} = 186$	$n_{24} + n_{34} = 28$
22	$n_{41} + n_{42} + n_{43} = 26$	$n_{44} = 66$

Do the same problems as Exercise 4.2.

4.4 Two dominant markers

If both markers are dominant, we have

	10	00
10	$n_{11}+n_{12}+n_{13}+n_{21}+n_{22}$ $+n_{23}+n_{31}+n_{32}+n_{33}=278$	$n_{14}+n_{24}+n_{34}=30$
00	$n_{41}+n_{42}+n_{43}=26$	$n_{44}=66$

Do the same as Exercise 4.2.

4.7 Notes

We have introduced a general framework for linkage analysis of any type of markers. For outcrossing species, the diplotypes among different markers are unknown for the parents used for the cross. We have also described joint models that incorporate the linkage, parental diplotype, and gene order. Lu et al. (2004) performed extensive simulation studies to demonstrate the advantages of joint modeling. In this section, we attempt to summarize the main results from Lu et al.'s (2004) studies.

4.7.1 Linkage Analysis

Suppose there are five markers of a known order, $\mathbf{M}_1$–$\mathbf{M}_2$–$\mathbf{M}_3$–$\mathbf{M}_4$–$\mathbf{M}_5$, on a chromosome. These five markers are segregating differently in order, 1:1:1:1, 1:2:1, 3:1, 1:1, and 1:1:1:1. Two parents, with diplotypes for the five markers given in Table 4.4, are crossed to generate a segregating full-sib family. This full-sib family is simulated with different degrees of linkage ($r = 0.05$ vs. 0.20).

As expected, more informative markers or more tightly linked markers display a greater estimation precision of linkage than less informative markers or less tightly linked markers (Table 4.4). Joint models for two-point analysis can provide an excellent estimation of the parental diplotype. For example, the MLE of the probability (p or q) of the parental diplotype is close to 1 or 0 (Table 4.4), suggesting that we can always accurately estimate parental diplotypes. But for two symmetrical markers (e.g., markers $\mathbf{M}_2$ and $\mathbf{M}_3$ in this example), two sets of MLEs, $\hat{p} = 1$, $\hat{q} = 0$ and $\hat{p} = 0$, $\hat{q} = 1$, give an identical likelihood ratio test statistic. Thus, two-point analysis cannot specify parental diplotypes for symmetrical markers even when the two parents have different diplotypes.

The estimation precision of linkage can be increased when a three-point analysis is performed (Table 4.5), but this depends on different marker types and different degrees of linkage. The advantage of three-point analysis over two-point analysis is more pronounced for partially informative markers than for fully informative ones, and for less tightly linked markers than for more tightly linked ones. For example, the sampling error of the MLE of the recombination fraction (assuming $r = 0.20$) between

Table 4.4. Estimation from two-point analysis of the recombination fraction ($\hat{r} \pm \text{SD}$) and the parental diplotype probability of parents P ($\hat{p}$) and Q ($\hat{q}$) for five markers in a full-sib family of $n = 100$.

	Parental Diplotype			$r = 0.05$			$r = 0.20$		
Marker	P[a]	×	Q[a]	$\hat{r}$	$\hat{p}$	$\hat{q}$	$\hat{r}$	$\hat{p}$	$\hat{q}$
	$\mid\ \mid$		$\mid\ \mid$						
$\mathbf{M}_1$	1 2		3 4						
	$\mid\ \mid$		$\mid\ \mid$	0.0530±0.0183			0.2097±0.0328		
$\mathbf{M}_2$	1 2		1 2		0.9960	0.9972		0.9882	0.9878
	$\mid\ \mid$		$\mid\ \mid$	0.0464±0.0303			0.2103±0.0848		
$\mathbf{M}_3$	1 0	×	0 1		$1(0^b)$	$0(1^b)$		$1(0^b)$	$0(1^b)$
	$\mid\ \mid$		$\mid\ \mid$	0.0463±0.0371			0.1952±0.0777		
$\mathbf{M}_4$	1 2		2 2		1	$1/0^c$		1	$1/0^c$
	$\mid\ \mid$		$\mid\ \mid$	0.0503±0.0231			0.2002±0.0414		
$\mathbf{M}_5$	1 2		3 4		1	$1/0^c$		1	$1/0^c$
	$\mid\ \mid$		$\mid\ \mid$						

The MLE of r is given between two markers under comparison, whereas the MLEs of p and q are given at the second marker. [a]Shown is the parental diplotype of each parent for the five markers hypothesized, where the vertical lines denote the two homologous chromosomes. [b]The values in the parentheses present a second possible solution. For any two symmetrical markers ($\mathbf{M}_2$ and $\mathbf{M}_3$), $\hat{p} = 1$, $\hat{q} = 0$ and $\hat{p} = 0$, $\hat{q} = 1$ give an identical likelihood ratio test statistic. Thus, when the two parents have different diplotypes for symmetrical markers, their parental diplotypes cannot be correctly determined from two-point analysis. [c]The parental diplotype of parent Q cannot be estimated in these two cases because marker 4 is homozygous in this parent.

markers $\mathbf{M}_2$ and $\mathbf{M}_3$ from a two-point analysis is 0.0848, whereas this value from a three-point analysis decreases to 0.0758 when combining fully informative marker $\mathbf{M}_1$ but increases to 0.0939 when combining partially informative marker $\mathbf{M}_4$. The three-point analysis can clearly determine the diplotypes of different parents as long as one of the three markers is asymmetrical. In our example, using either asymmetrical marker $\mathbf{M}_1$ or $\mathbf{M}_4$, the diplotypes of the two parents for two symmetrical markers ($\mathbf{M}_2$ and $\mathbf{M}_3$) can be determined. The model for three-point analysis can determine a most likely gene order. In the three-point analyses combining markers $\mathbf{M}_1$ and $\mathbf{M}_3$, markers $\mathbf{M}_2$ and $\mathbf{M}_4$, and markers $\mathbf{M}_3$ and $\mathbf{M}_5$, the MLEs of the probabilities of gene order are all almost equal to 1, suggesting that the estimated gene order is consistent with the order hypothesized.

4.7.2 The Diplotype Probability

Two linked dominant markers are simulated to show the advantage of joint modeling of the linkage and parental diplotype. In the two-point analysis, two different parental diplotype combinations are assumed:

Table 4.5. Estimation from three-point analysis of the recombination fraction ($\hat{r} \pm$SD) and the parental diplotype probabilities of parents P ($\hat{p}$) and Q ($\hat{q}$) for five markers in a full-sib family of $n = 100$.

	Parental Diplotype	$\hat{r}$				$\hat{r}$			
Marker	P × Q	Case 1	Case 2	$\hat{p}$	$\hat{q}$	Case 1	Case 2	$\hat{p}$	$\hat{q}$
Recombination fraction = 0.05									
	$\mid\mid$ $\mid\mid$								
$\mathbf{M}_1$	1 2 3 4								
	$\mid\mid$ $\mid\mid$	0.0511±0.0175							
$\mathbf{M}_2$	1 2 1 2		0.1008±0.0298	0.9978	0.9986				
	$\mid\mid$ $\mid\mid$	0.0578±0.0269				0.0557±0.0312			
$\mathbf{M}_3$	1 0 × 0 1			0.9977	0		0.0988±0.0277	1	0
	$\mid\mid$ $\mid\mid$	0.0512±0.0307				0.0476±0.0280			
$\mathbf{M}_4$	1 2 2 2		0.0932±0.0301	1	1/0			1	1/0
	$\mid\mid$ $\mid\mid$	0.0514±0.0229							
$\mathbf{M}_5$	1 2 3 4			1	1				
	$\mid\mid$ $\mid\mid$								
Recombination fraction = 0.20									
	$\mid\mid$ $\mid\mid$								
$\mathbf{M}_1$	1 2 3 4								
	$\mid\mid$ $\mid\mid$	0.2026±0.0348							
$\mathbf{M}_2$	1 2 1 2		0.3282±0.0482	0.9918	0.9916				
	$\mid\mid$ $\mid\mid$	0.2240±0.0758				0.2408±0.0939			
$\mathbf{M}_3$	1 0 × 0 1			0.9944	0		0.3241±0.0488	1	0
	$\mid\mid$ $\mid\mid$	0.1927±0.0613				0.1824±0.0614			
$\mathbf{M}_4$	1 2 2 2		0.3161±0.0502	1	1/0			1	1/0
	$\mid\mid$ $\mid\mid$	0.2017±0.0393							
$\mathbf{M}_5$	1 2 3 4			1	1				
	$\mid\mid$ $\mid\mid$								

Case 1 denotes the recombination fraction between two adjacent markers, whereas case 2 denotes the recombination fraction between the two markers separated by a third marker. See Table 4.4 for other explanations.

(1) $[aa][oo] \times [aa][oo]$ (*cis* × *cis*),
(2) $[ao][oa] \times [ao][oa]$ (*trans* × *trans*).

The MLE of the linkage under combination (2), in which two dominant alleles are in a repulsion phase, is not as precise as that under combination (1), in which two dominant nonalleles are in a coupling phase. For a given data set with unknown linkage phase, the traditional procedure for estimating the recombination fraction is to calculate the likelihood values under all possible linkage phase combinations (i.e., *cis* × *cis*, *cis* × *trans*, *trans* × *cis*, and *trans* × *trans*). The combinations *cis* × *cis*, and *trans* × *trans* have the same likelihood value, with the MLE of one combination being equal to 1–the MLE of the second combination (Table 4.6). The same relationship is true for *cis* × *trans* and *trans* × *cis*. A most likely phase combination is chosen that corresponds to the largest likelihood and a legitimate MLE of the recombination fraction ($r \leq 0.5$).

Table 4.6. Comparison of the estimation of the linkage and parental diplotype between two dominant markers in a full-sib family of $n = 100$ from the traditional and unifying model.

	Traditional Model				
	cis × *cis*	*cis* × *trans*	*trans* × *cis*	*trans* × *trans*	Unifying Model
Data simulated from *cis* × *cis*					
Correct diplotype combination	Correct	Incorrect	Incorrect	Incorrect	
Log-likelihood[a]	−46.2	−92.3	−92.3	−46.2	
$\hat{r}$ under each diplotype combination	0.1981±0.0446	0.5000±0.0000	0.5000±0.0000	0.8018±0.0446	
Estimated diplotype combination	Selected				
$\hat{r}$ under correct diplotype combination	0.1981±0.0446				0.1982±0.0446
Diplotype probability for parent P ($\hat{p}$)					1.0000±0.0000
Diplotype probability for parent Q ($\hat{q}$)					1.0000±0.0000
Data simulated from *trans* × *trans*					
Correct diplotype combination	Incorrect	Incorrect	Incorrect	Correct	
Log-likelihood[a]	−89.6	−89.6	−89.6	−89.6	
$\hat{r}$ under each diplotype combination	0.8573±0.1253	0.0393±0.0419	0.0393±0.0419	0.1426±0.1253	
Estimated diplotype combination		Selected	Selected		
$\hat{r}$ under correct diplotype combination				0.1426±0.1253	0.1428±0.1253
Diplotype probability for parent P ($\hat{p}$)					0.0000±0.0000
Diplotype probability for parent Q ($\hat{q}$)					0.0000±0.0000

[a]The log-likelihood values given here are those from one random simulation for each diplotype combination by the traditional model.

For the data set simulated from $[aa][oo] \times [aa][oo]$, one can easily select *cis* × *cis* as the best estimation of the phase combination because it corresponds to a larger likelihood and a smaller $\hat{r}$. The linkage model incorporating the parental diplotypes can provide a comparable estimation precision of the linkage for the data from $[aa][oo] \times [aa][oo]$ and precisely determine the parental diplotypes (see the MLEs of p and q; Table 4.6). The unifying model has a great advantage over the traditional model for the data derived from $[ao][oa] \times [ao][oa]$. For this data set, the same likelihood was obtained under all possible four–diplotype combinations (Table 4.6). In this case, one would select *cis* × *trans* or *trans* × *cis* because these two phase combinations are associated with a lower estimate of r. But this estimate of r (0.0393) is biased since it is far less than the value of 0.20 hypothesized. The unifying model gives the same estimation precision of the linkage for the data derived from $[ao][oa] \times [ao][oa]$ as obtained when the analysis is based on a correct diplotype combination (Table 4.6). Also, the unifying model can precisely determine the parental diplotypes ($\hat{p} = \hat{q} = 0$).

Table 4.7. Comparison of the estimation of the linkage and gene order among three dominant markers in a full-sib family of $n = 100$ from the traditional and unifying model.

	M_1–M_2–M_3	M_1–M_3–M_2	M_2–M_1–M_3	Unifying Model
Data simulated from $[aaa][ooo] \times [aaa][ooo]$				
Correct gene order	Correct	Incorrect	Incorrect	
Estimated best gene order (%[a])	100	0	0	
$\hat{r}_{12}$	0.2047±0.0422			0.2048±0.0422
$\hat{r}_{23}$	0.1980±0.0436			0.1985±0.0434
$\hat{r}_{13}$	0.3245±0.0619			0.3235±0.0618
Prob(M_1–M_2–M_3) ($\hat{o}_1$)				0.9860±0.0105
Prob(M_1–M_3–M_2) ($\hat{o}_2$)				0.0060±0.0071
Prob(M_2–M_1–M_3) ($\hat{o}_3$)				0.0080±0.0079
Data simulated from $[aao][ooa] \times [aao][ooa]$				
Correct gene order	Correct	Incorrect	Incorrect	
Estimated best gene order (%[a])	80	11	9	
$\hat{r}_{12}$	0.1991±0.0456	0.8165±0.1003	0.9284±0.0724	0.2104±0.0447
$\hat{r}_{23}$	0.1697±0.0907	0.8220±0.0338	0.1636±0.0608	0.2073±0.0754
$\hat{r}_{13}$	0.3218±0.0755	0.2703±0.0586	0.7821±0.0459	0.2944±0.0929
Prob(M_1–M_2–M_3) ($\hat{o}_1$)				0.9952±0.0058
Prob(M_1–M_3–M_2) ($\hat{o}_2$)				0.0045±0.0058
Prob($M_2 - M_1 - M_3$) ($\hat{o}_3$)				0.0003±0.0015

[a]The percentages of a total of 200 simulations that have the largest likelihoods for a given gene order estimated from the traditional approach. In this example used to examine the advantage of implementing gene orders, known linkage phases are assumed.

4.7.3 Gene Order

In three-point analysis, we examine the advantage of implementing linkage analysis with gene orders. Three dominant markers are assumed to have two different parental diplotype combinations:

(1) $[aaa][ooo] \times [aaa][ooo]$,
(2) $[aao][ooa] \times [aao][ooa]$.

The traditional approach is to calculate the likelihood values under three possible gene orders and choose one of a maximum likelihood to estimate the linkage. Under combination (1), a most likely gene order can be well determined and therefore the recombination fractions between the three markers well estimated, because the =likelihood value of the correct order is always larger than those of incorrect orders (Table 4.7). However, under combination (2), the estimates of linkage are not always precise because with a frequency of 20 percent gene orders are incorrectly determined. The estimates of r will largely deviate from their actual values based on a wrong gene order (Table 4.7). The unifying model incorporating gene order can provide a better estimation of linkage than the traditional approach, especially between those markers with dominant alleles in a repulsion phase. Furthermore, a most likely gene order can be determined from our model at the same time that the linkage is estimated.

4.7.4 M-Point Analysis

Three-point analysis considering the dependence of recombination events among different marker intervals can be extended to perform the linkage analysis of an arbitrary number of markers. Suppose there are m ordered markers on a linkage group. The joint genotype probabilities of the m markers form a $(4^{m-1} \times 4)$-dimensional matrix. There are $2^{m-1} \times 2^{m-1}$ such probability matrices given a particular parental diplotype combination. The reasonable estimates of the recombination fractions rely upon the characterization of a most likely phase combination based on the likelihood values calculated.

The m-marker joint genotype probabilities can be expressed as a function of the probability of whether or not there is a crossover occurring between two adjacent markers, $g_{l_1 l_2 \ldots l_{m-1}}$, where l_1, l_2, ..., l_{m-1} are the indicator variables denoting the crossover event between markers $\mathbf{M}_1$ and $\mathbf{M}_2$, markers $\mathbf{M}_2$ and $\mathbf{M}_3$, ..., and markers $\mathbf{M}_{m-1}$ and $\mathbf{M}_m$, respectively. An indicator is defined as 1 if there is a crossover and 0 otherwise. Because each indicator can be taken as one or zero, there are a total of 2^{m-1} g's.

The probability that an interval-specific crossover $g_{l_1 l_2 \ldots l_{m-1}}$ will occur can be estimated using the EM algorithm. In the E step, the expected number of interval-specific crossovers is calculated (see equation (4.22) for three-point analysis). In the M step, an explicit equation is used to estimate the probability $g_{l_1 l_2 \ldots l_{m-1}}$. The MLEs of $g_{l_1 l_2 \ldots l_{m-1}}$ are further used to estimate $m(m-1)/2$ recombination fractions between all possible marker pairs. By comparing the magnitudes of these recombination fractions, we can obtain essential information about marker ordering.

5

Linkage Analysis with Recombinant Inbred Lines

5.1 Introduction

Recombinant inbred lines (RILs) have proven powerful for QTL mapping. They can be derived either by repeated selfing or by repeated sibling (brother-sister) mating from the offspring of an F_1 cross between two inbred lines. Because of continuous inbreeding for a sufficiently number of generations (e.g., 7–10), RILs tend to be homozygous for all genes. With such fixed genotypes (ignoring mutations), RILs can be propagated eternally, allowing the replication of identical genotypes on the scale of time and space aimed to address many fundamental biological and genetic issues. Furthermore, RILs accumulate crossovers that occur at each meiosis with every generation, and thus the proportion of recombinant zygotes in RILs (i.e., the probability that two linked loci have different parental alleles) is higher than what it would be in the F_2. As a result, RILs that have been increasingly available due to community efforts by geneticists and breeders (Threadgill et al. 2002; Complex Trait Consortium 2004) provide powerful material for high-resolution mapping of QTLs.

Interest in genetic analysis with RILs can be traced back to the work of Jennings (1917) and Robbins (1918). It was Haldane and Waddington (1931) who laid a detailed foundation for linkage analysis in RILs generated by selfing or sibling mating. Today, the analysis of RIL data has been experiencing a renewal of interest for more theoretical and practical explorations (Broman 2005; Teuscher et al. 2005; Martin and Hospital 2006). In this chapter, we will describe some basic theory for linkage analysis in RILs generated by selfing and sibling mating. Statistical methods and algorithms for RIL analysis will also be explored.

5.2 RILs by Selfing

5.2.1 Two-Point Analysis

Consider a pair of markers **A** (with two alleles A and a) and **B** (with two alleles B and b) with the recombination fraction of r. Cross two inbred lines $AABB$ and *aabb* to generate a heterozygous F_1. The F_1 is selfed to generate the segregating F_2, and each of the F_2 genotypes is selfed again to generate the F_3. This selfing process is repeated for many generations. In generation t, the proportions of a total of ten zygotic types (diplotypes; that is, phase-known genotypes) in terms of five states are expressed as

State	Diplotype	Proportion
1	$AABB, aabb$	C_t
2	$AAbb, aaBB$	D_t
3	$AABb, AaBB, Aabb, aaBb$	E_t
4	$[AB][ab]$	F_t
5	$[Ab][aB]$	G_t

where we use double brackets to denote two different diplotypes for the double zygote *AaBb*. The genotypes above are selfed to form generation $t+1$, and have new diplotype proportions:

$$
\begin{aligned}
C_{t+1} &= C_t + \tfrac{1}{2}E_t + \tfrac{1}{4}(1-r)^2F_t + \tfrac{1}{4}r^2G_t, \\
D_{t+1} &= D_t + \tfrac{1}{2}E_t + \tfrac{1}{4}r^2F_t + \tfrac{1}{4}(1-r)^2G_t, \\
E_{t+1} &= \tfrac{1}{4}r(1-r)(F_t + G_t), \\
F_{t+1} &= \tfrac{1}{2}(1-r)^2F_t + \tfrac{1}{2}r^2G_t, \\
G_{t+1} &= \tfrac{1}{2}r^2F_t + \tfrac{1}{2}(1-r)^2G_t.
\end{aligned}
\tag{5.1}
$$

Because each genotype is composed of two gametes or haplotypes, we define $C_{t+1} + D_{t+1} + E_{t+1} + F_{t+1} + G_{t+1} = 2$, so that $C_1 = D_1 = E_1 = G_1 = 0$ and $F_1 = 2$. The derivations of the equations (5.1) can be explained as follows. When selfed, the homozygotes reproduce themselves only, so that C_t and D_t contribute only to C_{t+1} and D_{t+1}. But selfing the heterozygotes will generate segregation. For example, $AABb$ is selfed to give $\frac{1}{4}AABB : \frac{1}{2}AABb : \frac{1}{4}AAbb$. Therefore, the contribution of E_t to E_{t+1} is $\frac{1}{2}E_t$. But its contribution to C_{t+1} or D_{t+1} is doubled since there are twice as many classes in the proportions E_t as in C_{t+1} or D_{t+1}. The double heterozygote produces four haplotypes, AB, ab, aB, and ab with frequencies $\frac{1}{2}(1-r)$, $\frac{1}{2}r$, $\frac{1}{2}r$, and $\frac{1}{2}(1-r)$ for diplotype [AB][ab] or $\frac{1}{2}r$, $\frac{1}{2}(1-r)$, $\frac{1}{2}(1-r)$, and $\frac{1}{2}r$ for diplotype [Ab][aB]. Thus, the coefficients of the contribution of F_t or G_t to C_{t+1} and D_{t+1} must be the multiplication of the corresponding haplotype frequencies.

When t tends to be infinite, we will have $E_\infty = F_\infty = G_\infty = 0$, and D_∞ is the final proportion of crossover zygotes. Now let $C_t - D_t = c_t$ and $F_t - G_t = d_t$. Thus, by subtracting the equations for C_{t+1} and D_{t+1} as well as F_{t+1} and G_{t+1}, we have

$$
\begin{aligned}
c_{t+1} &= c_t + \tfrac{1}{4}(1-2r)d_t, \\
d_{t+1} &= \tfrac{1}{2}(1-2r)d_t.
\end{aligned}
\tag{5.2}
$$

Let us introduce a parameter λ, which makes $c_{t+1} + \lambda d_{t+1} \equiv c_t + \lambda d_t$ for all values of t. From equation (5.2), we have

$$c_t + \lambda d_t = c_t + \frac{1}{4}(1-2r)d_t + \frac{1}{2}\lambda(1-2r)d_t,$$

which leads to

$$\lambda = \frac{1-2r}{2(1+2r)}.$$

Based on the nature of RILs and the definition, we have $C_\infty + D_\infty = 1$ and $C_\infty - D_\infty = c_\infty$, which suggests $D_\infty = \frac{1}{2}(1-c_\infty)$. Note that

$$c_\infty = c_\infty + \lambda d_\infty = c_1 + \lambda d_1 = \frac{1-2r}{1+2r}$$

since $d_\infty = 0$ and $c_1 = 0$, $d_1 = 2$. Thus, it is easy to see that

$$D_\infty = \frac{1}{2}\left(1 - \frac{1-2r}{1+2r}\right) = \frac{2r}{1+2r}. \tag{5.3}$$

The proportion D_∞ is just the frequency of recombinant zygotes for two markers **A** and **B** in RILs denoted by R. Based on the relationship between two parameters R and r in equation (5.3), we solve the recombination fraction from the proportion of recombinant homozygotes as

$$r = \frac{R}{2(1-R)}. \tag{5.4}$$

Thus, by estimating R from a practical data set, equation (5.4) allows for the estimation of r.

Example 5.1. Zhang et al. (2004) reported a molecular linkage map for soybeans with 184 RILs derived from seven generations' selfing of the F_1 between varieties Kefeng No. 1 and Nannong 1138-2. The map covered 3596 cM of the soybean genome with 452 markers onto 21 linkage groups. We selected two pairs of markers, sat_300 (**A**) vs. sat_384 (**B**) and sat_384 (**C**) vs. sat_265 (**D**), for linkage analysis with observations given in Table 5.1.

Consider the pairs of markers above. The likelihood of marker observations for the proportion of recombinant homozygotes (R) can be constructed as

$$L(R) = \frac{n!}{n_{22}!n_{20}!n_{02}!n_{00}!} R^{n_{20}+n_{02}}(1-R)^{n_{22}+n_{00}}.$$

The maximization of the log-likelihood leads to the MLE of R as

$$\hat{R} = \frac{n_{20}+n_{02}}{n}. \tag{5.5}$$

Equation (5.5) is used to estimate the proportion of crossover zygotes as $\hat{R} = \frac{7+7}{175} = 0.080$ for the first (left panel) pair and $\hat{R} = \frac{0+6}{175} = 0.034$ for the second (right panel)

Table 5.1. Observations of nonrecombinant and recombinant homozygotes at two pairs of markers in soybean RILs by selfing.

Marker	Marker sat_384	
sat_300	BB	bb
AA	$n_{22} = 87$	$n_{20} = 7$
aa	$n_{02} = 7$	$n_{00} = 74$

Marker	Marker sat_265	
sat_384	DD	dd
CC	$n_{22} = 99$	$n_{20} = 0$
cc	$n_{02} = 6$	$n_{00} = 70$

Note: Total observations $n = n_{22} + n_{20} + n_{02} + n_{00}$.

pair. Equation (5.3) is further used to estimate the recombination fractions as $\hat{r} = \frac{0.080}{2(1-0.080)} = 0.043$ for the first marker pair and $\hat{r} = \frac{0.034}{2(1-0.034)} = 0.018$ for the second marker pair.

Robbins (1918) derived the proportions of different zygote types in each generation from equation (5.1), which are expressed as

$$C_n = \frac{1 - (\frac{1}{2} - r)^t}{1 + 2r} + \frac{1}{2}(\frac{1}{2} - r + r^2)^{t-1} - (\frac{1}{2})^{t-1},$$

$$D_n = \frac{2r + (\frac{1}{2} - r)^t}{1 + 2r} + \frac{1}{2}(\frac{1}{2} - r + r^2)^{t-1} - (\frac{1}{2})^{t-1}.$$

Using these two equations and estimated recombination fractions, we further estimate the proportions of nonrecombinant and recombinant homozygotes from generations 2 to 10 (Table 5.2). The dynamic patterns of these proportions will be useful for the prediction of genetic compositions over generations.

Table 5.2. Proportions of nonrecombinant and recombinant homozygotes in different generations for RILs derived from selfing.

	Generation								
Proportion	2	3	4	5	6	7	8	9	10
sat_300-sat_384									
C_t	0.457	0.677	0.778	0.824	0.845	0.854	0.859	0.861	0.862
D_t	0.001	0.012	0.018	0.020	0.021	0.022	0.022	0.022	0.022
sat_384-sat_265									
C_t	0.482	0.719	0.834	0.889	0.916	0.928	0.935	0.938	0.939
D_t	0.000	0.005	0.007	0.008	0.009	0.009	0.009	0.009	0.009

The proportion of double homozygotes increases dramatically in early generations. The degree and period of this increase depend on the magnitude of the recombination fraction. Compared to a loose linkage, a tight linkage is associated with a higher degree of increase and a smaller number of generations within which the homozygous proportions achieve a stable value.

5.2.2 Three-Point Analysis

Proportion of Recombinant Homozygotes

We turn to consider three markers in order **A**-**B**-**C**. In an RIL population, there are eight haplotypes, ABC, abc, ABc, abC, aBC, Abc, AbC, aBc, or genotypes, $AABBCC$, $aabbcc$, $AABBcc$, $aabbCC$, $aaBBCC$, $AAbbcc$, $AAbbCC$, $aaBBcc$. The frequencies of haplotypes are classified into four groups, $P(ABC) = P(abc)$, $P(ABc) = P(abC)$, $P(aBC) = P(Abc)$, $P(AbC) = P(aBc)$, which reflect the event that a haplotype has 0 crossover over marker intervals **A**-**B** and **B**-**C**, only one crossover in the second interval, only one crossover in the first interval, and two crossovers each in a different interval, respectively. Let $g_{(j_1 j_2)}$ $(\sum_{j_1,j_2} g_{(j_1 j_2)} = 1)$ be the probability that a haplotype has j_1 (0 or 1) crossovers in the first marker interval and j_2 (0 or 1) in the second marker interval. We use $n_{(j_1 j_2)}$ $(\sum_{j_1,j_2} n_{(j_1 j_2)} = n)$ to denote the corresponding genotype observations.

A likelihood function of marker genotype groups can be formulated as

$$L(n_{(j_1 j_2)}) = \frac{n!}{n_{(00)}! n_{(01)}! n_{(10)}! n_{(11)}!} g_{(00)}^{n_{(00)}} g_{(01)}^{n_{(01)}} g_{(10)}^{n_{(10)}} g_{(11)}^{n_{(11)}}, \tag{5.6}$$

with the MLE of $g_{(j_1 j_2)}$ obtained as

$$\hat{g}_{(j_1 j_2)} = \frac{n_{(j_1 j_2)}}{n}, \tag{5.7}$$

where

$$\begin{aligned} n_{(00)} &= n_{ABC} + n_{abc}, \\ n_{(01)} &= n_{ABc} + n_{abC}, \\ n_{(10)} &= n_{Abc} + n_{aBC}, \\ n_{(11)} &= n_{AbC} + n_{aBc}. \end{aligned} \tag{5.8}$$

We use allele-specific notation to denote marker observations. Table 5.3 presents the data structure of genotype observations for three linked markers **A**, **B**, and **C**.

More specifically, the probabilities of crossover occurrence can be expressed in terms of the proportion of offspring zygotes; that is,

$$\begin{aligned} \hat{g}_{(00)} &= \frac{n_{ABC} + n_{abc}}{n} = \hat{P}(ABC) + \hat{P}(abc) \\ \hat{g}_{(01)} &= \frac{n_{ABc} + n_{abC}}{n} = \hat{P}(ABc) + \hat{P}(abC) \\ \hat{g}_{(10)} &= \frac{n_{Abc} + n_{aBC}}{n} = \hat{P}(Abc) + \hat{P}(aBC) \\ \hat{g}_{(11)} &= \frac{n_{AbC} + n_{aBc}}{n} = \hat{P}(AbC) + \hat{P}(aBc). \end{aligned}$$

Table 5.3. Observations of nonrecombinant and recombinant homozygotes for three order-unknown markers **A**, **B**, and **C** in RILs by selfing.

Marker **A**	Marker **B**	Marker **C** CC	cc
AA	BB	n_{ABC}	n_{ABc}
AA	bb	n_{AbC}	n_{Abc}
aa	BB	n_{aBC}	n_{aBc}
aa	bb	n_{abC}	n_{abc}

Let $R_{\mathbf{AB}}$, $R_{\mathbf{BC}}$ and $R_{\mathbf{AC}}$ be the proportions of recombinant homozygotes between markers **A** and **B**, between **B** and **C**, and between **A** and **C**, respectively. Based on the definition of the recombination fraction, we have

$$\begin{aligned} R_{\mathbf{AB}} &= g_{(10)} + g_{(11)}, \\ R_{\mathbf{BC}} &= g_{(01)} + g_{(11)}, \cdot \\ R_{\mathbf{AC}} &= g_{(01)} + g_{(10)}. \end{aligned} \tag{5.9}$$

It is thus straightforward to derive the following equations:

$$\begin{aligned} g_{(00)} &= 2P(ABC) = 2P(abc) = \tfrac{1}{2}(2 - R_{\mathbf{AB}} - R_{\mathbf{BC}} - R_{\mathbf{AC}}), \\ g_{(01)} &= 2P(ABc) = 2P(abC) = \tfrac{1}{2}(-R_{\mathbf{AB}} + R_{\mathbf{BC}} + R_{\mathbf{AC}}), \\ g_{(10)} &= 2P(Abc) = 2P(aBC) = \tfrac{1}{2}(R_{\mathbf{AB}} - R_{\mathbf{BC}} + R_{\mathbf{AC}}), \\ g_{(11)} &= 2P(AbC) = 2P(aBc) = \tfrac{1}{2}(R_{\mathbf{AB}} + R_{\mathbf{BC}} - R_{\mathbf{AC}}), \end{aligned} \tag{5.10}$$

and

$$\begin{aligned} R_{\mathbf{AB}} &= P(AbC) + P(Abc) + P(aBC) + P(aBc) = 2P(AbC) + 2P(Abc), \\ R_{\mathbf{BC}} &= P(AbC) + P(abC) + P(ABc) + P(aBc) = 2P(AbC) + 2P(ABc), \\ R_{\mathbf{AC}} &= P(ABc) + P(Abc) + P(aBC) + P(abC) = 2P(ABc) + 2P(Abc). \end{aligned}$$

Thus, the proportion of recombinant homozygotes can be estimated from haplotype frequencies.

Gene Order Test in RILs

Three-point analysis can be used to find the most likely order of markers based on their proportions of recombinant homozygotes. Consider three unordered markers **A**, **B**, and **C**, whose observations are listed in Table 5.3. As shown above, the eight three-point genotypes are sorted into four groups, and the sample size for each group

Table 5.4. The expected frequencies of gametes under three different possible marker orders expressed in terms of the proportion of recombinant zygotes.

Gamete	Obser-	Expected Frequency under Order		
type	vation	**A-B-C**	**A-C-B**	**B-A-C**
ABC or *abc*	$n_{(00)}$	$(1-R_{\mathbf{AB}})(1-R_{\mathbf{BC}})$	$(1-R_{\mathbf{AC}})(1-R_{\mathbf{BC}})$	$(1-R_{\mathbf{AB}})(1-R_{\mathbf{AC}})$
ABc or *abC*	$n_{(01)}$	$(1-R_{\mathbf{AB}})R_{\mathbf{BC}}$	$R_{\mathbf{AC}}R_{\mathbf{BC}}$	$(1-R_{\mathbf{AB}})R_{\mathbf{AC}}$
Abc or *aBC*	$n_{(10)}$	$R_{\mathbf{AB}}(1-R_{\mathbf{BC}})$	$R_{\mathbf{AC}}(1-R_{\mathbf{BC}})$	$R_{\mathbf{AB}}R_{\mathbf{AC}}$
AbC or *aBc*	$n_{(11)}$	$R_{\mathbf{AB}}R_{\mathbf{BC}}$	$(1-R_{\mathbf{AC}})R_{\mathbf{BC}}$	$R_{\mathbf{AB}}(1-R_{\mathbf{AC}})$

is denoted by $n_{j_1j_2}$. By assuming three different possible marker orders, the expected frequencies of gametes under each order are given in Table 5.4.

Based on Table 5.4), we construct three order-specific likelihoods as

$$
\begin{aligned}
L_{\mathbf{ABC}} &\propto R_{\mathbf{AB}}^{n_{10}+n_{11}} R_{\mathbf{BC}}^{n_{01}+n_{11}} (1-R_{\mathbf{AB}})^{n_{00}+n_{01}} (1-R_{\mathbf{BC}})^{n_{00}+n_{10}},\\
L_{\mathbf{ACB}} &\propto R_{\mathbf{AC}}^{n_{01}+n_{10}} R_{\mathbf{BC}}^{n_{01}+n_{11}} (1-R_{\mathbf{AC}})^{n_{00}+n_{11}} (1-R_{\mathbf{BC}})^{n_{01}+n_{11}},\\
L_{\mathbf{BAC}} &\propto R_{\mathbf{AB}}^{n_{10}+n_{11}} R_{\mathbf{AC}}^{n_{01}+n_{10}} (1-R_{\mathbf{AB}})^{n_{00}+n_{01}} (1-R_{\mathbf{AC}})^{n_{00}+n_{11}}.
\end{aligned} \tag{5.11}
$$

Each individual likelihood yields the MLE of the proportion of recombinant homozygotes for each order, expressed as

$$
\begin{aligned}
\hat{R}_{\mathbf{AB}} &= \frac{n_{(10)}+n_{(11)}}{n},\\
\hat{R}_{\mathbf{BC}} &= \frac{n_{(01)}+n_{(11)}}{n},\\
\hat{R}_{\mathbf{AC}} &= \frac{n_{(01)}+n_{(10)}}{n}.
\end{aligned} \tag{5.12}
$$

These estimates are plugged in the likelihood (5.11), and we calculate the likelihood values for different orders. The largest likelihood value corresponds to the most marker order.

Genetic Distances and Mapping Function

Let $r_{\mathbf{AB}}$, $r_{\mathbf{BC}}$, and $r_{\mathbf{AC}}$ be the recombination fractions between markers **A** and **B**, between **B** and **C**, and between **A** and **C**, respectively. Under the assumption that there is no interference in crossover events during meiosis, the relationship among the three recombination fractions is determined by

$$
r_{\mathbf{AC}} = r_{\mathbf{AB}} + r_{\mathbf{BC}} - 2r_{\mathbf{AB}}r_{\mathbf{BC}} \tag{5.13}
$$

or

$$
(1-2r_{\mathbf{AC}}) = (1-2r_{\mathbf{AB}})(1-2r_{\mathbf{BC}}).
$$

Based on equation (5.4), we have as the relationship among the three R's

$$\left(\frac{1-2R_{\mathbf{AC}}}{1-R_{\mathbf{AC}}}\right)=\left(\frac{1-2R_{\mathbf{AB}}}{1-R_{\mathbf{AB}}}\right)\left(\frac{1-2R_{\mathbf{BC}}}{1-R_{\mathbf{BC}}}\right), \tag{5.14}$$

which can be changed for

$$(1-2R_{\mathbf{AC}})=\frac{(1-2R_{\mathbf{AB}})(1-2R_{\mathbf{BC}})}{1-2R_{\mathbf{AB}}R_{\mathbf{BC}}} \tag{5.15}$$

or

$$(1-R_{\mathbf{AC}})=\frac{(1-R_{\mathbf{AB}})(1-R_{\mathbf{BC}})}{1-2R_{\mathbf{AB}}R_{\mathbf{BC}}}.$$

Further, we have

$$R_{\mathbf{AC}}=\frac{R_{\mathbf{AB}}+R_{\mathbf{BC}}-3R_{\mathbf{AB}}R_{\mathbf{BC}}}{1-2R_{\mathbf{AB}}R_{\mathbf{BC}}}. \tag{5.16}$$

In case of no interference in each meiosis, the genetic distance (d) can be expressed in terms of the recombination fraction (r) by the Haldane map function $d=-\frac{1}{2}\ln(1-2r)$. Based on this, the genetic distance is calculated from the proportion of recombinant homozygotes as

$$d=-\tfrac{1}{2}\ln\left[\frac{1-2R}{1-R}\right]. \tag{5.17}$$

Nonindependence of Recombinations

When there is no interference between different marker intervals, the recombination fractions follow the relation of equation (5.13). It is obvious that this relation does not hold for the proportion of recombinant homozygotes; that is,

$$R_{\mathbf{AC}}\neq R_{\mathbf{AB}}+R_{\mathbf{BC}}-2R_{\mathbf{AB}}R_{\mathbf{BC}}.$$

This means that even if there is no interference at each meiosis, recombination in different intervals is still not independent in RILs. The degree of nonindependence in RIL data, in which there is no interference in each meiosis, is quantified by

$$\begin{aligned}\rho&=\frac{R_{\mathbf{AB}}+R_{\mathbf{BC}}-R_{\mathbf{AC}}}{2R_{\mathbf{AB}}R_{\mathbf{BC}}}\\&=\frac{2g_{(11)}}{2R_{\mathbf{AB}}R_{\mathbf{BC}}}.\end{aligned}$$

Together, we use the four ratios to quantify the degree of nonindependence for double recombinant zygotes ($\rho_{(00)}$), double nonrecombinant zygotes ($\rho_{(11)}$), and two single recombinant zygotes ($\rho_{(01)}$ and $\rho_{(10)}$), respectively:

$$
\begin{aligned}
\rho_{(11)} &= \frac{g_{(11)}}{R_{\mathbf{AB}}R_{\mathbf{BC}}} \geq 1, \\
\rho_{(10)} &= \frac{g_{(10)}}{R_{\mathbf{AB}}(1-R_{\mathbf{BC}})} \leq 1. \\
\rho_{(01)} &= \frac{g_{(01)}}{(1-R_{\mathbf{AB}})R_{\mathbf{BC}}} \leq 1, \\
\rho_{(00)} &= \frac{g_{(00)}}{(1-R_{\mathbf{AB}})(1-R_{\mathbf{BC}})} \geq 1.
\end{aligned}
\tag{5.18}
$$

Test of Meiosis Interference

In the derivations above, we assume no interference at each meiosis, but this may not be true for a real RIL data set. Martin and Hospital (2006) generalized Muller's (1916) coefficient of coincidence in individual meioses to examine true interference at each meiosis for RIL data using

$$
\begin{aligned}
I &= 1 - \frac{\hat{g}_{(11)}}{\breve{g}_{(11)}} \\
&= 1 - \frac{(1-2R_{\mathbf{AB}}R_{\mathbf{BC}})(R_{\mathbf{AB}}+R_{\mathbf{BC}}-R_{\mathbf{AC}})}{3-2R_{\mathbf{AB}}-2R_{\mathbf{BC}}},
\end{aligned}
\tag{5.19}
$$

where $\hat{g}_{(11)}$ is the observed frequency of double recombinant zygotes and $\breve{g}_{(11)}$ is the frequency of double recombinant zygotes if there is no interference. With no interference, we have shown the $R_{\mathbf{AC}}$ of double recombinant zygotes by equation (5.16), which is $\breve{g}_{(11)}$. If $I = 0$, this indicates that interference is absent.

Example 5.2. Zhang et al. (2004) reported a molecular linkage map for soybeans with 184 RILs derived from seven generations' selfing of the F_1 between varieties Kefeng No. 1 and Nannong 1138-2. The map covered 3596 cM of the soybean genome with 452 markers onto 21 linkage groups. A set of three markers, sat_300 (**A**), sat_384 (**B**), and sat_265 (**C**), with observations given in Table 5.5, are chosen for linkage analysis.

Based on the genotype observations, we calculate $n_{(00)} = n_{ABC} + n_{abc} = 148$, $n_{(01)} = n_{ABc} + n_{abC} = 6$, $n_{(10)} = n_{Abc} + n_{aBC} = 14$, and $n_{(11)} = n_{AbC} + n_{aBc} = 0$ with equation (5.8), which are used to calculate three proportions of recombinant homozygotes as $\hat{R}_{\mathbf{AB}} = 0.0833$, $\hat{R}_{\mathbf{BC}} = 0.0357$, and $\hat{R}_{\mathbf{AC}} = 0.1190$ with equation (5.12). Plugging in the likelihood (5.11) results in the likelihood values for three possible marker orders:

Table 5.5. Observations of nonrecombinant and recombinant homozygotes for three order-unknown markers **A**, **B**, and **C** in RILs by selfing in soybeans.

Marker **A**	Marker **B**	Marker **C** CC	Marker **C** cc
AA	BB	$n_{ABC} = 87$	$n_{ABc} = 0$
AA	bb	$n_{AbC} = 0$	$n_{Abc} = 7$
aa	BB	$n_{aBC} = 7$	$n_{aBc} = 0$
aa	bb	$n_{abC} = 6$	$n_{abc} = 61$

$$L_1 \propto (148+6)\log(1-0.08) + (148+14)\log(1-0.04) + (14+0)\log(0.08) + (6+0)\log(0.04) = -74.07, \quad \text{for order } \mathbf{A}\text{-}\mathbf{B}\text{-}\mathbf{C},$$

$$L_2 \propto (148+0)\log(1-0.12) + (148+14)\log(1-0.04) + (6+14)\log(0.12) + (6+0)\log(0.04) = -87.21, \quad \text{for order } \mathbf{A}\text{-}\mathbf{C}\text{-}\mathbf{B},$$

$$L_3 \propto (148+6)\log(1-0.08) + (148+0)\log(1-0.12) + (14+0)\log(0.08) + (6+14)\log(0.12) = -109.51, \quad \text{for order } \mathbf{B}\text{-}\mathbf{A}\text{-}\mathbf{C}.$$

According to the likelihood values above, we determine **A**-**B**-**C** as an optimal order for these three markers.

Using $\hat{g}_{(j_1 j_2)}$ calculated from equation (5.7) under order **A**-**B**-**C**, we calculate the proportions of recombinant homozygotes with equation (5.9) and recombination fractions and map distances by equations (5.4) and (5.17) as

	A-B	**B-C**	**A-C**
$R_{\mathbf{AB}}$	0.0833	0.0357	0.1190
			0.1108*
$r_{\mathbf{AB}}$	0.0454	0.0185	0.0676
			0.0623*
$d_{\mathbf{AB}}$	4.76	1.89	7.26
			6.65*

where the values indicated by asterisks were calculated from the relations of the three R's with equation (5.16). Based on the estimated R' values, we further perform the test of recombination nonindependence by equation (5.18) and meiotic interference by equation (5.19) as follows:

	Value
$\rho_{(11)}$	0
$\rho_{(10)}$	1.0370
$\rho_{(01)}$	1.0909
$\rho_{(00)}$	0.9267
I	1.0000

It can be seen that the adjacent marker intervals display some degree of nonindependence and interference among markers sat_300, sat_384, and sat_265 in soybean RILs.

5.3 RILs by Sibling Mating

5.3.1 Two-point Analysis

There are many species, such as the mouse, for which selfing is not possible. For these species, RILs can be produced by repeated sibling mating of the progeny initiated from the F_1 between two inbred lines $AABB\cdots$ and $aabb\cdots$. Analogous to selfing RILs, Haldane and Waddington (1931) also derived the relationships between the recombination fraction and the proportion of recombinant homozygotes in RILs by sibling mating. Considering two markers on autosomes, the brother–sister mating of the F_1 leads to ten different diplotypes in the F_2, for which there are a total of $\frac{1}{2}(10 \times 11) = 55$ sibling mating types to generate the F_3. Haldane and Waddington (1931) showed that when sibling mating was performed for a large number of generations, the relationship between the proportion of recombinant homozygotes (R) and recombination fraction (r) obeys the following forms:

$$R = \frac{4r}{1+6r}, \tag{5.20}$$

$$r = \frac{R}{2(2-3R)}. \tag{5.21}$$

In the case of sibling mating, sex-linked genes on chromosome X will inherit differently from autosomal genes. The original mating for sex-linked genes is $AABB\cdots \times ab\cdots$. Haldane and Waddington (1931) also derived similar relationships for two sex-linked genes, expressed as

$$R = \frac{8r}{3(1+4r)}, \tag{5.22}$$

$$r = \frac{3R}{4(2-3R)}. \tag{5.23}$$

5.3.2 Three-point Analysis

The procedure for three-point analysis in selfing RILs can be modified for RILs by sibling mating. Consider three markers **A**-**B**-**C**. Assuming no interference at meiosis, three proportions of recombinant homozygotes for markers **A**, **B**, and **C** in the case of sibling mating follow

$$\left(\frac{2-4R_{\mathbf{AC}}}{2-3R_{\mathbf{AC}}}\right) = \left(\frac{2-4R_{\mathbf{AB}}}{2-3R_{\mathbf{AB}}}\right)\left(\frac{2-4R_{\mathbf{BC}}}{2-3R_{\mathbf{BC}}}\right), \tag{5.24}$$

$$(1-2R_{\mathbf{AC}}) = \frac{(1-2R_{\mathbf{AB}})(1-2R_{\mathbf{BC}})}{1-3R_{\mathbf{AB}}R_{\mathbf{BC}}}, \tag{5.25}$$

or

$$R_{\mathbf{AC}} = \frac{2R_{\mathbf{AB}}+2R_{\mathbf{BC}}-7R_{\mathbf{AB}}R_{\mathbf{BC}}}{2(1-3R_{\mathbf{AB}}R_{\mathbf{BC}})}. \tag{5.26}$$

The genetic distance is calculated from the proportion of recombinant homozygotes by the Haldane map function as

$$d = -\frac{1}{2}\ln\left[\frac{2-4R}{2-3R}\right]. \tag{5.27}$$

5.4 Bias Reduction

5.4.1 RILs by Selfing

Martin and Hospital (2006) recognized that the estimate of the recombination fraction (r) from the estimated proportion of recombinant homozygotes (R) using equation (5.4), written as

$$\tilde{r} = \frac{\hat{R}}{2(1-\hat{R})}, \tag{5.28}$$

is biased because these two parameters are not linearly related. The unbiased estimate $\hat{R}$ by equation (5.5) can be obtained. As a result, $\hat{R}$ can be viewed as the sum of the true value and a random noise of zero mean (that is, $\hat{R} = R + e(0,\sigma^2)$), where $e(0,\sigma^2)$ denotes a random variable with mean 0 and variance σ^2.

Rewriting the estimator of r by equation (5.28) and then performing the Taylor series expansion in $\epsilon_R \equiv (\hat{R}-R)/(1-R)$, we have

$$
\begin{aligned}
\tilde{r} &= \frac{\hat{R}}{2(1-\hat{R})} \\
&= -\frac{1}{2} + \frac{1}{2(1-\hat{R})} \\
&= \frac{R + \epsilon_R + \epsilon_R^2 + \epsilon_R^3 + \ldots}{2(1-R)}.
\end{aligned}
$$

Since $\hat{R}$ is unbiased, the expected value of $\tilde{r}$ can be taken as

$$
E[\tilde{r}] = r + \frac{E[(\hat{R}-R)^2]}{2(1-R)^3} + \ldots,
$$

where the higher-order terms are associated with high-order moments of $(\hat{R} - R)$. Because the expectation $E[(\hat{R} - R)^2]$ is estimated by $\hat{R}(1-\hat{R})/n$, the estimator of r, for which most of the bias has been removed, can be approximated by

$$
\begin{aligned}
\hat{r} &= \frac{\hat{R}}{2(1-\hat{R})} - \frac{\hat{R}(1-\hat{R})}{2n(1-\hat{R})^3} \\
&= \tilde{r}\left[1 - \frac{1}{n-m}\right] \\
&= \frac{m(n-m-1)}{2(n-m)^2}, \qquad (5.29)
\end{aligned}
$$

where m is the number of recombinant zygotes in RILs. The estimator r is still biased (although the remaining bias is now only of order $1/n^2$) unless all order corrections are considered.

Example 5.3. Revisit Example 5.1. The recombination fractions for two pairs of markers were estimated for the nonadjusted ($\tilde{r}$) and adjusted values ($\hat{r}$) using equations (5.28) and (5.29), respectively, and are given below as

Marker Pair	$\tilde{r}$	$\hat{r}$
sat_300–sat_384	0.0435	0.0432
sat_384–sat_265	0.0178	0.0176

The nonadjusted equation slightly overestimates the recombination fraction for the two pairs of markers in this example. In practice, therefore, the adjusted equation is recommended, especially for a small sample size and a loose linkage.

5.4.2 RILs by Sibling Mating

Martin and Hospital (2006) also derived a procedure for reducing the bias of the estimator of r from R in RILs derived from sibling mating. The adjusted equation for removing a major bias for the estimate of the recombination fraction is given as

$$\hat{r} = \frac{\hat{R}}{4-6\hat{R}} - \frac{24\hat{R}(1-\hat{R})}{N(4-6\hat{R})^3}$$
$$= \tilde{r}\left[1 - \frac{6(n-m)}{(2n-m)^2}\right]$$
$$= \frac{m}{4n-m}\left[1 - \frac{6(n-m)}{(2n-3m)^2}\right]. \quad (5.30)$$

Equations (5.29) and (5.30) are derived for bias reduction when the recombination fraction is estimated from the proportion of recombinant homozygotes. In Note 5.7, we provide a general procedure for bias reduction with which interesting readers can derive any adjusted formula of interest.

5.5 Multiway RILs

Recombinant inbred strains are derived from repeated inbreeding of the progeny initiated with the F_1 between two inbred lines 1 and 2. The RILs derived from two inbred lines are called "two-way RILs" (1×2). Given the complexity of the genome, RILs have been produced from multiple crosses between many inbred lines. Currently, four-way RILs, $(1\times2)\times(3\times4)$, and eight-way RILs, $[(1\times2)\times(3\times4)]\times[(5\times6)\times(7\times8)]$, have been produced in model species, such as the mouse (Threadgill et al. 2002; Complex Trait Consortium 2004). Figure 5.1 describes the procedure for generating eight-way RILs in which the progeny initiated with eight parental inbred lines are repeatedly selfed or sibling-mated. This procedure will generate new inbred lines whose genome is a mosaic of the eight parental strains. Thus, higher-way RILs can be much more powerful to study the pattern of genetic variation in natural populations than current mapping resources and could be used to dissect virtually any complex human disease using mouse models.

Based on Haldane and Waddington's (1931) work, Broman (2005) proposed a general procedure for deriving the formulas for expressing the proportions of recombinant zygotes in terms of the recombination fraction. This procedure can be used for various types of RILs and also for linked genes on the X chromosome (Table 5.6).

5.6 Exercises

5.1 As two different mapping designs, discuss the similarities and differences between the RIL and backcross designs. What advantages do the RIL design have compared with the backcross design?

5.2 Show that the genetic distance calculated from R (that is, $d^* = -\frac{1}{2}\ln(1-2R)$), is not additive; that is, for three markers **A-B-C**, we have $d^*_{\mathbf{AC}} \neq d^*_{\mathbf{AB}} + d^*_{\mathbf{BC}}$.

5.3 For a backcross design, the estimates of the recombination fractions from two- and three-point analyses should be identical if there are no missing data. Show how different the estimates of the recombination fractions are from two- and three-point analyses for the RIL design.

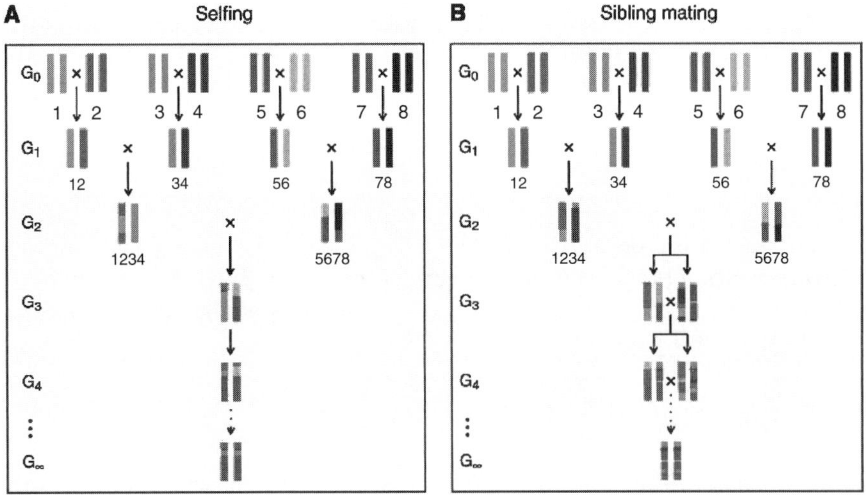

Fig. 5.1. A procedure for generating eight-way RILs by selfing (**A**) or sibling mating (**B**). Adapted from Broman (2005).

Table 5.6. Summary for the relationships between the recombination fraction and the proportion of recombinant homozygotes in different types of RILs.

		Sibling Mating	
	Selfing	Autosome	X Chromosome
2-way	$\frac{2r}{1+2r}$	$\frac{4r}{1+6r}$	$\frac{8r}{3(1+4r)}$
4-way	$\frac{3r}{1+2r}$	$\frac{6r}{1+6r}$	$\frac{4r}{1+4r}$
8-way	$\frac{r(4-r)}{1+2r}$	$\frac{7r}{1+6r}$	$\frac{14r}{3(1+4r)}$
16-way	$\frac{r(5-3r+r^2)}{1+2r}$	$\frac{r(8-r)}{1+6r}$	$\frac{r(16-r)}{3(1+4r)}$
2^t-way	$1-\frac{(1-r)^{t-1}}{1+2r}$	$1-\frac{(1-r)^{t-2}}{1+6r}$	–

5.4 If the same estimate for the proportion of recombinant homozygotes between two given markers is obtained between the RILs by selfing and RILs by sibling mating, show how different the respective recombination fractions are for these two different types of RILs.

5.7 Note

We provide a general procedure for deriving the equation for bias reduction. Consider an arbitrary relation between R and r, represented as $r = f(R)$. The estimator of the recombination fraction $\tilde{r} = f(\hat{R})$ is biased because f is nonlinear. The bias can be computed from the Taylor series expansion:

$$E[f(\hat{R})] = f(R) + f'(R)E[\hat{R} - R] + \frac{1}{2}f''(R)E[(\hat{R} - R)^2] + \ldots.$$

Since $\hat{R}$ is unbiased, the expectation $E[\hat{R} - R]$ vanishes and the leading bias in the estimator for r comes from the variance of $\hat{R}$, which is approximated by $\hat{R}(1 - \hat{R})/n$. Replacing $f(R)$ by $f(\hat{R})$, the correct estimator is

$$\hat{r} = f(\hat{R}) - \frac{f''(\hat{R})\hat{R}(1 - \hat{R})}{2n},$$

where $f''(\hat{R})$ is the second derivative of f evaluated at the point $\hat{R}$. Note that we have ignored the higher-order terms. Although this is usually not a problem, formally this needs to be justified.

6

Linkage Analysis for Distorted and Misclassified Markers

6.1 Introduction

We have described statistical methods for linkage analysis of different markers in a controlled cross. But these methods can only be appropriate for the markers whose segregation follows the Mendelian ratio (1:1 for the backcross or 1:2:1 for the F_2). In a practical molecular experiment, many markers may deviate from Mendelian segregation ratios due to some genetic or biological reason (e.g., differential *viability*, where viability is defined as an individual's ability to survive). If marker segregation is disturbed by viability effects, we can incorporate such effects into linkage analysis using modified methods. Bailey (1961) discussed several models for analyzing distorted markers in great detail. In this chapter we will describe the methods for estimating the recombination fraction between markers subject to differential viability.

The differential viability type of disturbance may arise in various ways. First, gametes bearing different genes may have unequal survival rates, so that fertilization is more likely to be brought about by one type than by another. Thus, an individual of genotype Aa produces two kinds of gametes, namely A and a, but these may not be equally available. Second, there may be differential survival of different types of zygotes. In a backcross, $Aa \times aa$, equal numbers of Aa and aa individuals can be expected, but they may survive unequally. For the backcross, gametic and zygotic differential viabilities have the same effect on linkage analysis. But, for the F_2, different models are needed to detect the influences of gametic and zygotic differential viability on linkage analysis.

6.2 Gametic Differential Viability

6.2.1 One-Gene Model

Suppose one of the two markers **A** and **B** subject to linkage analysis is affected by differential viability. Let r be the recombination fraction between the two markers and u be the viability of allele A relative to its alternative a for marker **A**.

Backcross $AaBb \times aabb$

For a double backcross $AaBb \times aabb$, we have observed numbers of individuals, $\mathbf{n} = (n_1, n_2, n_3, n_4)$, and expected frequencies for each of the four genotypes as follows:

Gamete	Expected	Observed
AB	$\frac{u}{1+u}(1-r)$	n_1
Ab	$\frac{u}{1+u}r$	n_2
aB	$\frac{1}{1+u}r$	n_3
ab	$\frac{1}{1+u}(1-r)$	n_4
Total	1	n

The log-likelihood of the data is

$$\begin{aligned}
&\log L(u, r|\mathbf{n}) \\
&= \log\left[\frac{n!}{n_1!n_2!n_3!n_4!}\left(\frac{u}{1+u}(1-r)\right)^{n_1}\left(\frac{u}{1+u}r\right)^{n_2}\left(\frac{1}{1+u}r\right)^{n_3}\left(\frac{1}{1+u}(1-r)\right)^{n_4}\right] \\
&= (n_1+n_2)\log u + (n_2+n_3)\log r + (n_1+n_4)\log(1-r) \\
&\quad -n\log(1+u) + \log\left[\frac{n!}{n_1!n_2!n_3!n_4!}\right],
\end{aligned}$$

whose scores are

$$\begin{cases} S_r = \dfrac{\partial}{\partial r}\log L(u, r|\mathbf{n}) = \dfrac{n_2+n_3}{r} - \dfrac{n_1+n_4}{1-r} \\ S_u = \dfrac{\partial}{\partial u}\log L(u, r|\mathbf{n}) = \dfrac{n_1+n_2}{u} - \dfrac{n}{1-u} \end{cases}$$

with information matrix

$$\begin{aligned}
\mathbf{I} &= -E\begin{bmatrix} \frac{\partial^2}{\partial r^2}\log L(u, r|\mathbf{n}) & \frac{\partial^2}{\partial r\partial u}\log L(u, r|\mathbf{n}) \\ \frac{\partial^2}{\partial u\partial r}\log L(u, r|\mathbf{n}) & \frac{\partial^2}{\partial u^2}\log L(u, r|\mathbf{n}) \end{bmatrix} \\
&= \begin{bmatrix} \frac{n}{r(1-r)} & 0 \\ 0 & \frac{n}{u(1+u)^2} \end{bmatrix}.
\end{aligned}$$

The MLEs of the recombination fraction and viability coefficient with their large-sample variances are thus

$$\begin{aligned}
\hat{r} &= \frac{n_2+n_3}{n}, \\
\hat{u} &= \frac{n_1+n_2}{n_3+n_4},
\end{aligned}$$

and

$$\text{var}(\hat{r}) = \frac{r(1-r)}{n},$$
$$\text{var}(\hat{u}) = \frac{u(1+u)^2}{n}.$$

It can be seen that the MLE of the recombination fraction for a backcross design is unchanged, regardless of the existence of viability disturbance for one of the two markers. The test of the linkage between markers **A** and **B** can be made using equation (3.8). The difference between the numbers n_1+n_2 and n_3+n_4 measures the departure from unity of the segregation ratio for A:a. The hypothesis about the existence of viability disturbance can be formulated as

$$H_0 : u = 1 \text{ vs. } H_1 \neq 1, \tag{6.1}$$

where H_0 corresponds to the reduced model (i.e., nonexistence of viability disturbance).

The test statistics for testing the hypotheses (6.1) are calculated as the log-likelihood ratio (LR) of the hypothesis under the full model over the reduced model:

$$\text{LR} = -2\log\left[\frac{L(u=1,\hat{r})}{L(\hat{u},\hat{r})}\right].$$

The test statistic LR can be viewed as being asymptotically χ^2-distributed with one degree of freedom.

Example 6.1. Differential viability of the genotypes is shown below in a double-backcross sample.

Gamete	AB	Ab	aB	ab
Observed	$n_1 = 50$	$n_2 = 2$	$n_3 = 1$	$n_4 = 32$

The MLEs of the recombination fraction and viability coefficient with their large-sample variances are thus

$$\hat{r} = \frac{2+1}{85} = 0.0353,$$
$$\hat{u} = \frac{50+2}{1+32} = 1.576,$$

and

$$\text{var}(\hat{r}) = \frac{0.0353(1-0.0353)}{85} = 0.00040,$$
$$\text{var}(\hat{u}) = \frac{1.576(1+1.576)^2}{85} = 0.1230.$$

The log-likelihood for no viability disturbance is

$$\begin{aligned}\mathrm{LL}_0 = 50\log\left[\frac{1-0.0353}{2}\right] + 2\log\left[\frac{0.0353}{2}\right] \\ +1\log\left[\frac{0.0353}{2}\right] + 32\log\left[\frac{1-0.0353}{2}\right] = -71.8961.\end{aligned}$$

Similarly, the log-likelihood for viability disturbance on the **A** locus is

$$\begin{aligned}LL_1 = 50\log\left[\frac{1.576}{1+1.576}\times(1-0.0353)\right] + 2\log\left[\frac{1.576}{1+1.576}\times 0.0353\right] \\ +1\log\left[\frac{1}{1+1.576}\times 0.0353\right] + 32\log\left[\frac{1}{1+1.576}\times(1-0.0353)\right] = -69.755.\end{aligned}$$

The LR statistic for viability disturbance on the **A** locus versus no disturbance is

$$LR = -2\log\left[\frac{L(u=1,\hat{r})}{L(\hat{u},\hat{r})}\right] = -2[-71.8961-(-69.7545)] = 4.283.$$

Compared with $\chi^2_{0.05}(1) = 3.841$, the viability disturbance on the **A** locus is significant.

Double Intercross $AaBb \times AaBb$

For a double intercross $AaBb \times AaBb$, the frequencies of nine F_2 genotypes for markers **A** and **B** can be expressed in matrix notation as

$$\begin{array}{c} \\ AA \\ Aa \\ aa \end{array}\begin{array}{c} \begin{array}{ccc} BB & Bb & bb \end{array} \\ \left[\begin{array}{ccc} \frac{u^2}{(1+u)^2}(1-r)^2 & \frac{2u^2}{(1+u)^2}r(1-r) & \frac{u^2}{(1+u)^2}r^2 \\ \frac{2u}{(1+u)^2}r(1-r) & \frac{2u}{(1+u)^2}[(1-r)^2+r^2] & \frac{2u}{(1+u)^2}r(1-r) \\ \frac{1}{(1+u)^2}r^2 & \frac{2}{(1+u)^2}r(1-r) & \frac{1}{(1+u)^2}(1-r)^2 \end{array}\right] \end{array}, \tag{6.2}$$

denoted by $P = \{p_{ij}, i = 2,1,0; j = 2,1,0\}$.

The data matrix is given by

$$\mathbf{n} = \begin{array}{c} \\ AA \\ Aa \\ aa \end{array}\begin{array}{c} \begin{array}{ccc} BB & Bb & bb \end{array} \\ \left[\begin{array}{ccc} n_{22} & n_{21} & n_{20} \\ n_{12} & n_{11} & n_{10} \\ n_{02} & n_{01} & n_{00} \end{array}\right] \end{array}.$$

The likelihood function of the data is

$$L(u,r|\mathbf{n}) = \frac{n!}{n_{22}!n_{21}!\cdots n_{00}!}\prod_{i=0}^{2}\prod_{j=0}^{2}p_{ij}^{n_{ij}}.$$

Based on matrix (6.2), it is not difficult to derive the MLE of the viability coefficient as

$$\hat{u} = \frac{2(n_{22}+n_{21}+n_{20})+n_{12}+n_{11}+n_{10}}{2(n-n_{22}-n_{21}-n_{20})-(n_{12}+n_{11}+n_{10})}.$$

The MLE of the recombination fraction in the F_2 can also be obtained from matrix (6.2). But the EM algorithm is necessary because no closed form for $\hat{r}$ can be derived. By looking closely at matrix (6.2), we found that the expected numbers of recombinants within the corresponding cells are actually unchanged when one marker displays viability disturbance. This is because the coefficients within each cell contain no information about the recombination fraction. Thus, we can estimate $\hat{r}$ directly using formulas in the M step,

$$\hat{r} = \frac{1}{2n}[n_{12}+n_{21}+n_{10}+n_{01}+2(n_{02}+n_{20})+\phi n_{11}],$$

where

$$\phi = \frac{2r^2}{(1-r)^2+r^2},$$

in the E step, is used to calculate the expected number of recombinants for the double heterozygote (see Chapter 3). In closing, linkage analysis in the F_2 is also not affected by viability effects of one marker, as in the backcross.

The sampling variances of the MLEs of the recombination fraction and viability coefficient in the F_2 can be derived as

$$\begin{aligned}\text{var}(\hat{r}) &= \frac{r(1-r)(1-2r+2r^2)^2(1+u)^2}{2n((2r^2-2r+1)u^2+(8r^2-8r+2)u+2r^2-2r+1)},\\ \text{var}(\hat{u}) &= \frac{u(1+u)^2}{2n}.\end{aligned}$$

Similarly, we can test the hypothesis about the existence of viability disturbance.

6.2.2 Two-Gene Model

We now turn to the more general case in which both genes are disturbed by viability effects. Suppose that, in addition to the parameter u defined above for the relative excess of A over a, we also have v for the viability of B relative to b. We also assume that the two viability effects operate independently of one another.

Backcross $AaBb \times aabb$

The observations and expectations for a double backcross are now as shown below:

Gamete	Expected	Observed
AB	$\frac{uv}{d}(1-r)$	n_1
Ab	$\frac{u}{d}r$	n_2
aB	$\frac{v}{d}r$	n_3
ab	$\frac{1}{d}(1-r)$	n_4
Total	1	n

where

$$d = uv(1-r) + ur + vr + (1-r).$$

The likelihood of the data is

$$L(u, r|\mathbf{n}) = \frac{n!}{n_1!n_2!n_3!n_4!}\left[\frac{uv}{d}(1-r)\right]^{n_1}\left[\frac{u}{d}r\right]^{n_2}\left[\frac{v}{d}r\right]^{n_3}\left[\frac{1}{d}(1-r)\right]^{n_4}.$$

The MLEs of the three parameters can be obtained in the usual way:

$$\hat{r} = \frac{\sqrt{n_2n_3}}{\sqrt{n_1n_4} + \sqrt{n_2n_3}},$$
$$\hat{u} = \sqrt{\frac{n_1n_2}{n_3n_4}},$$
$$\hat{v} = \sqrt{\frac{n_1n_3}{n_2n_4}}.$$

We further obtain the sampling variances

$$\text{var}(\hat{r}) = \frac{r^2(1-r)^2}{h},$$
$$\text{var}(\hat{u}) = \frac{u^2}{h},$$
$$\text{var}(\hat{v}) = \frac{v^2}{h},$$

where

$$h = \frac{4n_1n_2n_3n_4}{n_1n_2n_3 + n_1n_2n_4 + n_1n_3n_4 + n_2n_3n_4}.$$

We can formulate log-likelihood ratio test statistics to test for the existence of linkage and viability effects for two markers. From the analysis above, the test and detection of linkage between two distorted markers cannot be conducted using the formula derived for the two markers neither or only one of which has the deviation from the Mendelian segregation ratio.

Example 6.2. Consider the differential viability of both markers in the data in Example 6.1.

The MLEs of the recombination fraction and viability coefficient with their large-sample variances are thus

$$\hat{r} = \frac{\sqrt{2 \times 1}}{\sqrt{50 \times 32} + \sqrt{2 \times 1}} = 0.0341,$$
$$\hat{u} = \sqrt{\frac{50 \times 2}{1 \times 32}} = 1.7678,$$
$$\hat{v} = \sqrt{\frac{50 \times 1}{2 \times 32}} = 0.8839,$$

and

$$h = \frac{4n_1n_2n_3n_4}{n_1n_2n_3 + n_1n_2n_4 + n_1n_3n_4 + n_2n_3n_4} = 2.5786,$$

$$\text{var}(\hat{r}) = \frac{0.0341^2(1-0.0341)^2}{2.5786} = 0.000422,$$

$$\text{var}(\hat{u}) = \frac{1.7678^2}{2.5786} = 1.2119,$$

$$\text{var}(\hat{v}) = \frac{0.8839^2}{2.5786} = 0.3030.$$

The log-likelihood for viability disturbance on both the **A** and **B** loci is

$$\begin{aligned}\text{LL}_2 &= 50\log\left[(1-0.0341)1.7678 \times \frac{0.8839}{2.5655}\right] + 2\log\left[0.0341 \times \frac{1.7678}{2.5655}\right] \\ &+1\log\left[0.0341 \times \frac{0.8839}{2.5655}\right] + 32\log\left[\frac{1-0.0341}{2.5655}\right] = -69.734,\end{aligned}$$

where $\hat{d} = \hat{u}\hat{v}(1-\hat{r}) + \hat{u}\hat{r} + \hat{v}\hat{r} + (1-\hat{r}) = 2.5655$.

The LR statistic for viability disturbance on the **A** and **B** loci versus no disturbance is

$$\text{LR} = -2\log\left[\frac{L(u=1, v=1, \hat{r})}{L(\hat{u}, \hat{v}, \hat{r})}\right] = -2[-71.8961 - (-69.7344)] = 4.3234.$$

Compared with $\chi^2_{0.05}(2) = 5.991$, the viability disturbance is not significant.

Double Intercross $AaBb \times AaBb$ $\mathbf{F_2}$

When both markers are distorted, we have the frequencies of the nine genotypes in the F_2 population as

$$\mathbf{F} = \begin{matrix} \\ AA \\ Aa \\ aa \end{matrix} \begin{matrix} BB & Bb & bb \\ \left[\begin{matrix} \frac{u^2v^2}{d^2}(1-r)^2 \\ \frac{2uv^2}{d^2}r(1-r) \\ \frac{v^2}{d^2}r^2 \end{matrix}\right. & \begin{matrix} \frac{2u^2v}{d^2}r(1-r) \\ \frac{2uv}{d^2}[(1-r)^2+r^2] \\ \frac{2v}{d^2}r(1-r) \end{matrix} & \left.\begin{matrix} \frac{u^2}{d^2}r^2 \\ \frac{2u}{d^2}r(1-r) \\ \frac{1}{d^2}(1-r)^2 \end{matrix}\right] \end{matrix}, \tag{6.3}$$

$$= \begin{bmatrix} (1-r)^2 & 2r(1-r) & r^2 \\ 2r(1-r) & 2[(1-r)^2+r^2] & 2r(1-r) \\ r^2 & 2r(1-r) & (1-r)^2 \end{bmatrix} \circ \begin{bmatrix} \frac{u^2v^2}{d^2} & \frac{u^2v}{d^2} & \frac{u^2}{d^2} \\ \frac{uv^2}{d^2} & \frac{uv}{d^2} & \frac{u}{d^2} \\ \frac{v^2}{d^2} & \frac{v}{d^2} & \frac{1}{d^2} \end{bmatrix} \tag{6.4}$$

$$= \mathbf{F}_r \circ \mathbf{F}_{uvr},$$

where $\circ$ stands for the elementwise product between the two matrices, the first ($\mathbf{F}_r$) only associated with r and the second ($\mathbf{F}_{uvr}$) with u, v, and r.

The EM algorithm can be used to provide the MLE of r based on matrix (6.3), but this will be difficult to derive because the coefficients within each cell of this matrix contain r. By dividing the matrix (6.3) into two component matrices in matrix (6.4), however, we can simplify this derivation process. The MLE of r based on the first component matrix, $\mathbf{F}_r$, has been given, as can be seen in Section 3.5.2, which includes the E step to calculate the expected number of recombination events for the double F_2 heterozygote $AaBb$ using $\phi = \frac{2r^2}{(1-r)^2+r^2}$. The MLE of r based on the second component matrix, $\mathbf{F}_{uvr}$, can be obtained by maximizing the likelihood constructed by matrix (6.3) and the observation matrix $\mathbf{n}$. The likelihood is given by

$$\begin{aligned}\log L &= \mathbf{1}^T(\mathbf{n} \circ \log \mathbf{F})\mathbf{1} \\ &= \mathbf{1}^T[\mathbf{n} \circ \log(\mathbf{F} \circ \mathbf{F}_{uvr})]\mathbf{1} \\ &= \mathbf{1}^T(\mathbf{n} \circ \log \mathbf{F})\mathbf{1} + \mathbf{1}^T(\mathbf{n} \circ \mathbf{F}_{uvr})\mathbf{1},\end{aligned}$$

where $\mathbf{1} = (1, 1, 1)$.

Combining these two processes leads to the M step, as given below, to estimate r by

$$\begin{aligned}\hat{r} &= \frac{1}{2n}[n_{12} + n_{21} + n_{10} + n_{01} + 2(n_{02} + n_{20}) + \phi n_{11}] \\ &\quad + \frac{(u-1)(v-1)(1-r)r}{d},\end{aligned} \tag{6.5}$$

where the estimates of the viability coefficients are obtained by

$$\hat{u} = \frac{[2(n_{22} + n_{21} + n_{20}) + n_{12} + n_{11} + n_{10}][vr + (1-r)]}{[2(n - n_{22} - n_{21} - n_{20}) - (n_{12} + n_{11} + n_{10})][v(1-r) + r]},$$

$$\hat{v} = \frac{[2(n_{22} + n_{12} + n_{02}) + n_{21} + n_{11} + n_{01}][ur + (1-r)]}{[2(n - n_{22} - n_{12} - n_{02}) - (n_{21} + n_{11} + n_{01})][u(1-r) + r]}.$$

In the F_2, the estimation of the linkage will be affected when both markers are distorted. The existence of the linkage and viability effects can be tested using the likelihood ratio test.

The Fisher information matrix for r, u, and v can be derived as

$$2n \begin{bmatrix} \frac{u(1-2r+2r^2)(uv+(v+1)^2)-2uv+v+2rv(r-1)}{r(2r^2-2r+1)(1-r)d^2} & \frac{v^2-1}{d^2} & \frac{u^2-1}{d^2} \\ \frac{v^2-1}{d^2} & \frac{(vr+1-r)(rv-v-r)}{ud^2} & \frac{2r-1}{d^2} \\ \frac{u^2-1}{d^2} & \frac{2r-1}{d^2} & \frac{(ur+1-r)(ur-u-r)}{vd^2} \end{bmatrix}.$$

The sampling variances of the MLEs of the recombination fraction and viability coefficients in the F_2 can be derived from the inverse of the Fisher information matrix.

6.2.3 Simulation

We have described the equations for estimating the recombination fraction between distorted markers. But it is unclear how gametic viability influences the estimate of the recombination fraction. This can be examined by simulation studies. A marker genotype data set was simulated for a backcross of different sample sizes ($n = 80$ or 200) for two viability markers displaying tight ($r = 0.05$) or loose linkages ($r = 0.35$). The simulated data were analyzed with the viability model (1) proposed in Section 6.2.2 as well as the linkage model (2) used for normal markers.

Table 6.1. The MLEs of the recombination fraction between two distorted markers with different degrees of gametic viability disturbance in the backcross progeny. The sampling errors (SE) of the estimates are given from 100 simulation replicates.

	Model 1			Model 2
(u, v, r)	$\hat{r} \pm$ SE	$\hat{u} \pm$ SE	$\hat{v} \pm$ SE	$\hat{r}_0 \pm$ SE
$n = 80$				
(1.0, 1.0, 0.05)	0.053 ± 0.021	1.099 ± 0.438	1.067 ± 0.409	0.056 ± 0.022
(1.0, 1.0, 0.35)	0.359 ± 0.054	1.064 ± 0.226	1.014 ± 0.237	0.359 ± 0.052
(1.0, 1.0, 0.05)	0.053 ± 0.021	1.099 ± 0.438	1.067 ± 0.409	0.056 ± 0.022
(1.0, 1.0, 0.35)	0.359 ± 0.054	1.064 ± 0.226	1.014 ± 0.237	0.359 ± 0.052
(2.0, 2.0, 0.05)	0.062 ± 0.023	2.410 ± 1.112	2.119 ± 0.989	0.053 ± 0.020
(2.0, 2.0, 0.35)	0.362 ± 0.056	2.156 ± 0.596	2.045 ± 0.544	0.313 ± 0.047
$n = 200$				
(1.0, 1.0, 0.05)	0.047 ± 0.017	1.123 ± 0.453	1.054 ± 0.435	0.050 ± 0.016
(1.0, 1.0, 0.35)	0.354 ± 0.033	0.996 ± 0.144	1.017 ± 0.163	0.355 ± 0.033
(1.0, 1.0, 0.05)	0.053 ± 0.021	1.099 ± 0.438	1.067 ± 0.409	0.056 ± 0.022
(1.0, 1.0, 0.35)	0.359 ± 0.054	1.064 ± 0.226	1.014 ± 0.237	0.359 ± 0.052
(2.0, 2.0, 0.05)	0.049 ± 0.018	2.115 ± 0.857	2.230 ± 0.945	0.042 ± 0.014
(2.0, 2.0, 0.35)	0.349 ± 0.042	2.043 ± 0.365	1.989 ± 0.329	0.302 ± 0.034

As expected, the recombination fraction can be reasonably estimated by the two models when at least one marker is not distorted ($u = 1$ and/or $v = 1$) (Table 6.1). But when both markers are distorted by gametic viability, the estimate of the second model will be largely biased, especially when two markers are tightly linked, whereas the first model can always provide a reasonable estimate of r. The estimate of the second model is more precise (with smaller sampling errors), especially for a small sample size, than that of the first model, mainly because the former has fewer parameters being estimated (1) than the latter (3). But this advantage in precision comes with a large bias.

The coefficients of gametic viability (u and v) can be poorly estimated with a small sample size, although this does not affect the estimate of the recombination fraction (Table 6.1). To increase the estimation accuracy and precision of u and v, a large sample size is required. Similar findings have been obtained for simulation studies of the F_2 population.

6.3 Zygotic Differential Viability

6.3.1 One-Gene Model

The influence of zygotic differential viability can also occur at the genotype level. For the backcross, the models derived to study the effect of gametic viability disturbance can be used directly to examine zygotic viability disturbance. Here, we will focus our discussion on the modelling of zygotic differential viability in the F_2 progeny.

Consider an F_2 design in which there are three genotypes at each marker. If only one marker (say **A**) displays zygotic viability effects, we can assume that the viability of AA relative to Aa is u_1 and the viability of Aa relative to aa is u_2. Thus, the overall proportions of the three genotypes should be u_1u_2/U for AA, $2u_2/U$ for Aa, and $1/U$ for aa, where $U = u_1u_2 + 2u_2 + 1$. The expected frequencies of nine F_2 genotypes are a function of the viability coefficients and the recombination fraction, arrayed by

$$\tag{6.6} \begin{array}{c} \\ AA \\ Aa \\ aa \end{array} \begin{array}{c} \begin{array}{ccc} BB & Bb & bb \end{array} \\ \begin{bmatrix} \frac{u_1u_2}{U}(1-r)^2 & \frac{2u_1u_2}{U}r(1-r) & \frac{u_1u_2}{U}r^2 \\ \frac{2u_2}{U}r(1-r) & \frac{2u_2}{U}[(1-r)^2+r^2] & \frac{2u_2}{U}r(1-r) \\ \frac{1}{U}r^2 & \frac{2}{U}r(1-r) & \frac{1}{U}(1-r)^2 \end{bmatrix} \end{array}.$$

The MLE of r is obtained by the EM algorithm for a normal F_2 because the coefficients within each cell in matrix (6.6) are not dependent on the recombination fraction. It can be seen that the estimate of r is not affected by the zygotic differential viability of one marker. Parameters u_1 and u_2 are obtained by

$$\hat{u}_1 = \frac{2(n_{22}+n_{21}+n_{20})}{n_{12}+n_{11}+n_{10}},$$
$$\hat{u}_2 = \frac{n_{12}+n_{11}+n_{10}}{2(n_{02}+n_{01}+n_{00})}.$$

The Fisher information matrix is

$$I = n \begin{bmatrix} \frac{2(2r^2-2r+1)(u_1u_2+1)+4u_2(1-2r)^2}{Ur(1-r)(2r^2-2r+1)} & 0 & 0 \\ 0 & \frac{u_2(2u_2+1)}{u_1U^2} & -\frac{1}{U^2} \\ 0 & -\frac{1}{U^2} & \frac{u_1+2}{u_2U^2} \end{bmatrix}.$$

The sampling variances of the MLEs of the recombination fraction and zygotic viability coefficients in the F_2 can then be derived as

$$\text{var}(\hat{r}) = \frac{Ur(1-r)(2r^2-2r+1)}{2n[(2r^2-2r+1)(u_1u_2+1)+2u_2(1-2r)^2]},$$

$$\text{var}(\hat{u}_1) = \frac{U(u_1+2)u_1}{2nu_2},$$

$$\text{var}(\hat{u}_2) = \frac{U(2u_2+1)u_2}{2n}.$$

6.3.2 Two-Gene Model

Let us consider the situation in which both markers have viability effects. The viability effects of the second marker **B** can be similarly denoted using v_1 and v_2. If the two genes are subject to viability disturbances, the elements in matrix (6.6) should be changed to consider the influences of the second marker,

$$\mathbf{F} = \begin{matrix} \\ AA \\ Aa \\ aa \end{matrix} \begin{bmatrix} BB & Bb & bb \\ \frac{u_1u_2v_1v_2}{d}(1-r)^2 & \frac{2u_1u_2v_2}{d}r(1-r) & \frac{u_1u_2}{d}r^2 \\ \frac{2u_2v_1v_2}{d}r(1-r) & \frac{2u_2v_2}{d}[(1-r)^2+r^2] & \frac{2u_2}{d}r(1-r) \\ \frac{v_1v_2}{d}r^2 & \frac{2v_2}{d}r(1-r) & \frac{1}{d}(1-r)^2 \end{bmatrix},$$

$$= \begin{bmatrix} (1-r)^2 & 2r(1-r) & r^2 \\ 2r(1-r) & 2[(1-r)^2+r^2] & 2r(1-r) \\ r^2 & 2r(1-r) & (1-r)^2 \end{bmatrix} \circ \begin{bmatrix} \frac{u_1u_2v_1v_2}{d} & \frac{u_1u_2v_2}{d} & \frac{u_1u_2}{d} \\ \frac{u_2v_1v_2}{d} & \frac{u_2v_2}{d} & \frac{u_2}{d} \\ \frac{v_1v_2}{d} & \frac{v_2}{d} & \frac{1}{d} \end{bmatrix}$$

$$= \mathbf{F}_r \circ \mathbf{F}_{u_1u_2v_1v_2r}, \tag{6.7}$$

where

$$d = u_1u_2v_1v_2(1-r)^2 + 2u_2v_1v_2r(1-r) + v_1v_2r^2 + 2u_1u_2v_2r(1-r)$$
$$+2u_2v_2[(1-r)^2+r^2] + 2v_2r(1-r) + u_1u_2r^2 + 2u_2r(1-r) + (1-r)^2.$$

Two component matrices in (6.7) each contain the recombination fraction. The EM algorithm can be formulated to estimate r for the first component matrix, $\mathbf{F}_r$. In conjunction with the effect of the second component matrix, $\mathbf{F}_{u_1u_2v_1v_2r}$, on the estimate of r, we have the overall MLE of r expressed as

$$\hat{r} = \frac{1}{2n}(n_{12}+n_{21}+n_{10}+n_{01}+2(n_{02}+n_{20})+\phi n_{11}) - \frac{r(1-r)}{2d}\frac{\partial d}{\partial r}, \tag{6.8}$$

where

$$\frac{\partial d}{\partial r} = 2[u_1u_2v_1v_2(r-1) + u_2v_1v_2(1-2r) + v_1v_2r + u_1u_2v_2(1-2r)$$
$$+u_2v_2(4r-2) + v_2(1-2r) + u_1u_2r + u_2(1-2r) - 1 + r].$$

The four viability coefficients u_1, u_2, v_1, and v_2 can be estimated simultaneously:

$$\hat{u}_1 = \frac{(n_{22} + n_{21} + n_{20})d}{n[u_2v_1v_2(1-r)^2 + u_2r^2 + 2u_2v_2r(1-r)]},$$

$$\hat{u}_2 = \frac{(n_{22} + n_{21} + n_{20} + n_{12} + n_{11} + n_{10})d}{n[(u_1v_1v_2 + 2v_2)(1-r)^2 + (u_1 + 2v_2)r^2 + 2(v_1v_2 + u_1v_2 + 1)r(1-r)]},$$

$$\hat{v}_1 = \frac{(n_{22} + n_{12} + n_{02})d}{n[u_1u_2v_2(1-r)^2 + v_2r^2 + 2u_2v_2r(1-r)]},$$

$$\hat{v}_2 = \frac{(n_{22} + n_{12} + n_{02} + n_{21} + n_{11} + n_{01})d}{n[(u_1u_2v_1 + 2u_2)(1-r)^2 + (v_1 + 2u_2)r^2 + 2(u_1u_2 + u_2v_1 + 1)r(1-r)]}.$$

6.3.3 Simulation

The influence of zygotic viability on the estimate of the recombination fraction is examined through simulation studies. We simulated two markers that display different degrees of zygotic viability for the F_2 of sample sizes $n = 80$ and 200. These two markers can be tightly ($r = 0.05$) or loosely linked ($r = 0.35$). The simulated data were analyzed with the viability model (1) proposed in Section 6.3 as well as the linkage model (2) used for normal markers.

When only one marker is distorted, the estimation accuracy is not affected by viability disturbance (Table 6.2). However, if both markers are distorted, model 1 performs better than model 2 in terms of estimation accuracy. This is especially true for two loosely linked markers in which the estimate of r may be very downward biased. More importantly, the estimate by model 2 can be little improved if the sample size is more than doubled (from 80 to 200). It should be noted that model 1 may have low estimation accuracy and precision for a small sample size, but this can be improved dramatically when the sample size is increased (Table 6.2).

The model described in Section 6.3 also allows estimates of the coefficients of gametic viability disturbance (u_1, u_2 and v_1, v_2) for the two markers. The estimates of these two coefficients are affected by sample sizes and the degree of linkage. As expected, more sample sizes provide better estimates. But these two coefficients can be better estimated when the two markers are tightly linked rather than loosely.

6.4 Misclassification

6.4.1 One-gene Model

In practice, it is possible to misclassify one genotype as the other due to human errors. For example, in a double backcross $AaBb \times aabb$, there may be a proportion of λ of allele A that is misclassified as a (irrespective of whether these two alleles are associated with B or b). The appropriate expected frequencies and observed number of genotypes or gametes in a backcross population are shown below:

Table 6.2. The MLEs of the recombination fraction between two distorted markers with different degrees of zygotic viability disturbances in the backcross progeny. The sampling errors (SE) of the estimates are given from 100 simulation replicates.

	Model 1					Model 2
(u_1, u_2, v_1, v_2, r)	$\hat{r} \pm$ SE	$\hat{u}_1 \pm$ SE	$\hat{u}_2 \pm$ SE	$\hat{v}_1 \pm$ SE	$\hat{v}_2 \pm$ SE	$\hat{r}_0 \pm$ SE
$n = 80$						
(1.0, 1.0, 1.0, 1.0, 0.05)	0.045 ± 0.017	1.190 ± 0.699	1.332 ± 1.370	1.101 ± 0.613	1.090 ± 0.772	0.050 ± 0.017
(1.0, 1.0, 1.0, 1.0, 0.35)	0.359 ± 0.050	1.076 ± 0.298	1.027 ± 0.299	0.965 ± 0.273	1.040 ± 0.316	0.360 ± 0.049
(1.0, 1.0, 2.0, 2.0, 0.05)	0.044 ± 0.020	1.267 ± 1.181	1.548 ± 2.243	2.654 ± 2.346	2.534 ± 2.469	0.049 ± 0.016
(1.0, 1.0, 2.0, 2.0, 0.35)	0.349 ± 0.052	1.051 ± 0.291	1.018 ± 0.346	2.092 ± 0.575	2.457 ± 1.257	0.352 ± 0.048
(2.0, 2.0, 2.0, 2.0, 0.05)	0.044 ± 0.021	2.598 ± 2.435	4.982 ± 8.378	2.438 ± 2.313	2.456 ± 3.352	0.039 ± 0.015
(2.0, 2.0, 2.0, 2.0, 0.35)	0.347 ± 0.056	2.157 ± 0.584	4.545 ± 20.383	2.038 ± 0.523	2.346 ± 1.427	0.282 ± 0.036
$n = 200$						
(1.0, 1.0, 1.0, 1.0, 0.05)	0.047 ± 0.011	1.067 ± 0.435	1.108 ± 0.472	1.069 ± 0.409	1.049 ± 0.404	0.049 ± 0.010
(1.0, 1.0, 1.0, 1.0, 0.35)	0.352 ± 0.032	1.000 ± 0.186	1.008 ± 0.171	1.022 ± 0.210	1.024 ± 0.182	0.353 ± 0.032
(1.0, 1.0, 2.0, 2.0, 0.05)	0.048 ± 0.012	1.004 ± 0.292	1.037 ± 0.402	2.232 ± 1.001	2.192 ± 0.940	0.050 ± 0.011
(1.0, 1.0, 2.0, 2.0, 0.35)	0.351 ± 0.034	1.009 ± 0.167	1.057 ± 0.237	2.011 ± 0.314	2.051 ± 0.432	0.350 ± 0.030
(2.0, 2.0, 2.0, 2.0, 0.05)	0.051 ± 0.014	2.117 ± 0.637	2.258 ± 1.093	2.058 ± 0.660	2.151 ± 1.213	0.042 ± 0.011
(2.0, 2.0, 2.0, 2.0, 0.35)	0.351 ± 0.033	2.052 ± 0.362	2.101 ± 0.656	2.030 ± 0.318	2.038 ± 0.534	0.285 ± 0.024

Gamete	Expected	Observed
AB	$\frac{1}{2}(1-r)(1-\lambda)$	n_1
Ab	$\frac{1}{2}r(1-\lambda)$	n_2
aB	$\frac{1}{2}[r+\lambda(1-r)]$	n_3
ab	$\frac{1}{2}[(1-r)+\lambda r]$	n_4
Total	1	n

The expectations are derived because a proportion λ of each of the first two gamete classes involving A has been transferred to the corresponding one involving a but with the same B or b classification.

The maximum likelihood estimation of the two parameters r and λ is straightforward. We have

$$\hat{r} = \frac{n_2(n_1+n_3)}{n_2(n_1+n_3)+n_1(n_2+n_4)},$$

$$\hat{\lambda} = \frac{n_3n_4 - n_1n_2}{(n_1+n_3)(n_2+n_4)}.$$

The sampling variances of these estimators are

$$\text{var}(\hat{r}) = \frac{2r(1-r)}{n}\left[\frac{1}{1-\lambda} - 2r(1-r)\right],$$

$$\text{var}(\hat{\lambda}) = \frac{2(1-\lambda)^2}{n}\left[\frac{\lambda}{1-\lambda} + 2r(1-r)\right].$$

Although A is misclassified as a, it is possible to have the reverse misclassification; i.e., $a \rightarrow A$. The formulas for this alternative pattern can be derived by suitably changing the observational symbols n_1, n_2, n_3, and n_4.

When misclassification occurs in the formation of genotypes in the F_2, similar formulas can be derived to explore its influences on the estimates of the recombination fraction. In so doing, we assume that such misclassification arises from the ambiguity of individual alleles. Thus, the formulation for the backcross above can be extended to model misclassification in the F_2. The expected frequencies of the F_2 genotypes after gene **A** is misclassified are expressed as

$$\begin{array}{c} AA \\ Aa \\ aa \end{array} \overset{\begin{array}{ccc} BB & \qquad\qquad Bb & \qquad\qquad bb \end{array}}{\begin{bmatrix} \frac{1}{4}(1-\lambda)^2(1-r)^2 & \frac{1}{2}(1-\lambda)^2 r(1-r) & \frac{1}{4}(1-\lambda)^2 r^2 \\ \frac{1}{2}(1-\lambda)(1-r)(r+\lambda(1-r)) & \frac{1}{2}(1-\lambda)[r^2+(1-r)^2+2\lambda r(1-r)] & \frac{1}{2}(1-\lambda)r(1-r+\lambda r) \\ \frac{1}{4}[r+\lambda(1-r)]^2 & \frac{1}{2}[\lambda r^2+\lambda(1-r)^2+(1+\lambda^2)r(1-r)] & \frac{1}{4}[(1-r)+\lambda r]^2 \end{bmatrix}}. \tag{6.9}$$

We derive the EM algorithm to estimate the recombination fraction r. It is not difficult to find the expected number of recombinants within each cell in the matrix (6.9), expressed as

$$\begin{array}{c} \\ AA \\ Aa \\ aa \end{array} \begin{array}{c} \begin{array}{ccc} BB & Bb & bb \end{array} \\ \begin{bmatrix} 0 & 1 & 2 \\ \phi_1 & 2\phi_2 & \phi_3 \\ 2\phi_4 & \phi_5 & \phi_6 \end{bmatrix} \end{array},$$

where

$$\begin{aligned}
\phi_1 &= \frac{r(1-r)}{r(1-r)+\lambda(1-r)^2}, \\
\phi_2 &= \frac{r^2+\lambda r(1-r)}{r^2+(1-r)^2+2\lambda r(1-r)}, \\
\phi_3 &= 1+\frac{\lambda r^2}{r(1-r)+\lambda r^2}, \\
\phi_4 &= \frac{r}{r+\lambda(1-r)}, \\
\phi_5 &= \frac{2\lambda r^2+(1+\lambda^2)r(1-r)}{\lambda r^2+\lambda(1-r)^2+(1+\lambda^2)r(1-r)}, \\
\phi_6 &= \frac{2\lambda r}{(1-r)+\lambda r}.
\end{aligned} \tag{6.10}$$

The recombination fraction can be estimated using equation

$$\hat{r} = \frac{1}{2n}(n_{21}+2n_{20}+\phi_1 n_{12}+2\phi_2 n_{11}+\phi_3 n_{10}+2\phi_4 n_{02}+\phi_5 n_{01}+\phi_6 n_{00}). \tag{6.11}$$

In the E step, the expected number of recombinants for each genotype is calculated using equation (6.10). These expected numbers are used to update the estimate of r with equation (6.11) in the M step. These two steps are iterated until r converges to a stable value.

It can be seen that the estimate of r is affected by the degree of misclassification λ but not affected by the estimate of λ. The MLE of λ in the F_2 is given by solving the third-order polynomial equation

$$\frac{n_1}{1-\lambda} = \frac{n_2}{\lambda+\frac{r}{1-r}} + \frac{n_3}{\lambda+\frac{1-r}{r}} + \frac{n_{11}}{\lambda+\frac{r^2+(1-r)^2}{2r(1-r)}},$$

where

$$\begin{aligned}
n_1 &= 2(n_{22}+n_{21}+n_{20})+n_{12}+n_{11}+n_{10}, \\
n_2 &= n_{12}+2n_{02}+n_{01}, \\
n_3 &= n_{10}+2n_{00}+n_{01}.
\end{aligned}$$

The sampling variances of $\hat{r}$ and $\hat{\lambda}$ can be obtained by

$$\text{var}(\hat{r}) = \frac{\{r^2[12(1-\hat{\lambda})^2\hat{r}(\hat{r}-2) + 13\hat{\lambda}^2 - 34\hat{\lambda} + 21] - (\hat{\lambda}^2 - 10\hat{\lambda} + 9)r + 2\}\hat{r}(1-\hat{r})}{n(1-\hat{\lambda})[2-\hat{\lambda}-6(1-\hat{\lambda})\hat{r}(1-\hat{r})]},$$

$$\text{var}(\hat{\lambda}) = \frac{\hat{\lambda}(1-\hat{\lambda}) + 2(1-\hat{\lambda})^2\hat{r}(1-\hat{r})}{n}.$$

6.4.2 Two-Gene Model

When two genes are both misclassified, we should introduce an additional proportion for allele B misclassified as b. Let the misclassified proportions be λ_1 and λ_2 for markers **A** and **B**, respectively. Assuming that these two proportions are independent, we have the expected numbers of each of the four backcross genotypes, along with their observations, as follows

Gamete	Expected	Observed
AB	$\frac{1}{2}(1-\lambda_1)(1-\lambda_2)(1-r)$	n_1
Ab	$\frac{1}{2}(1-\lambda_1)[r+\lambda_2(1-r)]$	n_2
aB	$\frac{1}{2}(1-\lambda_2)[r+\lambda_1(1-r)]$	n_3
ab	$\frac{1}{2}[r(\lambda_1+\lambda_2)+(1-r)(1+\lambda_1\lambda_2)]$	n_4
Total	1	n

The MLE of the recombination fraction and the proportion of misclassifications can be derived as

$$\hat{r} = 1 - \frac{n_1 n}{2(n_1+n_2)(n_1+n_3)},$$

$$\hat{\lambda}_1 = \frac{(n_3+n_4)-(n_1+n_2)}{n},$$

$$\hat{\lambda}_2 = \frac{(n_2+n_4)-(n_1+n_3)}{n}.$$

The sampling variances of the MLEs of r, λ_1, and λ_2 are

$$\text{var}(\hat{r}) = \frac{nn_1[n(n_1^2+n_2n_3) - n_1(n_1+n_2)(n_1+n_3)]}{4(n_1+n_2)^3(n_1+n_3)^3},$$

$$\text{var}(\hat{\lambda}_1) = \frac{1-\lambda_1^2}{n},$$

$$\text{var}(\hat{\lambda}_2) = \frac{1-\lambda_2^2}{n}.$$

It is difficult to estimate the recombination fraction when both markers are misclassified in the F_2. With the assumption that misclassification arises from the ambiguity of individual alleles, we derive the expected frequencies of the F_2 genotypes after both genes **A** and **B** are misclassified, expressed as

Genes		Expected Frequencies	Obs.
AA	BB	$(1-\lambda_1)^2(1-\lambda_2)^2(1-r)^2/4$	n_{22}
AA	Bb	$(1-\lambda_1)^2(1-\lambda_2)(1-r)(r+\lambda_2(1-r))/2$	n_{21}
AA	bb	$(1-\lambda_1)^2(r+\lambda_2(1-r))^2/4$	n_{20}
Aa	BB	$(1-\lambda_1)(1-\lambda_2)^2(1-r)(r+\lambda_1(1-r))/2$	n_{12}
Aa	Bb	$(1-\lambda_1)(1-\lambda_2)[2r(1-r)(\lambda_1+\lambda_2)+r^2+(1-r)^2(2\lambda_1\lambda_2+1)]$	n_{11}
Aa	bb	$(1-\lambda_1)(r+\lambda_2(1-r))((1-r)(1+\lambda_1\lambda_2)+(\lambda_1+\lambda_2)r)/2$	n_{10}
aa	BB	$(1-\lambda_2)^2(r+\lambda_1(1-r))^2/4$	n_{02}
aa	Bb	$((1-r)(1+\lambda_1\lambda_2)+(\lambda_1+\lambda_2)r)(r+\lambda_1(1-r))(1-\lambda_2)/2$	n_{01}
aa	bb	$((1-r)(1+\lambda_1\lambda_2)+(\lambda_1+\lambda_2)r)^2/4$	n_{00}

The expected numbers of recombinants within each cell in the table above are expressed as

$$\begin{array}{c} \\ AA \\ Aa \\ aa \end{array} \begin{array}{c} \begin{array}{ccc} BB & Bb & bb \end{array} \\ \begin{bmatrix} 0 & \phi_1 & 2\phi_2 \\ \phi_3 & 2\phi_4 & \phi_5 \\ 2\phi_6 & \phi_7 & \phi_8 \end{bmatrix} \end{array},$$

where

$$\begin{aligned}
\phi_1 &= \frac{r(1-r)}{r(1-r)+\lambda_2(1-r)^2}, \\
\phi_2 &= \frac{r}{r+\lambda_2(1-r)}, \\
\phi_3 &= \frac{r(1-r)}{r(1-r)+\lambda_1(1-r)^2}, \\
\phi_4 &= \frac{r(1-r)(\lambda_1+\lambda_2)+r^2}{2r(1-r)(\lambda_1+\lambda_2)+r^2+(1-r)^2(2\lambda_1\lambda_2+1)}, \\
\phi_5 &= \frac{2r^2(\lambda_1+\lambda_2)+(\lambda_2^2+2\lambda_1\lambda_2+1)r(1-r)}{[(1-r)(1+\lambda_1\lambda_2)+(\lambda_1+\lambda_2)r](r+\lambda_2(1-r))}, \\
\phi_6 &= \frac{r}{r+\lambda_1(1-r)}, \\
\phi_7 &= \frac{2r^2(\lambda_1+\lambda_2)+(\lambda_1^2+2\lambda_1\lambda_2+1)r(1-r)}{[(1-r)(1+\lambda_1\lambda_2)+(\lambda_1+\lambda_2)r](r+\lambda_1(1-r))}, \\
\phi_8 &= \frac{2(\lambda_1+\lambda_2)r}{(1-r)(1+\lambda_1\lambda_2)+(\lambda_1+\lambda_2)r}.
\end{aligned} \tag{6.12}$$

The recombination fraction can be estimated using equation

$$\hat{r} = \frac{1}{2n}(\phi_1 n_{21} + 2\phi_2 n_{20} + \phi_3 n_{12} + 2\phi_4 n_{11} + \phi_5 n_{10} + 2\phi_6 n_{02} + \phi_7 n_{01} + \phi_8 n_{00}). \tag{6.13}$$

The EM algorithm is formulated to estimate r. In the E step, the expected number of recombinants for each genotype is calculated using equations (6.12). These expected numbers are used to update the estimate of r with equation (6.13) in the M step. These two steps are iterated until r converges to a stable value.

Again, for two misclassified markers, the estimate of r is not affected by the estimate of λ, although the former is affected by the degree of misclassification λ. The MLEs of λ_1 and λ_2 in the F_2 are given by solving the third-order polynomial equations

$$\frac{n_1}{1-\lambda_1} = \frac{n_2}{\lambda_1 + \frac{r}{1-r}} + \frac{n_3}{\lambda_1 + \frac{1-r+\lambda_2 r}{(1-r)\lambda_2 + r}} + \frac{n_{11}}{\lambda_1 + \frac{r^2+(1-r)^2+2\lambda_2 r(1-r)}{2r(1-r)+2\lambda_2(1-r)^2}},$$

$$\frac{m_1}{1-\lambda_2} = \frac{m_2}{\lambda_2 + \frac{r}{1-r}} + \frac{n_3}{\lambda_2 + \frac{1-r+\lambda_1 r}{(1-r)\lambda_1 + r}} + \frac{n_{11}}{\lambda_2 + \frac{r^2+(1-r)^2+2\lambda_1 r(1-r)}{2r(1-r)+2\lambda_1(1-r)^2}},$$

where

$$\begin{aligned}
n_1 &= 2(n_{22} + n_{21} + n_{20}) + n_{12} + n_{11} + n_{10}, \\
n_2 &= n_{12} + 2n_{02} + n_{01}, \\
n_3 &= n_{10} + 2n_{00} + n_{01}, \\
m_1 &= 2(n_{22} + n_{12} + n_{02}) + n_{21} + n_{11} + n_{01}, \\
m_2 &= n_{21} + 2n_{20} + n_{10}.
\end{aligned}$$

The sampling variances of the MLEs of r, λ_1, and λ_2 are

$$\begin{aligned}
\text{var}(\hat{r}) = \{&(1-r)[24(1 + \lambda_1^2\lambda_2^2 + \lambda_2^2 - 2\lambda_2^2\lambda_1 - 2\lambda_1 - 2\lambda_2 + \lambda_1^2 + 4\lambda_1\lambda_2 - 2\lambda_1^2\lambda_2)r^4 \\
&+(100\lambda_2 - 56\lambda_2^2 + 100\lambda_1 - 56\lambda_1^2 - 240\lambda_1\lambda_2 - 44 - 84\lambda_1^2\lambda_2^2 + 140\lambda_1^2\lambda_2 + 140\lambda_2^2\lambda_1)r^3 \\
&+(-144\lambda_2^2\lambda_1 + 42\lambda_1^2 + 42\lambda_2^2 + 110\lambda_1^2\lambda_2^2 - 144\lambda_1^2\lambda_2 + 34 - 76\lambda_2 + 212\lambda_1\lambda_2 - 76\lambda_1)r^2 \\
&+(59\lambda_2^2\lambda_1 - 12 + 59\lambda_1^2\lambda_2 - 64\lambda_1^2\lambda_2^2 + 27\lambda_1 - 11\lambda_1^2 - 82\lambda_1\lambda_2 - 11\lambda_2^2 + 27\lambda_2)r \\
&+(2 + \lambda_2^2 - 7\lambda_2^2\lambda_1 - 7\lambda_1^2\lambda_2 - 3\lambda_1 + 14\lambda_1^2\lambda_2^2 - 3\lambda_2 + 14\lambda_1\lambda_2 + \lambda_1^2)]\} / \\
&\{2n(1-\lambda_2)(1-\lambda_1)[6(\lambda_1\lambda_2 + 1 - \lambda_1 - \lambda_2)r^2 + 2(3\lambda_1 - 5\lambda_1\lambda_2 - 3 + 3\lambda_2)r \\
&+ 4\lambda_1\lambda_2 - \lambda_1 - \lambda_2 + 2]\},
\end{aligned}$$

$$\text{var}(\hat{\lambda}_1) = \frac{1-\lambda_1^2}{2n},$$

$$\text{var}(\hat{\lambda}_2) = \frac{1-\lambda_2^2}{2n}.$$

6.5 Simulation

The influence of marker misclassification on the estimate of the recombination fraction is investigated through simulation studies. Two markers are simulated for both the backcross and F_2 with different degrees of marker misclassification (λ_1, λ_2). The sample sizes considered are 80 and 200, and there are degrees of linkage ($r = 0.05$ and 0.35). The simulated data are analyzed by both the model (1) that incorporates marker misclassification and the model (2) that does not.

Table 6.3. The MLEs of the recombination fraction between two misclassified markers in the backcross progeny. The sampling errors (SE) of the estimates are given from 100 simulation replicates.

	Model 1			Model 2
$(\lambda_1, \lambda_2, r)$	$\hat{r} \pm$ SE	$\hat{\lambda}_1 \pm$ SE	$\hat{\lambda}_2 \pm$ SE	$\hat{r}_0 \pm$ SE
$n = 80$				
(0.0, 0.0, 0.05)	0.060 ± 0.107	-0.020 ± 0.119	0.019 ± 0.116	0.054 ± 0.026
(0.0, 0.0, 0.35)	0.349 ± 0.060	0.014 ± 0.098	0.003 ± 0.121	0.354 ± 0.052
(0.0, 0.1, 0.05)	0.040 ± 0.095	0.004 ± 0.109	0.109 ± 0.106	0.099 ± 0.032
(0.0, 0.1, 0.35)	0.351 ± 0.059	-0.015 ± 0.102	0.107 ± 0.108	0.366 ± 0.051
(0.2, 0.3, 0.05)	0.033 ± 0.124	0.198 ± 0.107	0.304 ± 0.096	0.215 ± 0.047
(0.2, 0.3, 0.35)	0.339 ± 0.090	0.208 ± 0.106	0.311 ± 0.103	0.382 ± 0.048
$n = 200$				
(0.0, 0.0, 0.05)	0.043 ± 0.069	0.002 ± 0.077	0.002 ± 0.075	0.050 ± 0.015
(0.0, 0.0, 0.35)	0.353 ± 0.040	0.001 ± 0.070	0.001 ± 0.071	0.354 ± 0.034
(0.0, 0.1, 0.05)	0.041 ± 0.069	0.005 ± 0.073	0.098 ± 0.072	0.093 ± 0.021
(0.0, 0.1, 0.35)	0.350 ± 0.040	-0.006 ± 0.062	0.090 ± 0.067	0.363 ± 0.034
(0.2, 0.3, 0.05)	0.043 ± 0.073	0.206 ± 0.064	0.310 ± 0.068	0.220 ± 0.028
(0.2, 0.3, 0.35)	0.347 ± 0.062	0.201 ± 0.069	0.302 ± 0.069	0.385 ± 0.035

When $\lambda_1 = \lambda_2 = 0$ (i.e., there is no marker misclassification), models 1 and 2 provide similarly good results (Tables 6.3 and 6.4). When one of the markers is misclassified to some extent, model 1 quickly ill-behaves, and its estimate is very biased, especially for a tight linkage. For example, when $\lambda_2 = 0.1$, model 1 estimates the true $r = 0.05$ as over 0.09. This is not improved when the size of the sample is increased. Model 1 can generally provide good estimates of r even when both

Table 6.4. The MLEs of the recombination fraction between two misclassified markers in the F_2 progeny. The sampling errors (SE) of the estimates are given from 100 simulation replicates.

	Model 1			Model 2
$(\lambda_1, \lambda_2, r)$	$\hat{r} \pm$ SE	$\hat{\lambda}_1 \pm$ SE	$\hat{\lambda}_2 \pm$ SE	$\hat{r_0} \pm$ SE
$n = 80$				
(0.0, 0.0, 0.05)	0.050 ± 0.049	-0.045 ± 0.183	-0.033 ± 0.169	0.048 ± 0.018
(0.0, 0.0, 0.35)	0.352 ± 0.049	-0.000 ± 0.082	-0.005 ± 0.073	0.354 ± 0.044
(0.0, 0.1, 0.05)	0.053 ± 0.052	-0.005 ± 0.064	0.100 ± 0.065	0.098 ± 0.024
(0.0, 0.1, 0.35)	0.369 ± 0.060	-0.014 ± 0.077	0.111 ± 0.078	0.385 ± 0.053
(0.2, 0.3, 0.05)	0.073 ± 0.063	0.196 ± 0.079	0.296 ± 0.071	0.225 ± 0.034
(0.2, 0.3, 0.35)	0.341 ± 0.085	0.193 ± 0.075	0.296 ± 0.084	0.372 ± 0.047
$n = 200$				
(0.0, 0.0, 0.05)	0.050 ± 0.040	-0.002 ± 0.045	-0.003 ± 0.046	0.049 ± 0.012
(0.0, 0.0, 0.35)	0.350 ± 0.030	-0.005 ± 0.048	-0.005 ± 0.046	0.349 ± 0.028
(0.0, 0.1, 0.05)	0.049 ± 0.036	0.001 ± 0.042	0.102 ± 0.041	0.097 ± 0.015
(0.0, 0.1, 0.35)	0.356 ± 0.034	0.001 ± 0.051	0.096 ± 0.045	0.371 ± 0.030
(0.2, 0.3, 0.05)	0.053 ± 0.045	0.201 ± 0.045	0.292 ± 0.048	0.215 ± 0.021
(0.2, 0.3, 0.35)	0.350 ± 0.056	0.200 ± 0.051	0.293 ± 0.045	0.373 ± 0.030

markers are misclassified. The estimation accuracy of r by model 1 can be dramatically increased with increasing sample size. Because model 2 has fewer parameters to be estimated than model 1, the former displays superior estimation precision over the latter. But this advantage of model 1 is not useful given its large biased estimate. The estimation precision of model 1 is increased with increasing sample size.

The backcross and F_2 display similar trends for parameter estimation. But it is observed that the F_2 (Table 6.4) is better in terms of estimation accuracy and precision than the backcross (Table 6.3). This may be because the F_2 contains a larger amount of information than the backcross.

6.6 Exercises

6.1 One-gene model in the F_2

In a mapping experiment, one wishes to estimate the linkage between two markers, **A** and **B**, in the F_2. One of the markers (**A**) is distorted due to gametic viability disturbance, but the other marker (**B**) is normal. The nine genotypes in this F_2 progeny are observed as follows:

$$\begin{array}{c} \\ AA \\ Aa \\ aa \end{array} \begin{array}{c} \begin{array}{ccc} BB & Bb & bb \end{array} \\ \left[\begin{array}{ccc} n_{22} = 111 & n_{21} = 32 & n_{20} = 1 \\ n_{12} = 56 & n_{11} = 123 & n_{10} = 13 \\ n_{02} = 7 & n_{01} = 32 & n_{00} = 25 \end{array}\right] . \end{array}$$

Derive the EM algorithm to calculate the recombination fraction, r, between the two markers based on the matrix above in the following cases.

(a) Markers **A** and **B** are both codominant. Show why the estimation of r is not dependent on the viability disturbance of marker **A**.

(b) The distorted marker **A** is dominant.

(c) The normal marker **B** is dominant.

(d) Both markers **A** and **B** are dominant.

(e) Show why the estimation of r is affected by the viability disturbance of a marker when at least one of the markers is dominant.

6.2 Two-gene model in the F_2

For two markers **A** and **B** distorted due to gametic viability, the observations of the nine genotypes are obtained for the F_2 as follows:

$$\begin{array}{c} \\ AA \\ Aa \\ aa \end{array} \begin{array}{c} \begin{array}{ccc} BB & Bb & bb \end{array} \\ \left[\begin{array}{ccc} n_{22} = 265 & n_{21} = 33 & n_{20} = 3 \\ n_{12} = 48 & n_{11} = 39 & n_{10} = 2 \\ n_{02} = 2 & n_{01} = 6 & n_{00} = 2 \end{array}\right] . \end{array}$$

Derive the EM algorithm to calculate the recombination fraction, r, between the two markers based on matrix (6.2) in the following cases.

(a) Markers **A** and **B** are both codominant.

(b) The distorted marker **A** is dominant.

(c) The normal marker **B** is dominant.

(d) Both markers **A** and **B** are dominant.

(e) Show why the estimation of r is affected by the viability disturbance of a marker when at least one of the markers is dominant.

6.3 Do the same things as required in Exercises 6.1 and 6.2 if one or two markers are distorted due to zygotic viability, respectively.

6.4 If the matrices in Exercises 6.1 and 6.2 contain the results for the misclassification of one marker (e.g., marker **A**), perform the following.

(a) Calculate the coefficient of marker misclassification and test its significance.

(b) Estimate the recombination fraction and test its significance.

(c) Based on the conclusions of simulation studies, analyze the possible accuracy and precision of your estimate of r.

6.5 Do the same thing as Exercise 6.3 if both markers are misclassified.

7

Special Considerations in Linkage Analysis

7.1 Introduction

In the preceding chapters, we discussed linkage analysis for a single controlled cross derived from inbred lines or outbred lines. We described a general framework for performing linkage analysis with any type of marker, fully or partially informative, that is qualitatively scored on the presence or absence of a DNA band for an allele. For an outbred cross, the procedure was given to simultaneously estimate the linkage, parental diplotype, and gene order. However, there are many situations in practice in which the strategies for linkage analysis described thus far cannot be appropriately used. In this chapter, we will consider two important issues that have been common or are becoming increasingly common in linkage analysis. The first concerns linkage analysis in a complicated nuclear pedigree, whereas the second deals with linkage analysis based on quantitative marker information.

7.2 Linkage Analysis with a Complicated Pedigree

The prerequisite for the use of a single cross in linkage analysis is the availability of a sufficiently large sample size. Although this can be easily met in plants, the generation of too many offspring from a single cross is difficult or even impossible for most animals and humans. For these organisms, the method for performing linkage analysis is to combine all possible individuals (including parents) derived from multiple related or unrelated families. In this section, we will introduce an approach for linkage analysis in a structured multigeneration pedigree initiated with multiple parents (or founders). Because the diplotypes of each founder are unknown, this approach should be implemented with the estimation of parental diplotype probabilities.

7.2.1 A Nuclear Family

To understand the principle of linkage analysis for a complicated pedigree, we start with a simple nuclear family. Assume that two parents generate three offspring. All these individuals, designated by 1–5, are genotyped for two markers **A** and **B** (Fig. 7.1). Marker **A** has two different alleles, whereas marker **B** has four alleles. Note that, although the genotypes of each individual can be known from a gel analysis, the linkage phase between the two markers cannot be measured. In this example, the linkage phases of both parents and offspring 5 are unknown, but they should be correctly estimated because different linkage phases for these individuals affect the inference and estimation of the recombination fraction (r) between the two markers.

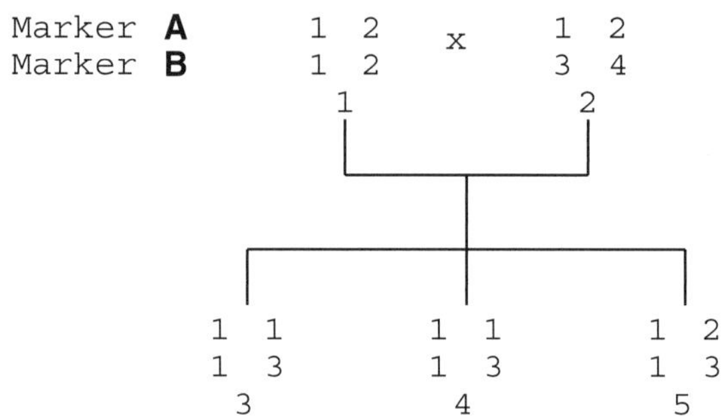

Fig. 7.1. A nuclear family with three offspring

Each of the three phase-unknown individuals has two possible phases; that is,

$$\Lambda_1 = \left.\begin{array}{c|c} 1 & 2 \\ & \\ 1 & 2 \end{array}\right| \quad \text{or} \quad \bar{\Lambda}_1 = \left.\begin{array}{c|c} 1 & 2 \\ & \\ 2 & 1 \end{array}\right| \quad \text{for parent 1,}$$

$$\Lambda_2 = \left.\begin{array}{c|c} 1 & 2 \\ & \\ 3 & 4 \end{array}\right| \quad \text{or} \quad \bar{\Lambda}_2 = \left.\begin{array}{c|c} 1 & 2 \\ & \\ 4 & 3 \end{array}\right| \quad \text{for parent 2,}$$

$$\Lambda_5 = \left.\begin{array}{c|c} 1 & 2 \\ & \\ 1 & 3 \end{array}\right| \quad \text{or} \quad \bar{\Lambda}_5 = \left.\begin{array}{c|c} 1 & 2 \\ & \\ 3 & 1 \end{array}\right| \quad \text{for offspring 5,}$$

which form eight possible diplotype combinations. Traditional linkage analysis assumes equal probability, i.e., 1/8, for these combinations. Table 7.1 tabulates possible numbers of recominants (R) and nonrecombinants (NR) for each of the three offspring and the resulting likelihood under different diplotype combination.

Based on the likelihood for each diplotype combination, we construct an overall likelihood for this family as

Table 7.1. Number of recombinants (R) and nonrecombinants (NR) for three offspring derived from a family shown in Fig. 7.1 under different diplotype combinations for parents 1 and 2 and offspring 5. The likelihoods for each diplotype combination are also given.

Diplotype Combination				Offspring			
Proba-bility	Parent 1	Parent 2	Offsping 5	3	4	5	Likelihood
$p_1 = \frac{1}{8}$	Λ_1	Λ_2	Λ_5	NR,NR	NR,NR	NR,R	$L_1 = r(1-r)^5$
$p_2 = \frac{1}{8}$	Λ_1	Λ_2	$\bar{\Lambda}_5$	NR,NR	NR,NR	NR,R	$L_2 = r(1-r)^5$
$p_3 = \frac{1}{8}$	Λ_1	$\bar{\Lambda}_2$	Λ_5	NR,NR	NR,R	NR,R	$L_3 = r^2(1-r)^4$
$p_4 = \frac{1}{8}$	Λ_1	$\bar{\Lambda}_2$	$\bar{\Lambda}_5$	NR,R	NR,R	R,R	$L_4 = r^4(1-r)^2$
$p_5 = \frac{1}{8}$	$\bar{\Lambda}_1$	Λ_2	Λ_5	R,NR	R,NR	R,R	$L_5 = r^4(1-r)^2$
$p_6 = \frac{1}{8}$	$\bar{\Lambda}_1$	Λ_2	$\bar{\Lambda}_5$	R,NR	R,NR	NR,NR	$L_6 = r^2(1-r)^4$
$p_7 = \frac{1}{8}$	$\bar{\Lambda}_1$	$\bar{\Lambda}_2$	Λ_5	R,R	R,R	R,NR	$L_7 = r^5(1-r)$
$p_8 = \frac{1}{8}$	$\bar{\Lambda}_1$	$\bar{\Lambda}_2$	$\bar{\Lambda}_5$	R,R	R,R	NR,R	$L_8 = r^5(1-r)$

$$\begin{aligned} L(r) &= \frac{1}{8}(L_1 + \ldots + L_8) \\ &= \frac{1}{8}\left[2r(1-r)^5 + r^2(1-r)^4 + r^4(1-r)^2 + r^5(1-r)\right] \\ &= \frac{1}{4}r(1-r)[1-3r(1-r)] \end{aligned} \tag{7.1}$$

By maximizing the likelihood (7.1), the MLE of the recombination fraction, r, is obtained as $\hat{r} = 0.21$. An approximate approach for estimating r with a complicated likelihood (7.1) is to calculate the likelihood for a grid of fixed r values, such as $r = (0.00, 0.01, 0.05, 0.10, 0.15, ...)$. The r value that leads to a maximal likelihood value is considered as the MLE of the recombination fraction (Fig. 7.2). Asymptotically, the likelihood curve approaches a Gaussian shape (Ott 1991).

The procedure for a linkage analysis described above assumes an equal probability for all eight possible diplotype combination. But this may not be true. The most likely diplotype combination can be determined from the data by calculating the MLE of r with the likelihood listed in Table 7.2. The results for each diplotype probability are presented below.

It can be seen that the optimal diplotype combination with the largest likelihood is Λ_1-Λ_2-Λ_5 or Λ_1-Λ_2-$\bar{\Lambda}_5$, whose probability is estimated as

$$\hat{p}_1 \text{ or } \hat{p}_2 = \frac{\hat{L}_1 \text{ or } \hat{L}_2}{\hat{L}_1 + \ldots + \hat{L}_8} = 0.2787.$$

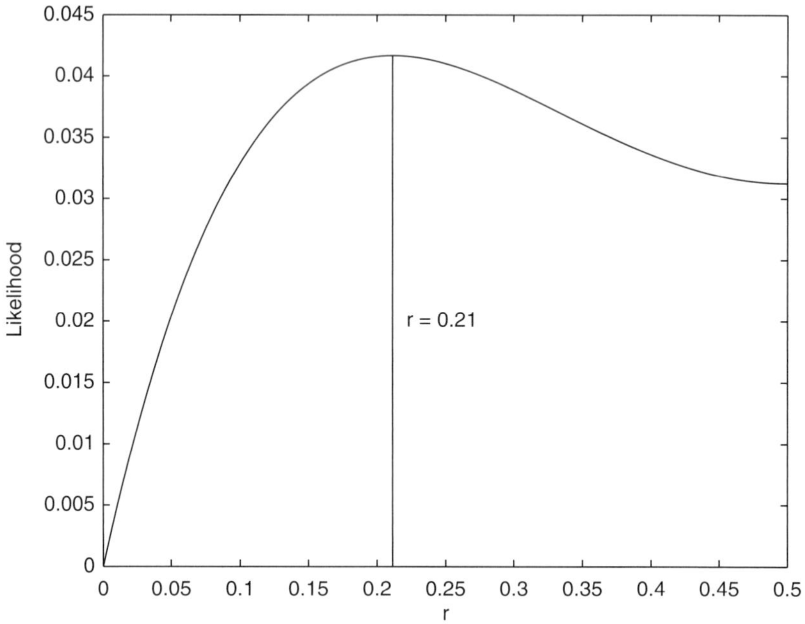

Fig. 7.2. The profile of the likelihood calculated for a nuclear family in Fig. 7.3

Table 7.2. The MLE of the recombination fraction and the plugged in likelihood values under different diplotype combinations.

Diplotype Combination					
Probability	Parent 1	Parent 2	Offspring 5	$\hat{r}$	Log-Likelihood
$\hat{p}_1 = 0.2787$	Λ_1	Λ_2	Λ_5	1/6	$\log \hat{L}_1 = -2.7034$
$\hat{p}_2 = 0.2787$	Λ_1	Λ_2	$\bar{\Lambda}_5$	1/6	$\log \hat{L}_2 = -2.7034$
$\hat{p}_3 = 0.0913$	Λ_1	$\bar{\Lambda}_2$	Λ_5	1/3	$\log \hat{L}_3 = -3.8191$
$\hat{p}_4 = 0.0650$	Λ_1	$\bar{\Lambda}_2$	$\bar{\Lambda}_5$	1/2	$\log \hat{L}_4 = -4.1589$
$\hat{p}_5 = 0.0650$	$\bar{\Lambda}_1$	Λ_2	Λ_5	1/2	$\log \hat{L}_5 = -4.1589$
$\hat{p}_6 = 0.0913$	$\bar{\Lambda}_1$	Λ_2	$\bar{\Lambda}_5$	1/3	$\log \hat{L}_6 = -3.8191$
$\hat{p}_7 = 0.0650$	$\bar{\Lambda}_1$	$\bar{\Lambda}_2$	Λ_5	1/2	$\log \hat{L}_7 = -4.1589$
$\hat{p}_8 = 0.0650$	$\bar{\Lambda}_1$	$\bar{\Lambda}_2$	$\bar{\Lambda}_5$	1/2	$\log \hat{L}_8 = -4.1589$

An alternative to determining the optimal diplotype combination is to construct a mixture likelihood of all diplotype combination weighted by the assumed probability. An algorithm can be derived to provide simultaneous estimation of r and p's. But this approach needs a larger sample size than currently used because more parameters are estimated.

7.2.2 Multipoint Estimation of Identical-By-Descent Sharing

Two alleles at a single locus are identical by descent (IBD) if they are identical copies of the same allele on the maternal and paternal chromosome, both copies that arose by DNA replication from the same ancestral sequence without any intervening mutation. The concept of IBD is of great use to understand population genetics and also shows a great utility for linkage analysis and genetic mapping. The estimation of IBD probabilities can be obtained from linkage analysis when genotypes at multiple linked markers are available. For a family as illustrated in Fig. 7.3, the mean IBD sharing for the siblings would be 0.25 at marker **A** and 0 at marker **B** if the two markers are unlinked.

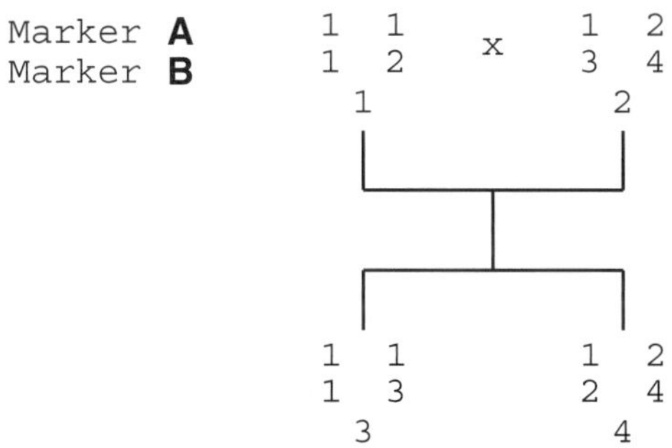

Fig. 7.3. Genotypes at two adjacent markers in a nuclear family

The IBD estimation for the less informative marker (**A**) can be obtained through the joint likelihood of the two markers if they are linked. Let r be the recombination fraction between the two markers. For marker **A**, it is not possible to determine the parent of origin of two identical alleles, 1, carried by parent 1. Thus, the linkage phase of parent 1 cannot be determined. We arbitrarily assume one allele derived from the paternal parent (denoted as 1^{p}) and the other from the maternal parent (denoted as 1^{m}). Two possible phases of parent 1 are expressed as

$$\Lambda_1 = \begin{array}{c|c|} 1^{\mathrm{p}} & 1^{\mathrm{m}} \\ 1 & 2 \end{array} \quad \text{or} \quad \bar{\Lambda}_1 = \begin{array}{c|c|} 1^{\mathrm{p}} & 1^{\mathrm{m}} \\ 2 & 1 \end{array} .$$

Similarly, parent 2 has two possible phases as follows:

$$\Lambda_2 = \begin{array}{c|c|} 1 & 2 \\ 3 & 4 \end{array} \quad \text{or} \quad \bar{\Lambda}_2 = \begin{array}{c|c|} 1 & 2 \\ 4 & 3 \end{array}.$$

Offspring 3 has two identical alleles for marker **A**, but one and only one of them, either $1^{\mathtt{m}}$ or $1^{\mathtt{p}}$, comes from parent 1. Thus, its possible linkage phase should be

$$\Lambda_3^{\mathtt{p}} = \begin{array}{c|c|} 1^{\mathtt{p}} & 1 \\ 1 & 3 \end{array} \quad \text{or} \quad \Lambda_3^{\mathtt{m}} = \begin{array}{c|c|} 1^{\mathtt{m}} & 1 \\ 1 & 3 \end{array}.$$

It is not possible for offspring 3 to have

$$\begin{array}{c|c|} 1^{\mathtt{p}} & 1 \\ 3 & 1 \end{array} \quad \text{or} \quad \begin{array}{c|c|} 1^{\mathtt{m}} & 1 \\ 3 & 1 \end{array},$$

because the gametes that form these genotype configurations do not exist. For the same reason, offspring 4 has only two possible phases as follows:

$$\Lambda_4^{\mathtt{p}} = \begin{array}{c|c|} 1^{\mathtt{p}} & 2 \\ 2 & 4 \end{array} \quad \text{or} \quad \Lambda_4^{\mathtt{m}} = \begin{array}{c|c|} 1^{\mathtt{m}} & 2 \\ 2 & 4 \end{array}.$$

Combining all the four individuals, we will have 16 diplotype combinations with likelihoods and IBD probabilities for each marker are given in Table 7.3. By assuming the same probability of the 16 diplotype combinations, we construct a joint likelihood to obtain the MLE of r. With $\hat{r}$, the IBD sharing for each marker can be calculated, which is

$$\frac{0.5[4r(1-r)^3 + 4r^3(1-r)]}{2(1-r)^4 + 4r(1-r)^3 + 4r^2(1-r)^2 + 4r^3(1-r) + 2r^4} = \frac{r(1-r)^3 + r^3(1-r)}{1 - 2r(1-r)},$$

for marker **A** and 0 for marker **B**.

When many markers are included for IBD analysis, the likelihood can be factored out in terms of a Markov model incorporating the IBD at a marker, along with the transition probabilities from marker to marker.

7.2.3 A Complex Pedigree

Because members from different families can be related by their common ancestors, the pedigree we are considering forms a complicated network. Figure 7.4 is a diagram illustrating such a complicated pedigree that can be used in many different species. Suppose there are a total of ten individuals that are genotyped at two multiallelic markers **A** and **B** and one biallelic marker **C**. Alleles at each of these markers are symbolized by Arabic numerals. Individuals 1, 2, and 3 are the founders whose parents

Table 7.3. Likelihoods and IBD probabilities for each marker under 16 possible diplotype combinations for the pedigree shown in Fig. 7.3.

Phase Combination					Offspring				
Probability	1	2	3	4	3	4	Likelihood$\propto$	IBD(**A**)	IBD(**B**)
$p_1 = \frac{1}{16}$	Λ_1	Λ_2	Λ_3^{p}	Λ_4^{p}	NR,NR	R,NR	$r(1-r)^3$	0.5	0.0
$p_2 = \frac{1}{16}$	Λ_1	Λ_2	Λ_3^{p}	Λ_4^{m}	NR,NR	NR,NR	$(1-r)^4$	0.0	0.0
$p_3 = \frac{1}{16}$	Λ_1	Λ_2	Λ_3^{m}	Λ_4^{p}	R,NR	R,NR	$r^2(1-r)^2$	0.0	0.0
$p_4 = \frac{1}{16}$	Λ_1	Λ_2	Λ_3^{m}	Λ_4^{m}	R,NR	NR,NR	$r(1-r)^3$	0.5	0.0
$p_5 = \frac{1}{16}$	Λ_1	$\bar{\Lambda}_2$	Λ_3^{p}	Λ_4^{p}	NR,R	R,R	$r^3(1-r)$	0.5	0.0
$p_6 = \frac{1}{16}$	Λ_1	$\bar{\Lambda}_2$	Λ_3^{p}	Λ_4^{m}	NR,R	NR,R	$r^2(1-r)^2$	0.0	0.0
$p_7 = \frac{1}{16}$	Λ_1	$\bar{\Lambda}_2$	Λ_3^{m}	Λ_4^{p}	R,R	R,R	r^4	0.0	0.0
$p_8 = \frac{1}{16}$	Λ_1	$\bar{\Lambda}_2$	Λ_3^{m}	Λ_4^{m}	R,R	NR,R	$r^3(1-r)$	0.5	0.0
$p_9 = \frac{1}{16}$	$\bar{\Lambda}_1$	Λ_2	Λ_3^{p}	Λ_4^{p}	R,NR	NR,NR	$r(1-r)^3$	0.5	0.0
$p_{10} = \frac{1}{16}$	$\bar{\Lambda}_1$	Λ_2	Λ_3^{p}	Λ_4^{m}	R,NR	R,NR	$r^2(1-r)^2$	0.0	0.0
$p_{11} = \frac{1}{16}$	$\bar{\Lambda}_1$	Λ_2	Λ_3^{m}	Λ_4^{p}	NR,NR	NR,NR	$(1-r)^4$	0.0	0.0
$p_{12} = \frac{1}{16}$	$\bar{\Lambda}_1$	Λ_2	Λ_3^{m}	Λ_4^{m}	NR,NR	R,NR	$r(1-r)^3$	0.5	0.0
$p_{13} = \frac{1}{16}$	$\bar{\Lambda}_1$	$\bar{\Lambda}_2$	Λ_3^{p}	Λ_4^{p}	R,R	NR,R	$r^3(1-r)$	0.5	0.0
$p_{14} = \frac{1}{16}$	$\bar{\Lambda}_1$	$\bar{\Lambda}_2$	Λ_3^{p}	Λ_4^{m}	R,R	R,R	r^4	0.0	0.0
$p_{15} = \frac{1}{16}$	$\bar{\Lambda}_1$	$\bar{\Lambda}_2$	Λ_3^{m}	Λ_4^{p}	NR,R	NR,R	$r^2(1-r)^2$	0.0	0.0
$p_{16} = \frac{1}{16}$	$\bar{\Lambda}_1$	$\bar{\Lambda}_2$	Λ_3^{m}	Λ_4^{m}	NR,R	R,R	$r^3(1-r)$	0.5	0.0

cannot be identified, whereas individuals 5–10 are the non-founders that are derived from parent-known individuals. The cross of individuals 1 and 2 leads to 4, which is crossed with 3 to produce 5. Individual 1 is crossed to 5 to produce 6, which is self-crossed to produce 7. Individual 8 is generated by crossing 7 and 3. The cross of 8 and 5 produces 9, which is crossed to 2 to produce 10. These pedigree relationships are further described in Table 7.4. Column 1 contains the identifications of all individuals studied, columns 2 and 3 are their parental origins (note that the parental information of the founders is missing), and columns 4–6 display genotypes of three markers **A**, **B** and **C**, respectively, for each individual. In this example, the linkage phase of each individual across these markers is unknown.

Likelihood Formulation

A prerequisite for linkage analysis is a known linkage phase between two markers considered for the founders that produce segregating progeny. In the example shown

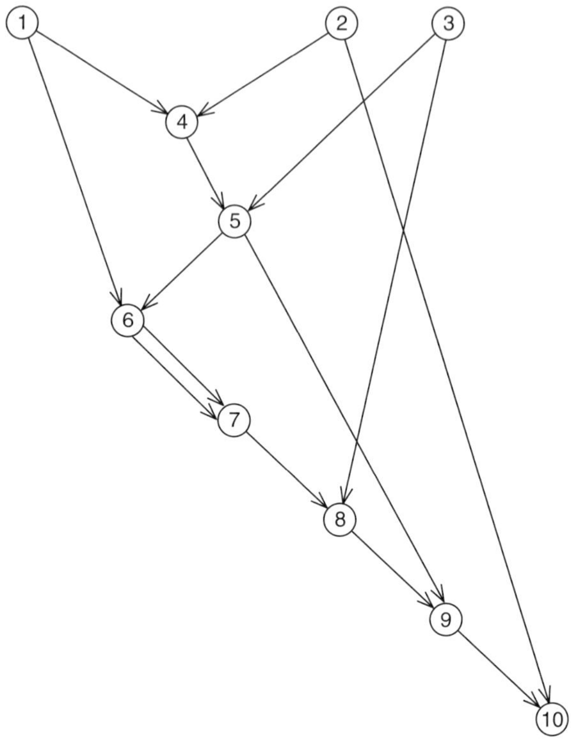

Fig. 7.4. A hypothesized pedigree structure.

in Table 7.4, which is likely to be an actual case in practice, marker data for non-inbred founders only present genotypes (designated as $\mathcal{G}$) but provide no information about linkage phase configuration or diplotype. However, one can always assume a possible diplotype for a founder. For the first two markers, each of the three founders has two alternative diplotypes (designated as Λ or $\bar{\Lambda}$), which forms eight different combinations, shown in Table 7.5.

Let us consider the first diplotype combination. Founders 1 and 2 are crossed to generate individual 4, which has genotype 13 at marker **A** and genotype 12 at marker **B** based on gel observations. The alleles of the two markers may be arranged in individual 4 in two linkage phases:

$$\Lambda_4 = \begin{array}{c|c|} 1 & 3 \\ \\ 2 & 1 \end{array} \quad \text{or} \quad \bar{\Lambda}_4 = \begin{array}{c|c|} 1 & 3 \\ \\ 1 & 2 \end{array}. \tag{7.2}$$

Yet, based on the assumed linkage phases of founders 1 and 2, the second arrangement does not exist because neither of the founders provides gamete 32. The only

Table 7.4. Observed genotypes for three hypothesized markers in a structured pedigree shown in Fig. 7.4.

Individual	Parent	Parent	Marker genotype		
ID	1	2	**A**	**B**	**C**
1	–	–	12	12	12
2	–	–	34	13	12
3	–	–	13	12	12
4	1	2	13	12	12
5	3	4	11	22	12
6	1	5	12	12	12
7	6	6	12	12	12
8	3	7	13	12	22
9	5	8	13	12	12
10	2	9	14	23	11

possible arrangement is the first one, whose formation is due to the combination of recombinant gamete 12 from founder 1 and recombinant gamete 31 from founder 2. Let r be the recombination fraction between these two markers, $P(\Lambda(t)|\Lambda(t-1))$ be the conditional probability of the diplotype of an individual (generation t) given its parents' diplotypes (generation $t-1$), and $P(\Lambda(t-1)|\mathcal{G}(t-1))$ be the conditional probability of the diplotype of a parent (generation $t-1$) given its observed genotype. Thus, the possibility of generating the observed genotype of individual 4 should be

$$\begin{aligned} P(\mathcal{G}_4|\Lambda_1,\Lambda_2) &= P(\Lambda_4|\Lambda_1,\Lambda_2) + P(\bar{\Lambda}_4|\Lambda_1,\Lambda_2) \\ &= P(12|\Lambda_1)P(31|\Lambda_2) \\ &= \tfrac{1}{4}r^2, \end{aligned} \tag{7.3}$$

where $P(\bar{\Lambda}_4|\Lambda_1,\Lambda_2) = 0$. For individual 5 derived from founder 3 and hybrid 4, its only possible diplotype is

$$\Lambda_5 = \begin{array}{c|c|} 1 & 1 \\ 2 & 2 \end{array}.$$

Thus, the genotype probability of individual 5 is expressed as

$$\begin{aligned} P(\mathcal{G}_5|\Lambda_3,\mathcal{G}_4) &= P(\mathcal{G}_5|\Lambda_3,\Lambda_4)P(\Lambda_4|\mathcal{G}_4) \\ &= P(\Lambda_5|\Lambda_3,\Lambda_4)P(\Lambda_4|\mathcal{G}_4) + P(\bar{\Lambda}_5|\Lambda_3,\Lambda_4)P(\bar{\Lambda}_4|\mathcal{G}_4) \\ &= \tfrac{1}{2}r(1-r), \end{aligned} \tag{7.4}$$

Table 7.5. Eight diplotype combinations for the founders.

Diplotype Combination	Founder 1	Founder 2	Founder 3	Probability
$1 = \Lambda_1\Lambda_2\Lambda_3$	$\begin{array}{c\|c\|} 1 & 2 \\ 1 & 2 \end{array}$	$\begin{array}{c\|c\|} 3 & 4 \\ 3 & 1 \end{array}$	$\begin{array}{c\|c\|} 1 & 3 \\ 1 & 2 \end{array}$	$p_1p_2p_3$
$2 = \Lambda_1\Lambda_2\bar{\Lambda}_3$	$\begin{array}{c\|c\|} 1 & 2 \\ 1 & 2 \end{array}$	$\begin{array}{c\|c\|} 3 & 4 \\ 3 & 1 \end{array}$	$\begin{array}{c\|c\|} 1 & 3 \\ 2 & 1 \end{array}$	$p_1p_2(1-p_3)$
$3 = \Lambda_1\bar{\Lambda}_2\Lambda_3$	$\begin{array}{c\|c\|} 1 & 2 \\ 1 & 2 \end{array}$	$\begin{array}{c\|c\|} 3 & 4 \\ 1 & 3 \end{array}$	$\begin{array}{c\|c\|} 1 & 3 \\ 1 & 2 \end{array}$	$p_1(1-p_2)p_3$
$4 = \Lambda_1\bar{\Lambda}_2\bar{\Lambda}_3$	$\begin{array}{c\|c\|} 1 & 2 \\ 1 & 2 \end{array}$	$\begin{array}{c\|c\|} 3 & 4 \\ 1 & 3 \end{array}$	$\begin{array}{c\|c\|} 1 & 3 \\ 2 & 1 \end{array}$	$p_1(1-p_2)(1-p_3)$
$5 = \bar{\Lambda}_1\Lambda_2\Lambda_3$	$\begin{array}{c\|c\|} 1 & 2 \\ 2 & 1 \end{array}$	$\begin{array}{c\|c\|} 3 & 4 \\ 3 & 1 \end{array}$	$\begin{array}{c\|c\|} 1 & 3 \\ 1 & 2 \end{array}$	$(1-p_1)p_2p_3$
$6 = \bar{\Lambda}_1\Lambda_2\bar{\Lambda}_3$	$\begin{array}{c\|c\|} 1 & 2 \\ 2 & 1 \end{array}$	$\begin{array}{c\|c\|} 3 & 4 \\ 3 & 1 \end{array}$	$\begin{array}{c\|c\|} 1 & 3 \\ 2 & 1 \end{array}$	$(1-p_1)p_2(1-p_3)$
$7 = \bar{\Lambda}_1\bar{\Lambda}_2\Lambda_3$	$\begin{array}{c\|c\|} 1 & 2 \\ 2 & 1 \end{array}$	$\begin{array}{c\|c\|} 3 & 4 \\ 1 & 3 \end{array}$	$\begin{array}{c\|c\|} 1 & 3 \\ 1 & 2 \end{array}$	$(1-p_1)(1-p_2)p_3$
$8 = \bar{\Lambda}_1\bar{\Lambda}_2\bar{\Lambda}_3$	$\begin{array}{c\|c\|} 1 & 2 \\ 2 & 1 \end{array}$	$\begin{array}{c\|c\|} 3 & 4 \\ 1 & 3 \end{array}$	$\begin{array}{c\|c\|} 1 & 3 \\ 2 & 1 \end{array}$	$(1-p_1)(1-p_2)(1-p_3)$

Note: Two possible linkage phases for each founder are designated by Λ or $\bar{\Lambda}$, whose subscripts stand for different individuals.

where $P(\Lambda_4|\mathcal{G}_4) = 1$ and $P(\bar{\Lambda}_5|\Lambda_3, \Lambda_4) = 0$. For individual 6 derived from individuals 1 and 5, we observe alleles 1 and 2 at marker **A** and alleles 1 and 2 at marker **B**. Although this individual may have allelic arrangements

$$\Lambda_6 = \begin{array}{c|c|} 2 & 1 \\ 2 & 1 \end{array} \quad \text{or} \quad \bar{\Lambda}_6 = \begin{array}{c|c|} 2 & 1 \\ 1 & 2 \end{array},$$

only the latter is possible given individual 5's genotype. Thus, the genotype probability of individual 6 is

$$P(\mathcal{G}_6|\Lambda_1, \mathcal{G}_5) = P(\bar{\Lambda}_6|\Lambda_1, \Lambda_5) = \tfrac{1}{2}r. \tag{7.5}$$

The selfing of individual 6 produces double heterozygous individual 7 with two possible diplotypes,

$$\Lambda_7 = \left.\begin{array}{c|c} 1 & 2 \\ 1 & 2 \end{array}\right| \quad \text{or} \quad \bar{\Lambda}_7 = \left.\begin{array}{c|c} 1 & 2 \\ 2 & 1 \end{array}\right| ,$$

whose overall genotype probability should be

$$\begin{aligned} P(\mathcal{G}_7|\mathcal{G}_6) &= P(\Lambda_7|\mathcal{G}_6) + P(\bar{\Lambda}_7|\mathcal{G}_6) \\ &= P(\Lambda_7|\Lambda_6)P(\Lambda_6|\mathcal{G}_6) + P(\Lambda_7|\bar{\Lambda}_6)P(\bar{\Lambda}_6|\mathcal{G}_6) \\ &\quad + P(\bar{\Lambda}_7|\Lambda_6)P(\Lambda_6|\mathcal{G}_6) + P(\bar{\Lambda}_7|\bar{\Lambda}_6)P(\bar{\Lambda}_6|\mathcal{G}_6) \\ &= P(\Lambda_7|\bar{\Lambda}_6)P(\bar{\Lambda}_6|\mathcal{G}_6) + P(\bar{\Lambda}_7|\bar{\Lambda}_6)P(\bar{\Lambda}_6|\mathcal{G}_6) \\ &= \tfrac{1}{2}r^2 + \tfrac{1}{2}(1-r)^2, \end{aligned} \tag{7.6}$$

where $P(\Lambda_6|\mathcal{G}_6) = 0$ and $P(\bar{\Lambda}_6|\mathcal{G}_6) = 1$. Individual 7 is crossed with founder 3 to produce individual 8 with observed alleles 1 and 3 at marker **A** and alleles 1 and 2 at marker **B**. It is possible that individual 8 has one of the two diplotypes

$$\Lambda_8 = \left.\begin{array}{c|c} 1 & 3 \\ 2 & 1 \end{array}\right| \quad \text{or} \quad \Lambda_8 = \left.\begin{array}{c|c} 1 & 3 \\ 1 & 2 \end{array}\right| .$$

The overall genotype probability of individual 8 given its parents' genotype is expressed as

$$\begin{aligned} P(\mathcal{G}_8|\Lambda_3,\mathcal{G}_7) &= P(\Lambda_8|\Lambda_3,\mathcal{G}_7) + P(\bar{\Lambda}_8|\Lambda_3,\mathcal{G}_7) \\ &= P(\Lambda_8|\Lambda_3,\Lambda_7)P(\Lambda_7|\mathcal{G}_7) + P(\Lambda_8|\Lambda_3,\bar{\Lambda}_7)P(\bar{\Lambda}_7|\mathcal{G}_7) \\ &\quad + P(\bar{\Lambda}_8|\Lambda_3,\Lambda_7)P(\Lambda_7|\mathcal{G}_7) + P(\bar{\Lambda}_8|\Lambda_3,\bar{\Lambda}_7)P(\bar{\Lambda}_7|\mathcal{G}_7) \\ &= \frac{r[r^3 + (1-r)^3 + (1-r)^2]}{4[r^2 + (1-r)^2]}, \end{aligned} \tag{7.7}$$

where

$$P(\Lambda_8|\Lambda_3,\Lambda_7) = \tfrac{1}{4}r^2,$$

with haplotypes 12 and 31 of Λ_8 derived from Λ_7 and Λ_3, respectively, each with a probability of $\frac{1}{2}r$ (note that Λ_7 is only derived from $\bar{\Lambda}_6$ so that the haplotype frequency of 12 is $\frac{1}{2}r$):

$$P(\Lambda_8|\Lambda_3,\bar{\Lambda}_7) = \tfrac{1}{4}r(1-r),$$

$$P(\bar{\Lambda}_8|\Lambda_3,\Lambda_7) = \tfrac{1}{4}(1-r)^2,$$

$$P(\bar{\Lambda}_8|\Lambda_3,\bar{\Lambda}_7) = \tfrac{1}{4}r(1-r),$$

$$P(\Lambda_7|\mathcal{G}_7) = \frac{r^2}{(1-r)^2 + r^2},$$

$$P(\bar{\Lambda}_7|\mathcal{G}_7) = \frac{(1-r)^2}{(1-r)^2 + r^2}.$$

Given its own genotype as well as the diplotypes of individuals 8 and 5, individual 9 should have only a possible diplotype like

$$\Lambda_9 = \begin{array}{c|c|} 1 & 3 \\ & \\ 2 & 1 \end{array},$$

with a probability of

$$\begin{aligned} P(\mathcal{G}_9|\mathcal{G}_5, \mathcal{G}_8) &= P(\Lambda_9|\mathcal{G}_5, \mathcal{G}_8) \\ &= P(\Lambda_9|\Lambda_5, \Lambda_8)P(\Lambda_5, \Lambda_8|\mathcal{G}_5, \mathcal{G}_8) + P(\Lambda_9|\bar{\Lambda}_5, \Lambda_8)P(\bar{\Lambda}_5, \Lambda_8|\mathcal{G}_5, \mathcal{G}_8) \\ &\quad + P(\Lambda_9|\Lambda_5, \bar{\Lambda}_8)P(\Lambda_5, \bar{\Lambda}_8|\mathcal{G}_5, \mathcal{G}_8) + P(\Lambda_9|\bar{\Lambda}_5, \bar{\Lambda}_8)P(\bar{\Lambda}_5, \bar{\Lambda}_8|\mathcal{G}_5, \mathcal{G}_8) \\ &= P(\Lambda_9|\Lambda_5, \Lambda_8)P(\Lambda_5|\mathcal{G}_5)P(\Lambda_8|\mathcal{G}_8) + P(\Lambda_9|\bar{\Lambda}_5, \Lambda_8)P(\bar{\Lambda}_5|\mathcal{G}_5)P(\Lambda_8|\mathcal{G}_8) \\ &\quad + P(\Lambda_9|\Lambda_5, \bar{\Lambda}_8)P(\Lambda_5|\mathcal{G}_5)P(\bar{\Lambda}_8|\mathcal{G}_8) + P(\Lambda_9|\bar{\Lambda}_5, \bar{\Lambda}_8)P(\bar{\Lambda}_5|\mathcal{G}_5)P(\bar{\Lambda}_8|\mathcal{G}_8) \\ &= P(\Lambda_9|\Lambda_5, \Lambda_8)P(\Lambda_5|\mathcal{G}_5)P(\Lambda_8|\mathcal{G}_8) + P(\Lambda_9|\Lambda_5, \bar{\Lambda}_8)P(\Lambda_5|\mathcal{G}_5)P(\bar{\Lambda}_8|\mathcal{G}_8) \\ &= \frac{r^3(1-r) + r(1-r)^3 + (1-r)^4}{2[r^3 + (1-r)^3 + (1-r)^2]}, \end{aligned} \tag{7.8}$$

where

$$\begin{aligned} P(\Lambda_5|\mathcal{G}_5) &= 1, \\ P(\bar{\Lambda}_5|\mathcal{G}_5) &= 0, \\ P(\Lambda_9|\Lambda_5, \Lambda_8) &= \frac{1}{2}(1-r), \\ P(\Lambda_9|\Lambda_5, \bar{\Lambda}_8) &= \frac{1}{2}r, \\ P(\bar{\Lambda}_8|\mathcal{G}_8) &= \frac{P(\Lambda_8|\Lambda_3, \mathcal{G}_7)}{P(\mathcal{G}_8|\Lambda_3, \mathcal{G}_7)} \\ &= \frac{P(\Lambda_8|\Lambda_3, \Lambda_7)P(\Lambda_7|\mathcal{G}_7) + P(\Lambda_8|\Lambda_3, \bar{\Lambda}_7)P(\bar{\Lambda}_7|\mathcal{G}_7)}{P(\mathcal{G}_8|\Lambda_3, \mathcal{G}_7)} \\ &= \frac{r^3 + (1-r)^3}{r^3 + (1-r)^3 + (1-r)^2}, \\ P(\Lambda_8|\mathcal{G}_8) &= 1 - P(\bar{\Lambda}_8|\mathcal{G}_8). \end{aligned}$$

For the last individual, we have only one possible phase,

$$\Lambda_{10} = \begin{array}{c|c|} 4 & 1 \\ & \\ 3 & 2 \end{array},$$

and its genotype probability is

$$P(\mathcal{G}_{10}|\Lambda_2, \mathcal{G}_9) = P(\mathcal{G}_{10}|\Lambda_2, \Lambda_9) = \tfrac{1}{4}r(1-r). \tag{7.9}$$

With the forming probability of each of the seven nonfounders under diplotype combination $\Lambda_1\Lambda_2\Lambda_3$, we can formulate the likelihood of the unknown recombination fraction given the markers ($\mathbf{M}$) by

$$\begin{aligned} L_{\Lambda_1\Lambda_2\Lambda_3}(r|\mathbf{M}) &= P(\mathcal{G}_4|\Lambda_1, \Lambda_2)P(\mathcal{G}_5|\Lambda_3, \mathcal{G}_4)P(\mathcal{G}_6|\Lambda_1, \mathcal{G}_5)P(\mathcal{G}_7|\mathcal{G}_6) \\ &\quad P(\mathcal{G}_8|\Lambda_3, \mathcal{G}_7)P(\mathcal{G}_9|\mathcal{G}_5, \mathcal{G}_8)P(\mathcal{G}_{10}|\Lambda_2, \mathcal{G}_9) \\ &= \frac{1}{2^{10}}r^6(1-r)^2[r^3(1-r) + r(1-r)^3 + (1-r)^4]. \end{aligned} \tag{7.10}$$

A grid approach over a range of r can be used to obtain the MLE of the recombination fraction. Based on equation (7.10), we attempt to draw a profile of likelihood from $r = 0$ to 0.5 (Fig. 7.5). But it turns out that the optimal estimate of r under the first diplotype combination among the three founders (Table 7.5) is 0.715. Obviously, this estimate is not the best given the data in Table 7.4 because it is beyond the limit of the recombination fraction. This also indicates that we need to search for the estimate of r under the seven other diplotype combinations. It should be noted that because only the first two markers were considered for this analysis, the conclusions will be completely identical for a pair of diplotype combinations 1 and 2, 3 and 4, 5 and 6, and 7 and 8. Each pair is different because of different diplotypes of the third marker, which was not considered. A final MLE of r was estimated as 0.285 under diplotype combination 3 or 4, which corresponds to the largest likelihood.

EM Algorithm

Based on the likelihood (7.10), we can derive a closed form for the EM algorithm to estimate the recombination fraction. In this likelihood, the last term contains a mixture of recombinants and nonrecombinants, in which the expected number of recombinants should be the sum of two probabilities,

$$\phi_1 = \frac{r^3(1-r)}{r^3(1-r) + r(1-r)^3 + (1-r)^4}, \tag{7.11}$$

and

$$\phi_2 = \frac{r(1-r)^2}{r^3(1-r) + r(1-r)^3 + (1-r)^4}. \tag{7.12}$$

The EM algorithm is implemented to obtain the MLE of r. In the E step, the expected number of recombination events is calculated by equations (7.11) and (7.12). In the M step, r is estimated using

$$\hat{r} = \tfrac{1}{12}(6 + 3\phi_1 + \phi_2). \tag{7.13}$$

Iterations among equations (7.11)–(7.13) are continued until a stable estimate of r is obtained. Initiated with $r = 0.5$, r converges to 0.715 after ten iterations, which is consistent with the estimate from the grid approach. The optimal MLE of r is found to be 0.285 under diplotype combination 3 or in Table 7.5.

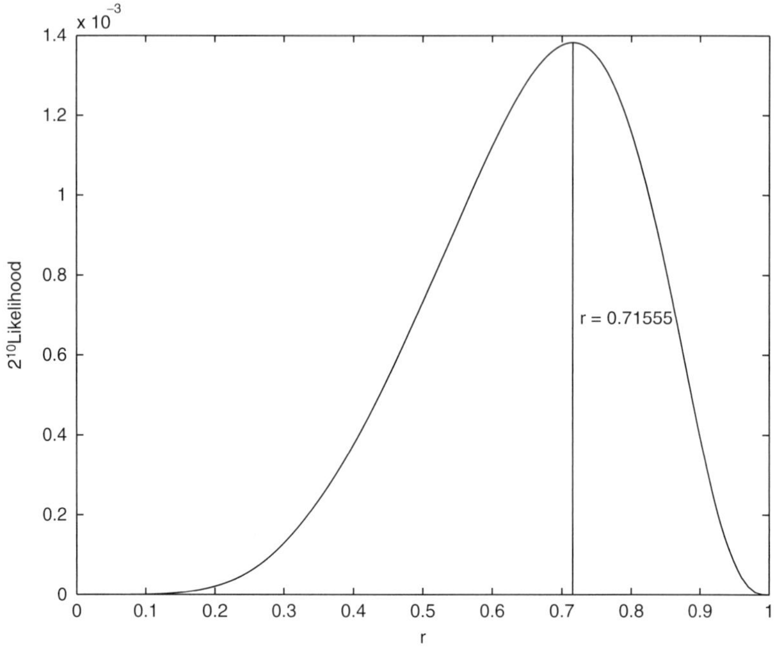

Fig. 7.5. The profile of the likelihood calculated for a complex family in Fig. 7.4.

Simultaneous Analysis of the Recombination Fraction and Diplotype Probability

A joint likelihood can be formulated to simultaneously estimate the recombination fraction and diplotype for each founder. Consider the marker data described in Table 7.4. Although the diplotypes of founders 1–3 are unknown, we can assume a probability for each of them to bear a particular diplotype. Let p_1 be the probability for founder 1 to have diplotype Λ_1, and thus the probability with diplotype Λ_2 is $1 - p_1$. Similarly, p_2 and p_3 are defined as the diplotype probabilities for founders 2 and 3, respectively. Thus, diplotype combination $\Lambda_1\Lambda_2\Lambda_3$ has a probability of $p_1p_2p_3$. The diplotype probabilities of the other combinations can also be defined (Table 7.5).

Individual 4, offspring of founders 1 and 2, has a two-marker genotype 13/12 for markers **A** and **B**. Individual 4 has two diplotypes, Λ_4 and $\bar{\Lambda}_4$ (7.2), whereas founders 1 and 2 may have the diplotypes defined in Table 7.5. The genotype probability of individual 4 given its parents, founders 1 and 2, is expressed as

$$\begin{aligned}
&P(\mathcal{G}_4|\mathcal{G}_1,\mathcal{G}_2)\\
&= P(\Lambda_4|\mathcal{G}_1,\mathcal{G}_2) + P(\bar{\Lambda}_4|\mathcal{G}_1,\mathcal{G}_2)\\
&= p_1p_2[P(\Lambda_4|\Lambda_1,\Lambda_2) + P(\bar{\Lambda}_4|\Lambda_1,\Lambda_2)]\\
&\quad +p_1(1-p_2)[P(\Lambda_4|\Lambda_1,\bar{\Lambda}_2) + P(\bar{\Lambda}_4|\Lambda_1,\bar{\Lambda}_2)]\\
&\quad +(1-p_1)p_2[P(\Lambda_4|\bar{\Lambda}_1,\Lambda_2) + P(\bar{\Lambda}_4|\bar{\Lambda}_1,\Lambda_2)]\\
&\quad +(1-p_1)(1-p_2)[P(\Lambda_4|\bar{\Lambda}_1,\bar{\Lambda}_2) + P(\bar{\Lambda}_4|\bar{\Lambda}_1,\bar{\Lambda}_2)],
\end{aligned} \tag{7.14}$$

where the probabilities of the two diplotypes, Λ_4 and $\bar{\Lambda}_4$, of individual 4 derived from each of four possible diplotype combinations of founders 1 and 2 are tabulated below:

Parental diplotype			Offspring 4	
1	2	Probability	Λ_4	$\bar{\Lambda}_4$
Λ_1	Λ_2	p_1p_2	$P(\Lambda_4\|\Lambda_1,\Lambda_2)=0$	$P(\bar{\Lambda}_4\|\Lambda_1,\Lambda_2)=\frac{1}{4}r^2$
Λ_1	$\bar{\Lambda}_2$	$p_1(1-p_2)$	$P(\Lambda_4\|\Lambda_1,\bar{\Lambda}_2)=0$	$P(\bar{\Lambda}_4\|\Lambda_1,\bar{\Lambda}_2)=\frac{1}{4}r(1-r)$
$\bar{\Lambda}_1$	Λ_2	$(1-p_1)p_2$	$P(\Lambda_4\|\bar{\Lambda}_1,\Lambda_2)=0$	$P(\bar{\Lambda}_4\|\bar{\Lambda}_1,\Lambda_2)=\frac{1}{4}r(1-r)$
$\bar{\Lambda}_1$	$\bar{\Lambda}_2$	$(1-p_1)(1-p_2)$	$P(\Lambda_4\|\bar{\Lambda}_1,\bar{\Lambda}_2)=0$	$P(\bar{\Lambda}_4\|\bar{\Lambda}_1,\bar{\Lambda}_2)=\frac{1}{4}(1-r)^2$

Unlike in equation (7.3), the genotype probability of individual 4 given its parents' genotypes, when all possible diplotype combinations are considered, is derived as

$$P(\mathcal{G}_4|\mathcal{G}_1,\mathcal{G}_2) = \tfrac{1}{4}[p_1p_2r^2 + (p_1+p_2-2p_1p_2)r(1-r) + (1-p_1)(1-p_2)(1-r)^2].$$

Using a similar approach, the genotype probabilities of the other offspring are also derived. Ultimately, we formulate a joint likelihood expressed as

$$\begin{aligned}
L(r|\mathbf{M}) = {} & P(\mathcal{G}_4|\mathcal{G}_1,\mathcal{G}_2)P(\mathcal{G}_5|\mathcal{G}_3,\mathcal{G}_4)P(\mathcal{G}_6|G_1,\mathcal{G}_5)P(\mathcal{G}_7|\mathcal{G}_6)\\
& P(\mathcal{G}_8|\mathcal{G}_3,\mathcal{G}_7)P(\mathcal{G}_9|\mathcal{G}_5,\mathcal{G}_8)P(\mathcal{G}_{10}|\mathcal{G}_2,\mathcal{G}_9).
\end{aligned} \tag{7.15}$$

By maximizing the likelihood (7.15), the recombination fraction r and diplotype probabilities of three founders (p_1, p_2, and p_3) can be estimated.

Three-Point Analysis

The idea of linkage analysis proposed for a complex nuclear family can be extended to three-point analysis. Recall Table 7.4 in which three markers **A**, **B** and **C** are genotyped. Let g_{00}, g_{01}, g_{10}, and g_{11} be the probabilities with which there are no recombinants over both marker intervals **A**–**B** and **B**–**C**, only one recombinant over **B**–**C**, only one recombinant over **A**–**C** and two recombinants each over a different interval, respectively. The three-marker genotype of individual 4 is 13/12/12, which may have four different diplotypes; that is,

$$\Lambda_4^1 = \left.\begin{array}{c|c} 1 & 3 \\ 1 & 2 \\ 1 & 2 \end{array}\right| , \quad \Lambda_4^2 = \left.\begin{array}{c|c} 1 & 3 \\ 1 & 2 \\ 2 & 1 \end{array}\right| , \quad \Lambda_4^3 = \left.\begin{array}{c|c} 1 & 3 \\ 2 & 1 \\ 1 & 2 \end{array}\right| , \quad \Lambda_4^4 = \left.\begin{array}{c|c} 1 & 3 \\ 2 & 1 \\ 2 & 1\,. \end{array}\right|$$

Because individual 4 is derived from founders 1 and 2, the diplotypes above must be determined by the diplotypes of the two founders. Each of the founders has 4 diplotypes, forming a total of 16 diplotype combinations. Assume one of the parental diplotype combinations is

$$\Lambda_1 = \left.\begin{array}{c|c} 1 & 2 \\ 1 & 2 \\ 1 & 2 \end{array}\right| \text{ (for founder 1)}, \quad \Lambda_2 = \left.\begin{array}{c|c} 3 & 4 \\ 1 & 3 \\ 1 & 2 \end{array}\right| \text{ (for founder 2)}.$$

It can be seen that the parental diplotypes above cannot generate diplotypes Λ_4^1 and Λ_4^2, but it is possible to generate Λ_4^3 and Λ_4^4, with a probability of $g_{11}g_{01}$ and $g_{10}g_{00}$, respectively. This can be expressed by a conditional probability

$$\begin{aligned} P(\mathcal{G}_4|\Lambda_1, \Lambda_2) &= P(\Lambda_4^3|\Lambda_1, \Lambda_2) + P(\Lambda_4^4|\Lambda_1, \Lambda_2) \\ &= g_{11}g_{01} + g_{10}g_{00}. \end{aligned}$$

Similar expressions for other offspring from 5 to 10 can be derived. All these are used to construct a likelihood from which parameters g_{00}, g_{01}, g_{10}, and g_{11} are estimated. The recombination fractions, $r_{\mathbf{AB}}$, $r_{\mathbf{BC}}$, and $r_{\mathbf{AC}}$, between the three markers can then be estimated by

$$\begin{aligned} \hat{r}_{\mathbf{AB}} &= \hat{g}_{10} + \hat{g}_{11}, \\ \hat{r}_{\mathbf{BC}} &= \hat{g}_{01} + \hat{g}_{11}, \\ \hat{r}_{\mathbf{AC}} &= \hat{g}_{01} + \hat{g}_{10}, \end{aligned}$$

under an assumed founder diplotype combination.

As in two-point analysis, the diplotype probabilities for the three founders can be incorporated into the likelihood and estimated simultaneously with the recombination fractions. The advantages of three-point analysis as compared with two-point analysis lie in the precise estimation of the recombination fractions and more power to detect significant linkage. Its drawback is that more computing resources are needed.

7.3 Information Analysis of Dominant Markers

7.3.1 Introduction

In many situations, codominant markers are preferable to dominant markers due to their larger information content. A codominant molecular marker allows the unequivocal distinction of homozygous and heterozygous genotypes on an electrophoretic gel.

By contrast, for dominant markers, dominant homozygous and heterozygous individuals cannot be distinguished on the basis of the presence or absence of bands on a gel. However, dominant markers, like AFLPs, continue to be very useful because of their low cost and simple molecular characterization. For this reason, the possibility of extracting codominant information from easily characterized dominant markers deserves exploration.

7.3.2 Segregation Analysis

Mixture Models

Suppose a PCR-based dominant marker system is used to score a population for their genotypes at a marker locus. Instead of scoring the presence or absence of a band, we use fluorescence techniques to quantify the band intensity of the degree of amplification of a fragment on an electrophoretic gel. Thus, for dominant homozygous individuals, the intensity is expected to be higher than for heterozygous individuals since for the latter the amount of PCR products is only that of homozygotes. However, because the distinction in band intensity among the genotypes is blurred, the band intensity measured varies among individuals from the same genotypes. This is mainly due to random variation that occurs during marker assays. A mixture model was developed to assign each individual to one of the genotype classes and estimate the segregation patterns of the genotypes in a progeny (Piepho and Koch 2000; Jansen et al. 2001).

Let us consider a family in which a dominant marker is segregating. The band intensity (y_{ij}) measured for an individual i from a genotype class j ($j = 1, ..., k$) can be expressed using a linear model as

$$y_{ij} = \mu_j + e_{ij}, \tag{7.16}$$

where μ_j is the expected mean band intensity for genotype j and e_{ij} is a residual error, assumed to be normally distributed with mean 0 and variance σ^2.

To further account for the genotype variation, we assume that each observation y_i comes from a mixture model,

$$y_i \sim \sum_{j=1}^{k} \gamma_j f(y_i|\mu_j, \sigma^2),$$

where $\gamma_1, ..., \gamma_k$ are the unknown (prior) mixing proportions satisfying $0 \leq \gamma_i \leq 1, \sum \gamma_j = 1$, and $f(\cdot|\mu_j, \sigma^2)$ is the normal density with mean μ_j and variance σ^2.

From a sample $y = (y_1, ..., y_n)$, the likelihood function is given by

$$L(\mathbf{\Omega}|y_{ij}) = \prod_{i=1}^{n} \sum_{j=1}^{k} \gamma_j f_j(y_i|\mu_j, \sigma^2), \tag{7.17}$$

where $\mathbf{\Omega} = (\mu_j, \sigma^2, \gamma_j)^T$, and the log-likelihood is

$$\log L(\mathbf{\Omega}) = \sum_{i=1}^{n} \log \left[\sum_{j=1}^{k} \gamma_j f_j(y_i|\mu_j, \sigma^2) \right]. \tag{7.18}$$

Differentiating with respect to an unknown of $\mathbf{\Omega}$, Ω_ℓ, we have

$$\begin{aligned}
\frac{\partial}{\partial \Omega_\ell} \log L(\mathbf{\Omega}) &= \sum_{i=1}^{n} \sum_{j=1}^{k} \frac{\gamma_j \frac{\partial}{\partial \Omega_\ell} f_j(y_i|\mu_j, \sigma^2)}{\sum_{j=1}^{k} \gamma_j f_j(y_i|\mu_j, \sigma^2)} \\
&= \sum_{i=1}^{n} \sum_{j=1}^{k} \frac{\gamma_j f_j(y_i|\mu_j, \sigma^2)}{\sum_{j=1}^{k} \gamma_j f_j(y_i|\mu_j, \sigma^2)} \frac{\partial}{\partial \Omega_\ell} \log f_j(y_i|\mu_j, \sigma^2) \\
&= \sum_{i=1}^{n} \sum_{j=1}^{k} \Gamma_{ij} \frac{\partial}{\partial \Omega_\ell} \log f_j(y_i|\mu_j, \sigma^2), \qquad (7.19)
\end{aligned}$$

where we define

$$\Gamma_{ij} = \frac{\gamma_j f_j(y_i|\mu_j, \sigma^2)}{\sum_{j=1}^{k} \gamma_j f_j(y_i|\mu_j, \sigma^2)}, \tag{7.20}$$

which could be thought of as a posterior probability that progeny i has marker genotype j. We then iterate between equations (7.19) and (7.20) with the expanded parameter set $\{\mathbf{\Omega}, \mathbf{\Gamma}\}$, where $\mathbf{\Gamma} = \{\Gamma_{ij}, j = 1, ..., k; i = 1, ..., n\}$. Conditional on $\mathbf{\Gamma}$, we solve for the zeros of $\frac{\partial}{\partial \Omega_\ell} \log L(\mathbf{\Omega})$ to get our estimates of $\mathbf{\Omega}$:

$$\begin{aligned}
\gamma_j &= \frac{1}{n} \sum_{i=1}^{n} \mathbf{\Gamma}_{ij}, \\
\mu_j &= \frac{\sum_{i=1}^{n} \Gamma_{ij} y_j}{\sum_{i=1}^{n} \Gamma_{ij}}, \\
\sigma^2 &= \frac{1}{n} \sum_{j=1}^{k} \sum_{i=1}^{n} \Gamma_{ij} (y_j - \mu_j)^2.
\end{aligned}$$

The estimates are then used to update $\mathbf{\Gamma}$, and the process is repeated until convergence. The values at convergence are the MLEs. The posterior probability of the marker genotype of an individual, given its phenotypic value y_i, can be evaluated by replacing parameters with their MLEs. On the basis of their estimated posterior probabilities, individuals may be assigned to one of the k genotypes.

For a segregating F_2 family, the mixing proportions will be known *a priori*, i.e., $\gamma_1 = 1/4$ for AA, $\gamma_2 = 1/2$ for Aa, and $\gamma_3 = 1/4$ for aa. In this case, the full model for estimating all six free parameters $\mathbf{\Omega}_F = (\mu_j, \sigma^2, \gamma_j)^{\mathrm{T}}$ is changed to a reduced model in which only four free parameters are estimated, $\mathbf{\Omega}_R = (\mu_j, \sigma^2)^{\mathrm{T}}$. Thus, to test for significant departure from the 1:2:1 segregation ratio, we calculate the log-likelihood ratio test statistic

$$\mathrm{LR} = -2[\log L(\mathbf{\Omega}_R) - \log L(\mathbf{\Omega}_F)],$$

which is asymptotically distributed as χ^2 with three degrees of freedom (the difference in the number of free parameters between $\boldsymbol{\Omega}_R$ and $\boldsymbol{\Omega}_F$). If the test is not significant, we conclude that the dominant marker follows 1:2:1. The departure from 1:2:1 in the F_2 initiated with two inbred lines may be due to many factors, such as more than one marker locus, differential survivals, etc.

The linear relation between the copy number of a dominant allele and the quantitative measurement can also be tested. The reduced model for this test corresponds to the mean values (μ_j) of three marker genotypes following $0\ (aa) : 1\ (Aa) : 2\ (AA)$, whereas the full model is formulated to estimate these mean values.

Correct Allocation Rate

The use of the estimated posterior probability to determine the genotype class of an individual is not error-free. In this section, we describe how to compute the correct allocation rate (CAR), i.e., the probability that a randomly selected individual is correctly classified (Piepho and Koch 2000). Suppose the classification limits are y_L and y_R so that individuals are classified as follows:

Classification	Condition for Value y
Genotype $= AA$	$y < y_L$
Genotype $= Aa$	$y_L \le y \le y_R$
Genotype $= aa$	$y_R < y$

The classification limit y_L is the point of intersection between $\gamma_1 f_1(y_i|\mu_1, \sigma)$ and $\gamma_2 f_2(y_i|\mu_2, \sigma)$ for $\mu_1 < y_L < \mu_2$; that is, the point at which $\gamma_1 f_1(y_i|\mu_1, \sigma) = \gamma_2 f_2(y_i|\mu_2, \sigma)$ for $\mu_1 < y_L < \mu_2$. The classification limit y_R is the point of intersection between $\gamma_2 f_2(y_i|\mu_2, \sigma_2)$ and $\gamma_3 f_3(y_i|\mu_3, \sigma)$ for $\mu_2 < y_R < \mu_3$. Because the residual variance is assumed homogeneous,

$$\begin{aligned} y_L &= \frac{\sigma^2 \log(\gamma_1/\gamma_2)}{\mu_2 - \mu_1} + \frac{\mu_2 + \mu_1}{2}, \\ y_R &= \frac{\sigma^2 \log(\gamma_2/\gamma_3)}{\mu_3 - \mu_2} + \frac{\mu_3 + \mu_2}{2}. \end{aligned}$$

Let $F(\cdot)$ denote the cumulative distribution function of the standard normal and CAR_j denote the correct allocation rate for the jth component. Then, we have

$$\begin{aligned} \mathrm{CAR}_1 &= F\left(\frac{y_L - \mu_1}{\sigma}\right), \\ \mathrm{CAR}_2 &= F\left(\frac{y_R - \mu_2}{\sigma}\right) - F\left(\frac{y_L - \mu_2}{\sigma}\right), \\ \mathrm{CAR}_3 &= 1 - F\left(\frac{y_R - \mu_3}{\sigma}\right). \end{aligned}$$

The overall correct classification rate is

$$\text{CAR} = \sum_{j=1}^{k} \gamma_j \text{CAR}_j.$$

We estimate the CAR by plugging in sample estimates for parameters. This approach provides a rough assessment of the true CAR.

Data Transformation

The mixture model derived above for classifying the marker genotypes of a dominant marker is based on the assumption of normality of the data. In practice, this assumption may not always be met, and the band intensity data may be brought closer to normality through transformations.

Typical transformations include the logarithm and the square root, where in most cases their actions are similar. They work to bring the data closer to symmetry if skewed and to produce more homogeneous variances. Specifically, the logarithm is appropriate when the mean is proportional to the standard deviation, and the square root is appropriate when the mean is proportional to the variance.

Gutierrez et al. (1995) discuss the application to normal mixtures of a more general family of transformations, the Box and Cox (1964) power transformations. This family of transformations is very flexible and includes the logarithmic transformation as a special case. It is given by

$$f(y,\lambda) = y^{\lambda} = \begin{cases} \dfrac{y^{\lambda}-1}{\lambda} & \text{if } \lambda \neq 0 \\ \log(y) & \text{if } \lambda = 0. \end{cases}$$

Note that taking $\lambda = 1$ is equivalent to not transforming the data. The simplest way to obtain an MLE of λ is by a grid search (Gutierrez et al. 1995). Other possible transformation approaches include square-roots, which can lessen the heterogeneity of residual variance due to scaling (Jansen et al. 2001). Once an appropriate transformation has been found, a normal mixture may be fitted to the transformed data.

Example 7.1. (**F_2 Progeny in Tomatoes**). Jansen et al. (2001) reported an example of codominant analysis of a dominant AFLP marker in 87 tomato plants from a segregating F_2 progeny. A histogram of band intensities of these progeny is shown in Fig. 7.6, in which a trimodal distribution can be seen. It appears that band intensity is linearly related to copy number and that variance increases linearly with the mean. A square-root transformation is used to remove the heterogeneity of the variance across different genotypes.

A normal mixture model is used to fit the square-root transformed data. The data can be well fit by three mixture distributions, as plotted over the histograms (Fig. 7.6). There is some overlap between the distributions of Aa and AA. The individuals with band intensities in the overlapping regions are not recommended for classification. The CAR for correctly classifying a marker genotype is more than 98 percent.

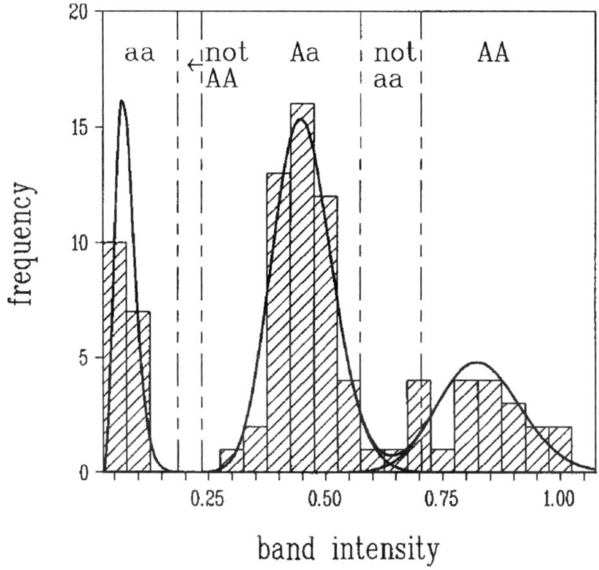

Fig. 7.6. Mixture distributions of band intensities in 87 F_2 tomato plants. The distributions of the three genotypes *aa*, *Aa* and *AA* are identified, but there is some overlap between the distributions of *aa* and *Aa* (not *AA*) and between the distributions of *Aa* and *AA* (not *aa*). Adapted from Jansen et al. (2001).

7.3.3 Linkage Analysis

Dominant markers, such as AFLPs and RAPDs, are usually analyzed based on the presence or absence of a band on an electrophoretic gel. This type of analysis does not allow a distinction among dominant homozygotes and heterozygotes. Such a distinction can be made possible if band intensities are quantitatively measured. In Section 7.3.2, a statistical mixture model implemented by the EM algorithm (Piepho and Koch 2000) was described to distinguish between dominant homozygotes and heterozygotes for single markers based on quantitative intensities. This model can extract more informative codominant information for linkage analysis from any dominant markers as long as their band intensities are quantified.

In this section, a similar normal mixture model is proposed for simultaneous estimation of quantitative codominance and the linkage between two dominant markers. Our analysis will be based on Piepho's (2001) work but extended for a three-point analysis.

Suppose there are n F_2 plants derived from a doubly heterozygous F_1 $AaBb$ from two contrasting inbred lines. Two markers, **A** and **B**, generate a total of nine genotypes in the F_2 population, $AABB$, $AABb$, $AAbb$, $AaBB$, $AaBb$, $Aabb$, $aaBB$, $aaBb$, and $aabb$, with respective frequencies as a function of the recombination fraction r between the two markers that can be described in matrix form. For dominant markers, AA and Aa, as well as BB and Bb, cannot be distinguished because they both produce a band on the gel. However, if the band intensities of these two markers are measured, we can expect that AA or BB will have a larger value than Aa or Bb. However, such a difference may be blurred due to random errors that can be assumed to follow a normal distribution.

The quantitative value of band intensity at the j_1th genotype of marker **A** for the ith F_2 plant includes two components, mean value μ_{1j_1} and variance $\sigma^2_{1j_1}$. The marginal distribution of $y_i = (y_{1i}\ y_{2i})$ for the two markers **A** and **B** is a mixture of nine bivariate normal distributions,

$$f(y_i|\boldsymbol{\Omega}) = \sum_{j_0=1}^{2}\sum_{j_2=0}^{2} \gamma_{j_1j_2} f(y_{1i}|\mu_{1j_1}, \sigma^2_{1j_1}) f(y_{2i}|\mu_{2j_2}, \sigma^2_{2j_2}), \tag{7.21}$$

where $\boldsymbol{\Omega} = (\gamma_{j_1j_2},\ \mu_{1j_1},\ \sigma^2_{1j_1},\ \mu_{2j_2},\ \sigma^2_{2j_2})$ are the unknown parameters to be estimated, $j_1, j_2 = 0$ for aa or bb, 1 for Aa or Bb, and 2 for AA or BB, respectively, and $\gamma_{j_1j_2}$ is the genotype frequency of the j_1th genotype at marker **A** and the j_2th genotype at marker **B**. The model can be simplified by assuming variance homogeneity at different levels; e.g.,

$$\sigma^2_{1j_1} = \sigma^2_1, \sigma^2_{2j_2} = \sigma^2_2, \tag{7.22}$$

or

$$\sigma^2_{1j_1} = \sigma^2_{2j_2} = \sigma^2. \tag{7.23}$$

Whereas model (7.23) is simpler in computation, model (7.22) may be closer to reality because the type and quantity of product measured at a band position vary among markers.

Let $z_{j_1j_2i}$ be a random variable with $z_{j_1j_2i} = 1$ for an observed marker genotype and $z_{j_1j_2i} = 0$ otherwise. We have that $\mathbf{z}_i = (z_{00i}, ..., z_{22i})$ follows a multinomial distribution with a constant 1 and cell probabilities of $\gamma_{j_1j_2}$. While y_i is observed, z_i is not observed (missing). Both y_i and z_i constitute the complete data. Thus, the EM algorithm can be used to provide the estimation of $\boldsymbol{\Omega}$ (McLachlan and Krishnan 1997). The likelihood for the complete data is

$$\log L(\boldsymbol{\Omega}) = \sum_{i=1}^{n}\sum_{j_1=0}^{2}\sum_{j_2=0}^{2} z_{j_1j_2i} \log\left[\gamma_{j_1j_2} f(y_{1i}|\mu_{1j_1}, \sigma^2_{1j_1}) f(y_{2i}|\mu_{2j_2}, \sigma^2_{2j_2})\right].$$

The conditional expectation of the complete-data log-likelihood, given the observed data y_i, using the current estimates for the parameter $\boldsymbol{\Omega}^{(t)}$, may be expressed as

$$Q(\boldsymbol{\Omega}; \boldsymbol{\Omega}^{(t)}) = E_{\boldsymbol{\Omega}^{(t)}}\left[\log L(\boldsymbol{\Omega}^{(t)})|y_i\right]. \tag{7.24}$$

Since $Q(\boldsymbol{\Omega}; \boldsymbol{\Omega}^{(t)})$ is linear in $z_{j_1j_2i}$, $Q(\boldsymbol{\Omega}; \boldsymbol{\Omega}^{(t)})$ is computed by replacing $z_{j_1j_2i}$ by its expectation, given y_i, in the complete-data log-likelihood evaluated at $Q(\boldsymbol{\Omega}; \boldsymbol{\Omega}^{(t)})$; i.e., $z_{j_1j_2i}$ is replaced by

$$\Gamma_{j_1j_2i}^{(t)} = \frac{\gamma_{j_1j_2} f\left(y_{1i}|\mu_{1j_1}^{(t)}, \sigma_{1j_1}^{2(t)}\right) f\left(y_{2i}|\mu_{2j_2}^{(t)}, \sigma_{2j_2}^{2(t)}\right)}{\sum_{j_1=0}^{2}\sum_{j_2=0}^{2} \gamma_{j_1j_2} f\left(y_{1i}|\mu_{1j_1}^{(t)}, \sigma_{1j_1}^{2(t)}\right) f\left(y_{2i}|\mu_{2j_2}^{(t)}, \sigma_{2j_2}^{2(t)}\right)}. \tag{7.25}$$

In the E step, $\Gamma_{j_1j_2i}^{(t)}$ is updated using the current parameter estimates, and, in the M step, $Q(\mathbf{\Omega};\mathbf{\Omega}^{(t)})$ is maximized with respect to the parameters. The estimating equations for the means and variances have explicit solutions, whereas the equation for r is a third-degree polynomial, which may be solved numerically or by explicit formulas. The equations for estimating the means and variances in the M step are given as follows:

$$\begin{aligned}
\mu_{1j_1}^{(t)} &= \frac{\sum_{i=1}^{n}\sum_{j_2=0}^{2}\Gamma_{j_1j_2i}^{(t)} y_{1i}}{\sum_{i=1}^{n}\sum_{j_2=0}^{2}\Gamma_{j_1j_2i}^{(t)}}, \\
\mu_{2j_2}^{(t)} &= \frac{\sum_{i=1}^{n}\sum_{j_1=0}^{2}\Gamma_{j_1j_2i}^{(t)} y_{2i}}{\sum_{i=1}^{n}\sum_{j_1=0}^{2}\Gamma_{j_1j_2i}^{(t)}}, \\
\sigma_1^{2(t)} &= \frac{1}{n}\sum_{i=1}^{n}\sum_{j_1=0}^{2}\sum_{j_2=0}^{2}\Gamma_{j_1j_2i}^{(t)}\left(y_{1i}-\mu_{1j_1}^{(t)}\right)^2, \\
\sigma_2^{2(t)} &= \frac{1}{n}\sum_{i=1}^{n}\sum_{j_1=0}^{2}\sum_{j_2=0}^{2}\Gamma_{j_1j_2i}^{(t)}\left(y_{2i}-\mu_{2j_2}^{(t)}\right)^2.
\end{aligned} \tag{7.26}$$

Update $r^{(t)}$ by the noncomplex root of

$$\frac{\partial Q(\mathbf{\Omega};\mathbf{\Omega}^{(t)})}{\partial r} = \sum_{i=1}^{n}\sum_{j_1=0}^{2}\sum_{j_2=0}^{2}\Gamma_{j_1j_2i}^{(t)} S_{j_1j_2} = 0,$$

where

$$S_{j_1j_2} = \frac{1}{\Gamma_{j_1j_2}}\frac{\partial \Gamma_{j_1j_2}}{\partial r},$$

that maximizes $Q(\mathbf{\Omega};\mathbf{\Omega}^{(t)})$. If the solution of this equation is larger than 0.5, set $r^{(t+1)} = 0.5$. If the solution is smaller than 0, set $r^{(t+1)} = 0$.

Piepho (2001) performed a simulation study to compare the estimator of the recombination fraction by treating dominant markers as quantitative codominant or band presence/absence. It was found that the quantitative method displayed more precise estimates of r than the qualitative method consistently for different r values, different means and variances, and different sample sizes.

The EM method described above can be used to simultaneously estimate marker codominance and the linkage among three dominant markers, **A**, **B**, and **C**, whose band intensities are quantified. Simultaneous consideration of three markers may increase the precision of linkage analysis and provide information about marker order. Assuming that the three markers are ordered as **A-B-C**, the frequencies of 27 different marker genotypes $AABBCC, AABBCc, ..., aabbcc$ in an F_2 population can be expressed as a function of the recombination fractions, $r_{\mathbf{AB}}$, $r_{\mathbf{AC}}$, and $r_{\mathbf{BC}}$, between the three markers; that is,

$$\mathbf{H}_{\mathbf{ABC}} =$$

$$\begin{array}{l} \\ BBCC \\ BBCc \\ BBcc \\ BbCC \\ BbCc \\ Bbcc \\ bbCC \\ bbCc \\ bbcc \end{array} \begin{array}{c} \\ \left[\begin{array}{ccc} \overset{AA}{\tfrac{1}{4}(1-r_{\mathbf{AB}})^2(1-r_{\mathbf{BC}})^2} & \overset{Aa}{\tfrac{1}{2}r_{\mathbf{AB}}(1-r_{\mathbf{AB}})r_{\mathbf{BC}}(1-r_{\mathbf{BC}})} & \overset{aa}{\tfrac{1}{4}r_{\mathbf{AB}}^2 r_{\mathbf{BC}}^2} \\ \tfrac{1}{2}(1-r_{\mathbf{AB}})^2 r_{\mathbf{BC}}(1-r_{\mathbf{BC}}) & \tfrac{1}{2}r_{\mathbf{AB}}(1-r_{\mathbf{AB}})\phi_{\mathbf{BC}} & \tfrac{1}{2}r_{\mathbf{AB}}^2 r_{\mathbf{BC}}(1-r_{\mathbf{BC}}) \\ \tfrac{1}{4}(1-r_{\mathbf{AB}})^2 r_{\mathbf{BC}}^2 & \tfrac{1}{2}r_{\mathbf{AB}}(1-r_{\mathbf{AB}})r_{\mathbf{BC}}(1-r_{\mathbf{BC}}) & \tfrac{1}{4}r_{\mathbf{AB}}^2(1-r_{\mathbf{BC}})^2 \\ \tfrac{1}{2}r_{\mathbf{AB}}(1-r_{\mathbf{AB}})(1-r_{\mathbf{BC}})^2 & \tfrac{1}{2}\phi_{\mathbf{AB}} r_{\mathbf{BC}}(1-r_{\mathbf{BC}}) & \tfrac{1}{4}r_{\mathbf{AB}}(1-r_{\mathbf{AB}})r_{\mathbf{BC}}^2 \\ r_{\mathbf{AB}}(1-r_{\mathbf{AB}})r_{\mathbf{BC}}(1-r_{\mathbf{BC}}) & \tfrac{1}{2}\phi_{\mathbf{AB}}\phi_{\mathbf{BC}} & r_{\mathbf{AB}}(1-r_{\mathbf{AB}})r_{\mathbf{BC}}(1-r_{\mathbf{BC}}) \\ \tfrac{1}{2}r_{\mathbf{AB}}(1-r_{\mathbf{AB}})r_{\mathbf{BC}}^2 & \tfrac{1}{2}\phi_{\mathbf{AB}} r_{\mathbf{BC}}(1-r_{\mathbf{BC}}) & \tfrac{1}{2}(r_{\mathbf{AB}})(1-r_{\mathbf{AB}})(1-r_{\mathbf{BC}})^2 \\ \tfrac{1}{4}r_{\mathbf{AB}}^2(1-r_{\mathbf{BC}})^2 & \tfrac{1}{2}r_{\mathbf{AB}}(1-r_{\mathbf{AB}})r_{\mathbf{BC}}(1-r_{\mathbf{BC}}) & \tfrac{1}{4}(1-r_{\mathbf{AB}})^2 r_{\mathbf{BC}}^2 \\ \tfrac{1}{2}r_{\mathbf{AB}}^2 r_{\mathbf{BC}}(1-r_{\mathbf{BC}}) & \tfrac{1}{2}r_{\mathbf{AB}}(1-r_{\mathbf{AB}})\phi_{\mathbf{BC}} & \tfrac{1}{2}(1-r_{\mathbf{AB}})^2 r_{\mathbf{BC}}(1-r_{\mathbf{BC}}) \\ \tfrac{1}{4}r_{\mathbf{AB}}^2 r_{\mathbf{BC}}^2 & \tfrac{1}{2}r_{\mathbf{AB}}(1-r_{\mathbf{AB}})r_{\mathbf{BC}}(1-r_{\mathbf{BC}}) & \tfrac{1}{4}(1-r_{\mathbf{AB}})^2(1-r_{\mathbf{BC}})^2 \end{array}\right] \end{array}, \tag{7.27}$$

where $\phi_{\mathbf{AB}} = 1 - 2r_{\mathbf{AB}} + 2r_{\mathbf{AB}}^2$ and $\phi_{\mathbf{BC}} = 1 - 2r_{\mathbf{BC}} + 2r_{\mathbf{BC}}^2$. Two similar matrices of the genotype frequencies $\mathbf{H}_{\mathbf{ACB}}$ (in terms of $r_{\mathbf{AC}}$ and $r_{\mathbf{BC}}$) and $\mathbf{H}_{\mathbf{BAC}}$ (in terms of $r_{\mathbf{AB}}$ and $r_{\mathbf{AC}}$) can also be derived for marker orders **A-C-B** and **B-A-C**.

Since a correct marker order is not known, an algorithm should be formulated to estimate the most likely marker order for linkage analysis. Let q_1 and q_2 be the probabilities of order **A-B-C** and **A-C-B**. Thus, the probability of order **B-A-C** is $(1 - q_1 - q_2)$. In the 27-normal mixture model, the frequency of each component is a weighted mean of the frequency of a genotype under three different orders. In other words, the matrix for the frequencies of 27 genotypes at the three markers can be written as

$$\mathbf{H} = q_1\mathbf{H}_{\mathbf{ABC}} + q_2\mathbf{H}_{\mathbf{ACB}} + (1 - q_1 - q_2)\mathbf{H}_{\mathbf{BAC}}.$$

The elements in this matrix represent the *a priori* genotype frequencies $\pi_{j_1j_2j_3}$, where j_1, j_2, j_3 are the genotypes of each of the three markers.

Using a similar principle for two-point analysis, we can formulate the marginal distribution of $y_i = (y_{1i}, y_{2i}, y_{3i})$ in terms of a mixture of 27 trivariate normal distributions. The EM algorithm is developed to estimate the recombination fractions, $r_{\mathbf{AB}}$, $r_{\mathbf{AC}}$, and $r_{\mathbf{BC}}$, the probabilities of marker order, q_1 and q_2, and the three means and residual variance at each marker. The estimates corresponding to the highest probabilities of marker orders are regarded as the optimal estimates.

7.4 Exercises

7.1 Based on the genotypic information provided by Table 7.4, prove equation (7.5); that is, $P(\mathcal{G}_6|\Lambda_1, \mathcal{G}_5) = P(\bar{\Lambda}_6|\Lambda_1, \Lambda_5) = \frac{1}{2}r$.

7.2 Recall Table 7.4. We describe the procedure for linkage analysis between markers **A** and **B**. Show a similar procedure with which linkage analysis can be performed between markers **B** and **C**.

(a) Write the likelihood and estimate the recombination fraction, $r_{\mathbf{AC}}$, under an optimal founder diplotype combination.

(b) Write a joint likelihood that incorporates the founder diplotype probability and the recombination fractions.

7.3 Morton (1956) reported a complicated pedigree for linkage analysis of familial elliptocytosis (Fig. 7.7). This is extremely rare, with the population frequency of the disease allele being very low. Investigations of this pedigree and others show that the inheritance of this disease is consistent with fully penetrant autosomal dominance inheritance. The low disease-allele frequency and its dominant inheritance allow scoring each affected person as *Dd* and each unaffected person as *dd* and taking each *D* allele as being identical by descent (IBD). Show that the likelihood function of this family can be expressed as

$$\begin{aligned}
L \propto\; & 810r(1-r)^{19} + 324r(1-r)^{18} + 180r(1-r)^{17} + 72r(1-r)^{16} + 90r^3(1-r)^{17} \\
& + 72r^3(1-r)^{16} + 40r^3(1-r)^{15} + 24r^3(1-r)^{14} + 90r^4(1-r)^{15} + 20r^4(1-r)^{13} \\
& + 90r^5(1-r)^{15} + 432r^5(1-r)^{14} + 20r^5(1-r)^{13} + 104r^5(1-r)^{12} \\
& + 1800r^6(1-r)^{14} + 558r^6(1-r)^{13} + 440r^6(1-r)^{12} + 176r^6(1-r)^{11} \\
& + 90r^7(1-r)^{13} + 324r^7(1-r)^{12} + 120r^7(1-r)^{10} + 360r^8(1-r)^{12} \\
& + 378r^8(1-r)^{11} + 80r^8(1-r)^{10} + 76r^8(1-r)^9 + 4r^8(1-r)^4 + 180r^9(1-r)^{11} \\
& + 522r^9(1-r)^{10} + 80r^9(1-r)^9 + 100r^9(1-r)^8 + 10r^9(1-r)^3 + 180r^{10}(1-r)^{10} \\
& + 846r^{10}(1-r)^9 + 40r^{10}(1-r)^8 + 216r^{10}(1-r)^7 + 18r^{10}(1-r)^4 + 4r^{10}(1-r)^2 \\
& + 1170r^{11}(1-r)^9 + 378r^{11}(1-r)^8 + 260r^{11}(1-r)^7 + 72r^{11}(1-r)^6 \\
& + 45r^{11}(1-r)^3 + 180r^{12}(1-r)^8 + 396r^{12}(1-r)^7 + 40r^{12}(1-r)^5 + 18r^{12}(1-r)^2 \\
& + 270r^{13}(1-r)^7 + 234r^{13}(1-r)^6 + 40r^{13}(1-r)^5 + 52r^{13}(1-r)^4 + 180r^{14}(1-r)^6 \\
& + 108r^{14}(1-r)^5 + 80r^{14}(1-r)^4 + 16r^{14}(1-r)^3 + 90r^{15}(1-r)^5 + 162r^{15}(1-r)^4 \\
& + 20r^{15}(1-r)^3 + 180r^{16}(1-r)^4 + 72r^{16}(1-r)^3 + 90r^{17}(1-r)^3.
\end{aligned}$$

The peak of the likelihood profile over a range of r values is found as 3.31 at the recombination fraction of 0.05.

7.4 We have introduced two approaches for linkage analysis of dominant markers. The first is based on the "qualitative" observations of band presence or absence, whereas the second makes use of the "quantitative" measurement of bands based on fluorescence techniques. It is important to compare how these two different approaches work in a real example, which can be investigated through simulation studies as follows.

We simulate normally distributed "quantitative" values of band intensity by assuming that two linked markers are each segregating 1:2:1 in an F_2 population. The statistical method described in this chapter is used to estimate the recombination fraction between two hypothesized markers. The simulated "quantitative" markers are then treated as (1/0)-dominant markers. The EM algorithm introduced in Chapter 4 is employed to estimate the recombination fraction of dominant markers. Compare the results of the estimates from these "quantitative" and "qualitative" treatments.

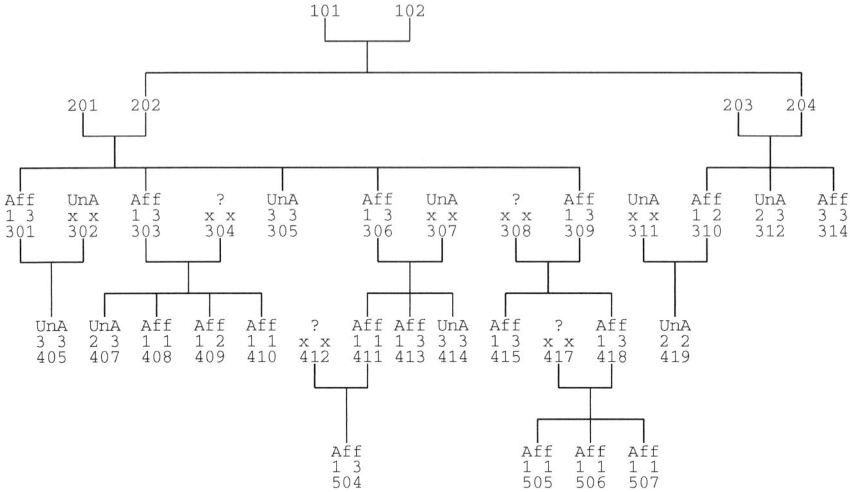

Fig. 7.7. A complicated pedigree in humans. Adapted from Morton (1956).

8

Marker Analysis of Phenotypes

8.1 Introduction

In the preceding chapters, we described numerous statistical issues related to linkage analysis of molecular markers and the construction of genetic linkage maps. One of the most important aims of these marker analyses is to provide ordered hallmarks on chromosomes with which one can map functional quantitative trait loci (QTLs) determining complex phenotypic variation to particular genomic regions. The genome-wide identification of QTLs, their locations and effects, is of fundamental importance for agricultural, evolutionary, and biomedical genetics.

A variety of methods have been developed for QTL mapping (Hoeschele et al. 1997; Lynch and Walsh 1998). These methods can be classified as t–tests and analysis of variance, least–squares analysis (LS), maximum–likelihood analysis (ML), and Bayesian analysis. These methods differ in computational requirements, efficiency in terms of extracting information, flexibility with regard to handling different data structures, and ability to map multiple QTLs. The simple LS method is efficient in terms of computational speed but cannot extract all information from the data and is restricted to specific mating designs. The technique of ML interval mapping (Lander and Botstein 1989) is one of the most widely used methods for QTL analysis in controlled crosses or structured pedigrees. The interval mapping method has been extended to composite interval mapping (Zeng 1994) and multiple interval mapping (Kao et al. 1999).

In this chapter, we will discuss t–test, analysis of variance, and regression analysis of multiple markers and perform statistical tests based solely on single DNA marker information. For single-marker analysis, no genetic map is required and the calculations are based on phenotypic means and variances within each of the genotypic classes. Marker analysis can be extended to include all markers of the genome (Xu 2003). Although single marker analyses, as shown later in this chapter, confound the QTL effect and the QTL location, they provide preliminary results that facilitate the use of more advanced interval mapping to detect QTLs within a genomic interval bracketed by two linked markers (Routman and Cheverud 1997). ML interval mapping and its extension, composite interval mapping, will be presented in Chapters 11 and 13.

8.2 QTL Regression Model

Prior to the introduction of an advanced statistical method for QTL mapping, we first consider a hypothetical example of a backcross design for mice shown in Table 8.1. This example contains ten mice phenotyped for body weight, y, and a QTL with two known genotypes, Qq (indicated by 1) and qq (indicated by 0). It appears that the mice that carry QTL genotype Qq tend to be heavier than those that carry genotype qq, although the mice that carry the same genotype do not have exactly the same body weight. To test whether this is actually the case and estimate the effect of a QTL on body weight, we formulate a simple regression model as

$$y_i = \mu + z_i a + e_i, \tag{8.1}$$

where y_i is the phenotypic value for mouse i, μ is the overall mean, z_i is the indicator variable that specifies the QTL genotype of mouse i and is defined as

$$z_i = \begin{cases} 1 & \text{if QTL genotype is } Qq \\ 0 & \text{if QTL genotype is } qq, \end{cases}$$

a is the additive effect of the QTL, and e_i is the random error, typically assumed to be normally distributed as $N(0, \sigma^2)$.

The linear model (8.1) can be extended to estimate and test the genetic effects of a QTL in an F_2 population with three QTL genotypes, QQ (indicated by 2), Qq (indicated by 1), and qq (indicated by 0). The model for the F_2 is written as

$$y_i = \mu + z_{1i} a + z_{2i} d + e_i, \tag{8.2}$$

with an additional parameter, d, that is the dominance effect of the QTL, and indicator variables z_{1i} and z_{2i} expressed as

$$z_{1i} = \begin{cases} 1 & \text{if QTL genotype is } QQ \\ -1 & \text{if QTL genotype is } qq \end{cases}$$

and

$$z_{2i} = \begin{cases} 0 & \text{if QTL genotype is } QQ \text{ or } qq \\ 1 & \text{if QTL genotype is } Qq. \end{cases}$$

Standard least squares (LS) approaches can be used to estimate the unknown model intercept, μ, and regression coefficients, a and/or d. Thus, by directly testing the significance of a and/or d, one can determine whether this QTL triggers an effect on body weight.

Example 8.1. Assume a small population of ten backcross mice. Each mouse was genotyped for two markers and measured for body weight as well. The marker and phenotypic data are given in Table 8.1. With this table, we provide a procedure for the estimation and test of the genetic effect, a, of the QTL on mouse body weight. Let

$$\mathbf{y} = \begin{pmatrix} 30 \\ 32 \\ 28 \\ 29 \\ 29 \\ 22 \\ 20 \\ 21 \\ 20 \\ 21 \end{pmatrix}, \quad \mathbf{X} = \begin{pmatrix} 1 & 1 \\ 1 & 1 \\ 1 & 1 \\ 1 & 1 \\ 1 & 1 \\ 1 & 0 \\ 1 & 0 \\ 1 & 0 \\ 1 & 0 \\ 1 & 0 \end{pmatrix}, \quad \mathbf{e} = \begin{pmatrix} e_1 \\ e_2 \\ e_3 \\ e_4 \\ e_5 \\ e_6 \\ e_7 \\ e_8 \\ e_9 \\ e_{10} \end{pmatrix},$$

and $\mathbf{b} = (\mu, a)^{\mathrm{T}}$. We then have $\mathbf{y} = \mathbf{X}\mathbf{b} + \mathbf{e}$. The LS estimates of the parameters are

$$\hat{\mathbf{b}} = (\mathbf{X}^{\mathrm{T}}\mathbf{X})^{-1}(\mathbf{X}^{\mathrm{T}}\mathbf{y}) = (20.8, 8.8)^{\mathrm{T}}$$

and

$$\hat{\sigma}^2 = \frac{1}{n}(\mathbf{y} - \mathbf{X}\mathbf{b})^{\mathrm{T}}(\mathbf{y} - \mathbf{X}\mathbf{b}) = 1.2.$$

Table 8.1. Data structure for two genotyped markers and mouse body weight in a backcross design.

	Marker		Body
Sample	**A**	**B**	Weight
1	1	1	30
2	1	1	32
3	1	1	28
4	1	1	29
5	1	0	29
6	0	1	22
7	0	0	20
8	0	0	21
9	0	0	20
10	0	0	21

With these estimates, we calculate the total sum of squares (SST),

$$\sum_{i=1}^{10}(y_i - \mu)^2 = 205.6,$$

and the residual sum of squares (SSE),

$$\sum_{i=1}^{10}(y_i - \mu - z_i a)^2 = 12.0.$$

The significance of the QTL genetic effect (a) is then tested by calculating the F-value,

$$F = \frac{(\mathrm{SST} - \mathrm{SSE})/(2-1)}{\mathrm{SSE}/(10-2)} = 129.07.$$

Compared with the critical value $F_{0.05,(2,8)} = 4.46$, we conclude that this QTL exerts a significant effect on body weight in the backcross population of mice.

8.3 Analysis at the Marker

A QTL statistical model assumes that the QTL genotypes can be observed in a mapping population. This is not possible in practice. What we can do is to use observable markers to predict such unobservable QTLs through the linkage between markers and QTLs. Thus, by performing the association analysis between the markers and phenotypes, we can still infer the effect of a putative QTL on phenotypic variation.

The use of a single-marker is limited for QTL identification since it cannot determine at which side of the marker, left or right, the QTL is located. However, single marker analyses are useful for a preliminary test of the existence of a QTL, although they cannot estimate the QTL location. Below, we introduce two testing approaches for marker analysis based on the t and F test statistics.

8.3.1 Two-Sample t Test

The mouse backcross data in Table 8.1 are genotyped for two linked molecular markers **A** and **B** and are given in Table 8.1. Two genotypes at each marker are denoted by 1 and 0. The linkage between these two markers can be seen from the consistency of their genotypes among the samples, except for mice 5 and 6. The recombination fraction between the two markers is $r = 2/10 = 0.2$. We will analyze these two markers separately.

Looking at marker **A**, it seems that the two groups of marker genotypes differ in body weight. A question arises naturally about whether this difference in body weight between genotypes 1 and 0 at marker **A** is statistically significant. This can be tested by a two-sample t test. Let μ_1 and μ_0 be the true trait means of two different groups of mice with genotypes 1 and 0, respectively, and let m_1 and m_0 be the corresponding sample means. The hypotheses for the test can be formulated as

$$H_0 : \mu_1 = \mu_0,$$
$$H_1 : \mu_1 \neq \mu_0.$$

The t test statistic used to test for the significance of the difference between the two means is

$$t = \frac{m_1 - m_0}{\sqrt{s^2(\frac{1}{n_1} + \frac{1}{n_0})}}, \tag{8.3}$$

where s^2 is the pooled sampling variance given by

$$s^2 = \frac{(n_1 - 1)s_1^2 + (n_0 - 1)s_0^2}{n_1 + n_0 - 2},$$

with n_1, n_0 and s_1^2, s_0^2 being the sample sizes and variances in two different marker groups, respectively.

The null hypothesis H_0 will be rejected if the t test statistic calculated is larger than or equal to the critical value to be obtained from the t–distribution. If we denote the upper α critical point by $t_{(\alpha,\nu)}$, we reject the hypothesis at $\alpha = 0.05$ if $t > t_{(0.025,\nu)}$, the two-tailed t value for the 0.05 significance level, with $\nu = n_1 + n_0 - 2$ degrees of freedom.

Example 8.2. In the example with $n_1 = n_0 = 5$ provided in Table 8.1, we calculate $m_1 = 29.6$, $m_0 = 20.8$, $s_1 = 0.8367$, $s_0 = 1.5166$, and $s = 1.50$ for marker **A**. We further calculate $t = \frac{29.6-20.8}{\sqrt{1.5^2(\frac{1}{5}+\frac{1}{5})}} = 11.3608$. Compared with the critical value of $t_{(0.025,5+5-2=8)} = 2.3060$, we conclude that marker **A** is significantly associated with body weight.

In this example, we also find that genotype 1 for each of the two markers tends to be heavier than genotype 2. Since the conclusion from the t–test is only that there is a difference, we cannot make a formal statement about the direction of the difference (we could have if a one-tailed test were done, but typically a two-tailed test is carried out). However, most would be comfortable with the informal conclusion that genotype 1 at each marker tends to be heavier than genotype 2.

Example 8.3. (**Tomato Plant Heights**). We again look at the data of Example 2.1, but now introduce marker data. Suppose that we have the following data y on heights (in cm) of 12 tomato plants of a particular species grouped into two marker classes

$$M_1M_1 : y = (79, 82, 100, 102, 124)$$
$$M_1M_2 : y = (85, 87, 101, 103, 125, 126, 127).$$

The observed means and variances of the marker classes are

$$\bar{y}_{M_1M_1} = 97.4 \quad \bar{y}_{M_1M_2} = 101.71$$
$$s_{M_1M_1} = 18.10 \quad s_{M_1M_2} = 18.33$$

To test H_0, is would seem natural to use a two-sample t-test, and this has been one of the first approaches. The t-statistic to test H_0 is

$$t = \frac{y_{M_1M_1} - y_{M_1M_2}}{\sqrt{s^2\left(\frac{1}{n_1} + \frac{1}{n_2}\right)}},$$

where s^2 is the pooled variance of the two groups and n_1 and n_2 are the sample sizes in each group.

A problem with using this statistic to test H_0 is that, due to the underlying mixture distribution, this statistic does not have the usual Student-t null distribution. Hence, calibrating this statistic, that is, computing a p-value, will not give a valid inference. One way around this is to use a permutation test, and to get the null distribution of the test statistic based on permutations.

In this situation, the permutation test is quite simple. The data are assigned at random into one of the two groups, and the resulting t-statistic is calculated. The premise is that under H_0 there is no difference in the two marker groups, and hence the observation could have just as likely been from either group. This random assignment is repeated many times, and the resulting t-statistics and made into a histogram which serves as the null distribution.

Example 8.4. (**Tomato Plant Heights-Continued**). For the data of Example 8.3, the t statistic of the observed data is -.965, and a histogram of 5000 permuted t-statistics is given in Figure 8.1 with the program that generated the histogram given in Appendix B.2. The upper and lower 5 percent cutoff points from the permutation distribution are 1.97 and −2.11, respectively. Based on this null distribution, we make the conclusion that there is no QTL near this marker.

8.3.2 Analysis of Variance

For an F_2 population, there are three different groups of marker genotypes, which can be denoted by 2, 1, and 0, respectively, at each marker (see Table 8.2). To test the overall difference among the three genotypes, a traditional analysis of variance (ANOVA) can be used. The mean square due to the difference among the three marker genotypes reflects the degree to which the marker is associated with a putative QTL for a particular trait, while the mean square due to the difference within the genotypes reflects the residual variance. The ratio of these two mean squares, the F-value, is a test statistic used to test for the significance of the difference among the three marker genotypes.

The calculated F-value is compared with the critical value obtained from the F distribution, $F_{0.05,(2,n-3)}$. The genetic variance due to a significant marker can be estimated by equating the expected mean squares (Table 8.3) to the mean squares (MS) and solving the resulting equation:

$$\sigma_g^2 = \frac{\text{MS}_1 - \text{MS}_2}{k}. \tag{8.4}$$

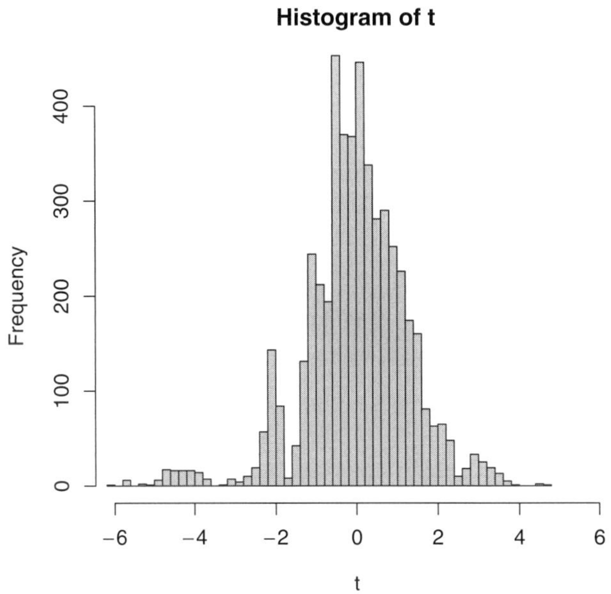

Fig. 8.1. Null distribution of the t-statistic of Example 8.4 based on permutations.

Table 8.2. Data structure for two genotyped markers and mouse body weight in an F_2 design.

	Marker		Body
Sample	**A**	**B**	Weight
1	2	2	30
2	2	1	32
3	2	0	28
4	1	2	29
5	1	1	29
6	1	0	22
7	1	0	20
8	0	2	21
9	0	1	20
10	0	0	21

The proportion of the phenotypic variance in a quantitative trait explained by the marker, the broad-sense *heritability*, is estimated by

$$R^2 = \frac{\sigma_g^2}{\sigma_g^2 + \sigma_e^2}. \tag{8.5}$$

This proportion is widely used as a parameter to assess the contribution of the **A** marker to the phenotypic variation.

Table 8.3. Summary of ANOVA for the difference among three genotype groups in an F_2 population.

Source of Variation	df	Mean Square	F-value	Expected Mean Square
Among marker genotypes	2	MS_1	MS_1/MS_2	$\sigma_e^2 + k\sigma_q^2$
Within marker genotypes	$n-3$	MS_2		σ_e^2

Note: $k = 3/(\frac{1}{n_2} + \frac{1}{n_1} + \frac{1}{n_0})$ is a harmonic mean, with n_2, n_1 and n_0 standing for sample sizes of the three different marker genotypes in an F_2 population.

The overall difference among the three marker genotypes in the F_2 population may be due to either the additive or dominance effect, or both. The significance of these two effects can also be tested by using the t test. To test the marker's additive effect, we have the test statistic

$$t_1 = \frac{m_2 - m_0}{\sqrt{s^2(\frac{1}{n_2} + \frac{1}{n_0})}}, \tag{8.6}$$

with

$$s^2 = \frac{(n_2 - 1)s_2^2 + (n_0 - 1)s_0^2}{n_2 + n_0 - 2},$$

and to test the marker's dominance effect, we have the test statistic

$$t_2 = \frac{m_1 - \frac{1}{2}(m_2 + m_0)}{\sqrt{s^2(\frac{1}{4n_2} + \frac{1}{n_1} + \frac{1}{4n_0})}}, \tag{8.7}$$

with

$$s^2 = \frac{(n_2 - 1)s_2^2 + (n_1 - 1)s_1^2 + (n_0 - 1)s_0^2}{n_2 + n_1 + n_0 - 3},$$

where s_1^2, s_1^2, and s_0^2 are the sample variances in three different marker groups of the F_2, respectively.

Example 8.5. Table 8.2 provides an example for the F_2 population with ten mice, each measured for body weight and genotyped for two codominant markers **A** and **B**. We compute the mean squares for among- and within-genotype differences $MS_1 = 65.47$ and $MS_2 = 10.67$ for marker **A**, from which the F-value is calculated as 6.14. Compared with the critical $F_{0.05,(2,10-3\ =\ 7)} = 4.7374$ value, this marker is thought to be significantly associated with body weight. The genetic variance due to this marker is calculated as $\sigma_g^2 = 5.0233$ with equation (8.4), and the heritability is then estimated as 0.3202.

For marker **A**, three genotype groups contain $n_2 = 3$, $n_1 = 4$, and $n_0 = 3$, and the three sampling means are calculated as $m_2 = 30$, $m_1 = 25$, and $m_0 = 20.67$ and three sampling variances calculated as $s_2^2 = 4$, $s_1^2 = 22$, and $s_0^2 = 0.3333$, respectively. We calculate the t test statistics for the additive and dominant effects, respectively, with equations (8.6) and (8.7), as $t_1 = 7.7658$ and $t_2 = -0.1220$. Compared with the critical values $t_{(0.025,\nu\ =\ 3+3-2\ =\ 4)} = 2.1318$ for the additive test and $t_{(0.025,\nu\ =\ 3+4+3-3\ =\ 7)} = 1.8946$ for the dominance effect, we conclude that marker **A** displays a significant additive effect, but an insignificant dominance effect on body weight.

A similar computing procedure is taken for marker **B**. This marker has the F-value 0.85, suggesting it has no significant association with body weight in mice. The t–values for testing additive and dominance effects are calculated as $t_1 = 1.2264$ and $t_2 = 0.5635$, respectively. It can be seen that both the additive effect and dominance effects are nonsignificant.

8.3.3 Genetic Analysis

Why can we infer the existence of an underlying QTL for a quantitative trait via a simple t test or ANOVA on marker means? Consider a putative QTL linked to a marker with a recombination fraction of r. The conditional expected genotypic values associated with each marker genotype are calculated from the conditional probabilities of the QTL genotypes given a marker genotype and from the genotypic values of different QTL genotypes. Given known marker genotypes, Aa (1) and aa (0), we can derive the conditional probabilities of two QTL genotypes, Qq (1) and qq (0), for the backcross as

Marker	QTL Genotypic Value	
Genotypic Value	μ_1	μ_0
m_1	$1-r$	r
m_0	r	$1-r$

The genetic values of these two backcross QTL genotypes can be denoted by

$$\mu_1 = \mu + \tfrac{1}{2}a \text{ and } \mu_0 = \mu - \tfrac{1}{2}a,$$

respectively. For each marker genotype, two different QTL genotypes are mixed, weighted by the conditional probabilities. Thus, the conditional expected genotypic values associated with different marker genotypes can be calculated as

$$m_1 = (1-r)\mu_1 + r\mu_0 = (1-r)\left(\mu + \tfrac{1}{2}a\right) + r\left(\mu - \tfrac{1}{2}a\right) = \mu + \tfrac{1}{2}(1-2r)a,$$

$$m_0 = r\mu_1 + (1-r)\mu_0 = r\left(\mu + \tfrac{1}{2}a\right) + (1-r)\left(\mu - \tfrac{1}{2}a\right) = \mu - \tfrac{1}{2}(1-2r)a.$$

Thus, the difference of the two marker means is

$$\mu_1 - \mu_0 = (1-2r)a. \tag{8.8}$$

If a is not significantly different from zero, the t test statistic based on equation (8.3) will be smaller than the critical value. In this sense, the t test can provide information about the significance of the QTL effect. But a nonsignificant t value may also be due to nonlinkage between the marker and QTL ($r = 0.5$) according to equation (8.8). Therefore, the t test only gives a composite test for the QTL effect and QTL–marker linkage.

For an F_2 population, three marker means can be similarly derived by conditional probabilities expressed as

Marker Genotypic Value	QTL Genotypic Value μ_2	μ_1	μ_0
m_2	$(1-r)^2$	$2r(1-r)$	r^2
m_1	$r(1-r)$	$(1-r)^2 + r^2$	$r(1-r)$
m_0	r^2	$2r(1-r)$	$(1-r)^2$

and the assigned QTL genotype values. These marker means are written as

$$\begin{aligned} m_2 &= (1-r)^2\mu_2 + 2r(1-r)\mu_1 + r^2\mu_0 \\ &= \mu + (1-2r)a + 2r(1-r)d, \end{aligned}$$

$$\begin{aligned} m_1 &= r(1-r)\mu_2 + [(1-r)^2 + r^2]\mu_1 + r(1-r)\mu_0 \\ &= \mu + (1-2r+2r^2)d, \end{aligned}$$

$$\begin{aligned} m_0 &= r^2\mu_2 + 2r(1-r)\mu_1 + (1-r)^2\mu_0 \\ &= \mu - (1-2r)a + 2r(1-r)d. \end{aligned}$$

The tests for the additive effect (a) from equation (8.6) and dominance effect (d) from equation (8.7) are equivalent to testing whether composite parameters

$$\tfrac{1}{2}(m_2 - m_0) = (1-2r)a = a - 2ra$$

and

$$m_1 - \tfrac{1}{2}(m_2 + m_0) = (1-2r)^2 d = d - 4r(1-r)d$$

are equal to zero, respectively.

From the analysis above, although the t test and ANOVA can be used to test the significance of marker differences, they cannot separate QTL genotypic means and the recombination fraction between a single marker and a QTL. If the marker difference is significant, as shown for the two markers in mouse body weight, we still do not know whether this difference is due to a tight linkage (small r) between the marker and a QTL of small effect or a loose linkage (large r) between the marker and a QTL of large effect. In fact, the additive and dominance effects of QTLs are underestimated by $2r$ and $4r(1-r)$, respectively, from a simple comparison of marker means. Also, the t test and ANOVA cannot separate the effects of individual QTLs on the phenotype if there are two or more QTL on the same chromosome. The two confounded parameters, QTL genetic means and the recombination fraction, can be separated using the approaches explained below.

8.4 Moving Away from the Marker

We now illustrate a single-marker analysis where we do not assume that the QTL is at the marker. Following Doerge et al. (1997), we illustrate the technique with a backcross design in which there are two genotypes at each marker or QTL.

8.4.1 Likelihood

Realize that the observed genotype will either be M_1M_1 or M_1M_2, but given this observed genotype, the QTL genotype will either be Q_1Q_1 or Q_1Q_2 with probabilities given below

	M_1M_1	M_1M_2
Q_1Q_1	$\frac{1-r}{2}$	$\frac{r}{2}$
Q_1Q_2	$\frac{r}{2}$	$\frac{1-r}{2}$

,

where r is the recombination fraction between the marker and QTL. Assume that a phenotypic trait (y) follows a normal distribution. Relating the phenotypes of the trait to the respective phenotypic means for the marker genotypes, we have the following mixture model:

$$\begin{aligned} &\text{observe } M_1M_1 \rightarrow y \sim (1-r)N(\mu_1, \sigma^2) + rN(\mu_2, \sigma^2), \\ &\text{observe } M_1M_2 \rightarrow y \sim rN(\mu_1, \sigma^2) + (1-r)N(\mu_2, \sigma^2), \end{aligned} \tag{8.9}$$

where μ_1 and μ_2 are the phenotypic means (or genotypic values) of the trait for QTL genotypes Q_1Q_1 and Q_1Q_2, respectively.

Under model (8.9), the mean and variance of the distributions are (Exercise 8.3)

$$\begin{aligned} \mu_{M_1M_1} &= (1-r)\mu_1 + r\mu_2 \\ \mu_{M_1M_2} &= r\mu_1 + (1-r)\mu_2 \\ \sigma^2_{M_1M_1} = \sigma^2_{M_1M_2} &= \sigma^2 + r(1-r)(\mu_1 - \mu_2)^2 \end{aligned} \tag{8.10}$$

Note that is there is no linkage between the markers and the QTL, that is, if $r = \frac{1}{2}$, then $\mu_{M_1M_1}$ and $\mu_{M_1M_2}$ are equal. Thus, the hypothesis of no linkage is

$$H_0 : r = \frac{1}{2} \text{ or } H_0 : \mu_{M_1M_1} - \mu_{M_1M_2} = (1-2r)(\mu_1 - \mu_2) = 0,$$

and it is important to note that under H_0, we cannot tell whether $r = \frac{1}{2}$ or $\mu_1 = \mu_2$. In either case, however, there is no practical phenotypic variation detectable.

If we assume that $y_1, \ldots, y_{n_1}$ are from marker group M_1M_1 and that $y_{n_1+1}, \ldots, y_n$ are from marker group M_1M_2, then the likelihood function based on model (8.9) is

$$\begin{aligned} L(\mu_1, \mu_2, \sigma^2, r|y) &= \prod_{i=1}^{n_1} (1-r) f(y_i|\mu_1, \sigma^2) + r f(y_i|\mu_2, \sigma^2) \\ &\times \prod_{i=n_1+1}^{n} r f(y_i|\mu_1, \sigma^2) + (1-r) f(y_i|\mu_2, \sigma^2). \end{aligned} \tag{8.11}$$

To test the null hypothesis of no linkage, H_0 : no QTL, we could use the likelihood ratio statistic

$$\lambda = \frac{\max_{\mu_1=\mu_{12},\sigma^2,r} L(\mu_1, \mu_2, \sigma^2, r|y)}{\max_{\mu_1,\mu_2,\sigma^2,r} L(\mu_1, \mu_{12}, \sigma^2, r|y)}. \tag{8.12}$$

The test statistic λ would reject H_0 if it is too small; alternatively we could transform to $-2\log\lambda$, which would reject if it is big and often has an approximate χ^2 distribution. However, the mixture model invalidates the χ^2 assumption, and what is typically done is a permutation test on $-2\log\lambda$ or its variant, the LOD score.

Likelihood Ratio Test

To calculate the test statistic λ, we have to maximize both the numerator and denominator of (8.12). The numerator is easy since under the null hypothesis $\mu_1 = \mu_{12} = \mu$, the likelihood (8.11) becomes

$$L(\mu_1, \mu_{12}, \sigma^2, r|y) = L(\mu, \sigma^2, r|y) = \prod_{i=1}^{n} f(y_i|\mu, \sigma^2), \tag{8.13}$$

with MLEs $\hat{\mu} = \bar{y}$ and $\hat{\sigma}^2 = (1/n)\sum_i (y_i - \bar{y})^2$ (Exercise 8.5).

To maximize the denominator of equation (8.12), we need to maximize the likelihood (8.11). To do this, we differentiate the log, set it equal to zero, and solve. We show some of the details here and leave the rest to Exercise 8.6. Differentiating with respect to μ_1 gives

$$\begin{aligned}\frac{\partial}{\partial\mu_1}\log L(\mu_1,\mu_2,\sigma^2,r|y) &= \frac{\partial}{\partial\mu_1}\sum_{i=1}^{n_1}\log\left((1-r)f(y_i|\mu_1,\sigma^2)+rf(y_i|\mu_2,\sigma^2)\right)\\ &\quad+\frac{\partial}{\partial\mu_1}\sum_{i=n_1+1}^{n}\log\left(rf(y_i|\mu_1,\sigma^2)+(1-r)f(y_i|\mu_2,\sigma^2)\right)\\ &= \sum_{i=1}^{n_1}\frac{(1-r)\frac{\partial}{\partial\mu_1}f(y_i|\mu_1,\sigma^2)}{(1-r)f(y_i|\mu_1,\sigma^2)+rf(y_i|\mu_2,\sigma^2)}\\ &\quad+\sum_{i=n_1+1}^{n}\frac{r\frac{\partial}{\partial\mu_1}f(y_i|\mu_1,\sigma^2)}{rf(y_i|\mu_1,\sigma^2)+(1-r)f(y_i|\mu_2,\sigma^2)}.\end{aligned} \tag{8.14}$$

Now take the derivative using Exercise 8.4 and define

$$\begin{aligned}P_1(y_i) &= \frac{(1-r)f(y_i|\mu_1,\sigma^2)}{(1-r)f(y_i|\mu_1,\sigma^2)+rf(y_i|\mu_2,\sigma^2)},\\ P_2(y_i) &= \frac{rf(y_i|\mu_1,\sigma^2)}{rf(y_i|\mu_1,\sigma^2)+(1-r)f(y_i|\mu_2,\sigma^2)},\end{aligned} \tag{8.15}$$

to get

$$\frac{\partial}{\partial\mu_1}\log L(\mu_1,\mu_2,\sigma^2,r|y) = \sum_{i=1}^{n_1}P_1(y_i)(y_i-\mu_1)+\sum_{i=n_1+1}^{n}P_2(y_i)(y_i-\mu_1). \tag{8.16}$$

Setting this equal to 0 and solving for μ_1 yields

$$\hat{\mu}_1 = \frac{\sum_{i=1}^{n_1}P_1(y_i)y_i+\sum_{i=n_1+1}^{n}P_2(y_i)y_i}{\sum_{i=1}^{n_1}P_1(y_i)+\sum_{i=n_1+1}^{n}P_2(y_i)}. \tag{8.17}$$

We can similarly solve for μ_2 and σ^2 to get

$$\begin{aligned}\hat{\mu}_2 &= \frac{\sum_{i=1}^{n_1}[1-P_1(y_i)]y_i+\sum_{i=n_1+1}^{n}[1-P_2(y_i)]y_i}{\sum_{i=1}^{n_1}[1-P_1(y_i)]+\sum_{i=n_1+1}^{n}[1-P_2(y_i)]}\\ \hat{\sigma}^2 &= \frac{1}{n}\left(\sum_{i=1}^{n_1}[P_1(y_i)(y_i-\hat{\mu}_1)^2+(1-P_1(y_i))(y_i-\hat{\mu}_2)^2]\right.\\ &\quad\left.+\sum_{i=n_1+1}^{n}[P_2(y_i)(y_i-\hat{\mu}_1)^2+(1-P_2(y_i))(y_i-\hat{\mu}_2)^2]\right).\end{aligned} \tag{8.18}$$

Of course, equations (8.17) and (8.18) do not solve the likelihood for all its parameters because $P_1(y_i)$ and $P_2(y_i)$ depend on the parameters and also depend on r. We have part of an iteration scheme to find the parameters. After estimating μ_1, μ_2, and σ^2 we use the current values to update r and then P_1 and P_2. We iterate until convergence. Specifically:

(1) Fix r.

(2) Use equations (8.17) and (8.18) to estimate μ_1, μ_2 and σ^2.
(3) Using the current estimates $\hat{\mu}_1, \hat{\mu}_2$, and $\hat{\sigma}^2$, maximize

$$\log L(\mu_1, \mu_2, \sigma^2, r|y) = \sum_{i=1}^{n_1} \log\left((1-r)f(y_i|\hat{\mu}_1, \hat{\sigma}^2) + rf(y_i|\hat{\mu}_2, \hat{\sigma}^2)\right) + \sum_{i=n_1+1}^{n} \log\left(rf(y_i|\mu_1, \sigma^2) + (1-r)f(y_i|\mu_2, \sigma^2)\right)$$

(4) Iterate between (2) and (3) until convergence.

For the maximization of (3) see Exercise 8.7. We illustrate it in the following example.

Example 8.6. (**MLEs for Tomato Plant Heights**). Suppose that we have the following data y on heights (in cm) of 12 tomato plants of a particular species grouped into two marker classes

$$M_1M_1 : y = (79, 82, 100, 102, 124)$$
$$M_1M_2 : y = (85, 87, 101, 103, 125, 126, 127).$$

The observed means and variances of the marker classes are $\bar{y}_{M_1M_1} = 97.4, \bar{y}_{M_1M_2} = 101.71$, $s_{M_1M_1} = 18.10$, and $s_{M_1M_2} = 18.33$. We further calculate the MLEs of μ_1, μ_2, σ^2, and r:

$$\hat{\mu}_1 = 92.169, \quad \hat{\mu}_2 = 124.561, \quad \hat{\sigma} = 8.08, \quad \hat{r} = .409.$$

See Fig. 8.2 for the convergence of the estimates, and see Appendix B.2 for the R program. The test statistics is calculated as $-2\log\lambda = 3.539$.

To assess the significance of the hypothesis test, we do a permutation test. We ran 5000 permuted samples and calculated $-2\log\lambda$ for each. The distribution is shown in Fig. 8.3. From the 5000 permutations, the .95 cutoff is 5.645, so the statistic is not significant and we do not have linkage.

8.5 Power Calculation

It is practically important to calculate the minimum sample size required to achieve a predetermined significance level. The calculation of the power to detect significant QTLs depends on the type of mapping population. For example, the backcross can only estimate the additive effect of a QTL, whereas the F_2 can estimate both the additive and dominant effects. Thus, different approaches should be used for power calculation for these two designs. In this section, we will introduce a general procedure for calculating the power for QTL analysis in the F_2. From this, interested readers can consider more complicated designs.

Assume that there is an F_2 population in which individual markers are genotyped to detect their associations with the underlying QTL for a quantitative trait. Three

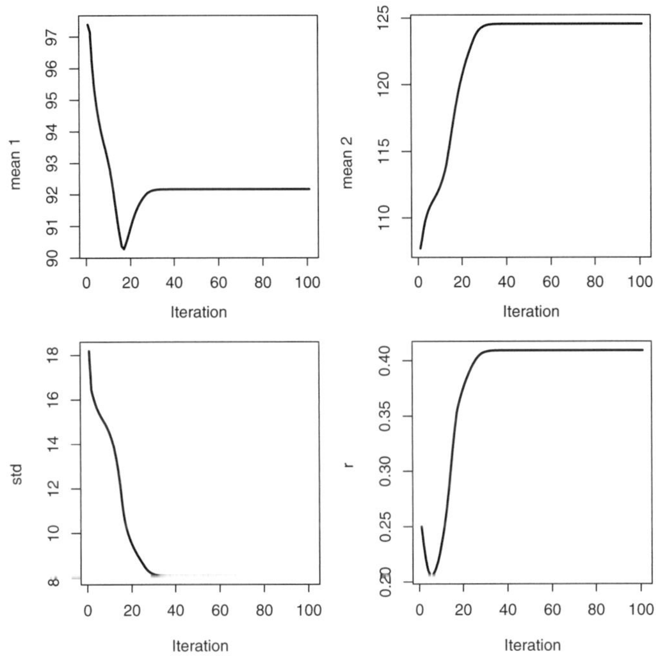

Fig. 8.2. MLEs for the tomato data of Example 8.6.

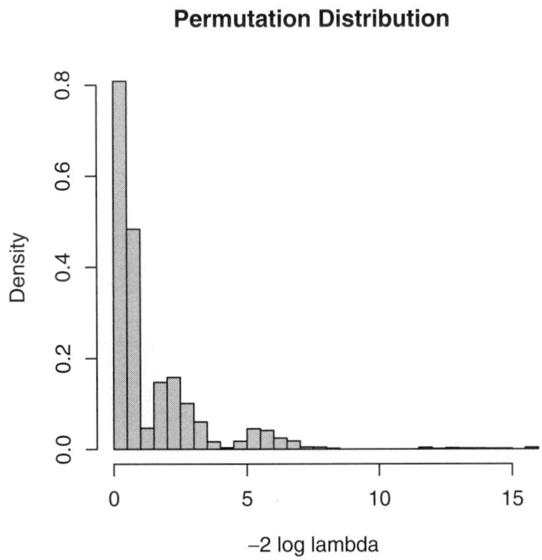

Fig. 8.3. Permutation distribution of the likelihood ratio statistic $-2\log\lambda$ under H_0 based on 5000 permutations from Example 8.6.

genotypes at a marker are segregating in the $1(AA) : 2(Aa) : aa$ ratio, which implies that a sample size of n is allocated into $n/4$, $n/2$, and $n/4$ for these three genotypes. Based on analyses in Section 8.3.3, the difference between the two homozygous marker genotypes can reflect the additive effect of a putative QTL, although the QTL is confounded with the QTL location. We rewrite equation (8.6) to test the QTL additive effect as

$$\begin{aligned} t_1 &= \frac{m_2 - m_0}{\sqrt{s^2(\frac{4}{n} + \frac{4}{n})}} \\ &= \frac{(1-2r)2a}{\sqrt{8s^2/n}}. \end{aligned}$$

When n is large, the calculated test statistic, $\hat{t}_1$, follows an approximately normal distribution, $\hat{t}_1 \sim N(t_1, 1)$. Thus, the power to detect the difference $(m_2 - m_0)$ for a two-tailed test is expressed as

$$\begin{aligned} 1 - \beta &= \text{Prob}(\hat{t}_1 > z_{\alpha/2}) \\ &= 1 - \Phi(z_\alpha - t), \end{aligned} \tag{8.19}$$

where z_α is the critical value of the test with $(1-\alpha)$ confidence under the null hypothesis $t_1 = 0$, and $\Phi(z_\alpha - t)$ is the standard normal cumulative distribution function.

For a given type I error (α) and type II error (β) in the t test, the sample size n required for detecting the additive effect of a QTL is determined by

$$n = 8\left[\frac{z_{\alpha/2} + z_\beta}{(1-2r)2a/s}\right]^2. \tag{8.20}$$

If the QTL search is performed over the entire genome, then the type I error α for each test should be substantially lower to account for the increased false-positive probability for the genome-wide test.

Similarly, based on equation (8.7), we rewrite the test statistic for the dominant effect as

$$\begin{aligned} t_2 &= \frac{m_1 - (m_2 + m_0)/2}{\sqrt{s^2(\frac{2}{n} + \frac{1}{n} + \frac{1}{n})}} \\ &= \frac{(1-2r)^2 d}{\sqrt{4s^2/n}}. \end{aligned}$$

From this, the sample size required for detecting the dominance effect is determined by

$$n = 4\left[\frac{z_{\alpha/2} + z_\beta}{(1-2r)^2 d/s}\right]^2. \tag{8.21}$$

It can be seen from equations (8.20) and (8.21) that the sample size for detecting a QTL depends on many factors, which are (1) the magnitude of the (additive or dominance) effect of a QTL (the larger the magnitude, the smaller the sample size required), (2) the degree of linkage between the marker considered and a putative QTL (the stronger the linkage, the smaller the sample size required), and (3) the residual variance within a QTL genotype class (the smaller the residual variance, the smaller the sample size required).

Example 8.7. A mouse geneticist plans to launch a molecular mapping study for the identification of QTLs that affect mouse body weight. Although the exact effect of a QTL is unknown, $\phi_a = a/s$ or $\phi_d = d/s$ is used to define the genetic effects relative to the residual standard deviation. Assume that the significance level for detecting a QTL is $\alpha = .01$. The power of $(1 - \beta) = .90$ is hoped for the QTL experiment. Using equations (8.20) and (8.21), the sample sizes required are calculated for a range of recombination fractions under different levels of the additive and dominance effects of the QTL, respectively (Figs. 8.4A and 8.4B).

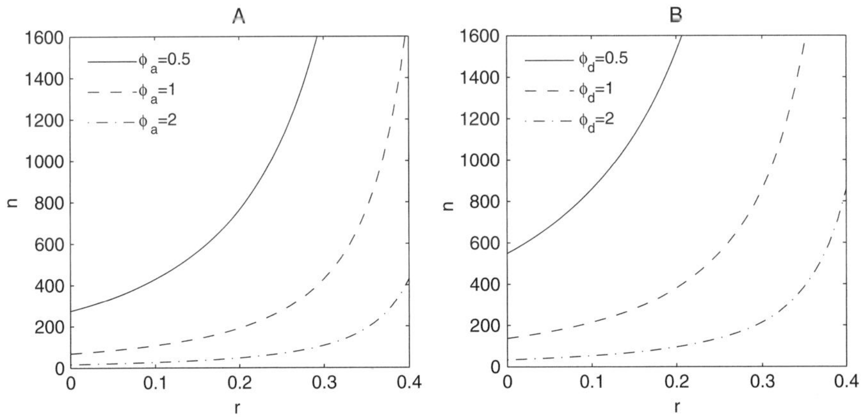

Fig. 8.4. Sample sizes required under different recombination fractions.

Only a small sample size is needed for a large QTL with $(\phi_a, \phi_d = 2)$ if the QTL is located exactly at the marker. The required sample size increases exponentially with the recombination fraction between the marker and QTL, with the extent of the increase being markedly greater for a small QTL $(\phi_a, \phi_d = 0.5)$ than for a large QTL $(\phi_a, \phi_d = 2)$. The sample size required to detect the dominance effect is larger than that to detect the additive effect of the same size if there is the same level of QTL-marker linkage.

8.6 Marker Interaction Analysis

8.6.1 ANOVA

Genetic interactions between different QTLs play an important role in trait control and expression. A two-way ANOVA based on different markers can be performed to estimate and test the main effect of the markers and their interaction effects (Routman and Cheverud 1997). Consider a mapping population in which there are l genotypes at each marker. A standard two-way ANOVA approach is used to analyze the two markers simultaneously, with the results summarized in Table 8.4.

Based on the structure of the expected mean squares due to the main and interaction effects (assuming that they are all random), we can calculate the F test statistics for each effect compared with the critical values $F_{0.05,(l-1,(l-1)^2)}$ and $F_{0.05,((l-1)^2,n-l^2-1)}$, respectively. The genetic variances for the main effect of markers **A** and **B** and their interaction effect are calculated as

$$\sigma^2_{\mathbf{A}} = \frac{\text{MS}_1 - \text{MS}_3}{k_1},$$
$$\sigma^2_{\mathbf{B}} = \frac{\text{MS}_2 - \text{MS}_3}{k_2},$$
$$\sigma^2_{\mathbf{AB}} = \frac{\text{MS}_3 - \text{MS}_4}{k_3},$$

with harmonic means calculated differently for the backcross and the F_2.

Table 8.4. Summary of two-way ANOVA aimed at detecting the main effects and interaction effect between different markers in a mapping population.

Source of Variation	df	Mean Square	F-value	Expected Mean Square
Main effect due to marker **A**	$l-1$	MS_1	MS_1/MS_3	$\sigma^2_e + k_3\sigma^2_{\mathbf{AB}} + k_1\sigma^2_{\mathbf{A}}$
Main effect due to marker **B**	$l-1$	MS_2	MS_2/MS_3	$\sigma^2_e + k_3\sigma^2_{\mathbf{AB}} + k_2\sigma^2_{\mathbf{B}}$
Interaction effect	$(l-1)^2$	MS_3	MS_3/MS_4	$\sigma^2_e + k_3\sigma^2_{\mathbf{AB}}$
Residual error	$n-l^2-1$	MS_4		σ^2_e

Note: $l = 2$ for the backcross and 3 for the F_2.

For the backcross, we have

$$k_1 = \frac{2}{\frac{1}{n_1} + \frac{1}{n_0}},$$

where n_1 and n_0 are the observations of the two genotypes at marker **A**;

$$k_2 = \frac{2}{\frac{1}{n'_1} + \frac{1}{n'_0}},$$

where n'_1 and n'_0 are the observations of the two genotypes at marker **B**; and

$$k_3 = \frac{4}{\frac{1}{n_{11}} + \frac{1}{n_{10}} + \frac{1}{n_{01}} + \frac{1}{n_{00}}},$$

where n_{11}, n_{10}, n_{01}, and n_{00} are the observations of the four genotypes at markers **A** and **B**. For the F_2, we have

$$k_1 = \frac{3}{\frac{1}{n_2} + \frac{1}{n_1} + \frac{1}{n_0}},$$

where n_2, n_1, and n_0 are the observations of the three genotypes at marker **A**;

$$k_2 = \frac{3}{\frac{1}{n'_2} + \frac{1}{n'_1} + \frac{1}{n'_0}},$$

where n'_2, n'_1, and n'_0 are the observations of the three genotypes at marker **B**; and

$$k_3 = \frac{9}{\frac{1}{n_{22}} + \frac{1}{n_{21}} + \frac{1}{n_{20}} + \frac{1}{n_{12}} + \frac{1}{n_{11}} + \frac{1}{n_{10}} + \frac{1}{n_{02}} + \frac{1}{n_{01}} + \frac{1}{n_{00}}},$$

where $n_{22}, \ldots, n_{00}$ are the observations of the nine genotypes at markers **A** and **B**.

The proportions of each of these genetic variances over the total phenotypic variance, the marker-specific heritability, can be calculated by

$$(8.22) \qquad \begin{aligned} R^2_{\mathbf{A}} &= \frac{\sigma^2_{\mathbf{A}}}{\sigma^2_{\mathbf{A}} + \sigma^2_{\mathbf{B}} + \sigma^2_{\mathbf{AB}} + \sigma^2_e}, \\ R^2_{\mathbf{B}} &= \frac{\sigma^2_{\mathbf{B}}}{\sigma^2_{\mathbf{A}} + \sigma^2_{\mathbf{B}} + \sigma^2_{\mathbf{AB}} + \sigma^2_e}, \\ R^2_{\mathbf{AB}} &= \frac{\sigma^2_{\mathbf{AB}}}{\sigma^2_{\mathbf{A}} + \sigma^2_{\mathbf{B}} + \sigma^2_{\mathbf{AB}} + \sigma^2_e}. \end{aligned}$$

For the F_2, the main effect for each marker, **A** and **B**, can be partitioned into the additive (a) and dominance (d) effect components, whereas the interaction effect between different markers in the F_2 can be partitioned into the additive $\times$ additive (I_{aa}), additive $\times$ dominance (I_{ad}), dominance $\times$ additive (I_{da}) and dominance $\times$ dominance (I_{dd}) components. All these component effects can be tested by a t test statistic, which, along with the corresponding critical value, is expressed sequentially as

$$(8.23) \quad t_{\mathbf{A}a} = \frac{(m_{22} + m_{20}) - (m_{02} + m_{00})}{\sqrt{s^2(\frac{1}{n_{22}} + \frac{1}{n_{20}} + \frac{1}{n_{02}} + \frac{1}{n_{00}})}}, \quad t_{(0.025,\nu=n_{22}+n_{20}+n_{02}+n_{00}-4)},$$

with

$$s^2 = \frac{(n_{22}-1)s_{22}^2 + (n_{20}-1)s_{20}^2 + (n_{02}-1)s_{02}^2 + (n_{00}-1)s_{00}^2}{n_{22}+n_{20}+n_{02}+n_{00}-4};$$

(8.24)
$$t_{\mathbf{A}d} = \frac{(m_{12}+m_{10}) - \frac{1}{2}(m_{22}+m_{20}+m_{02}+m_{00})}{\sqrt{s^2(\frac{1}{4n_{22}} + \frac{1}{4n_{20}} + \frac{1}{n_{12}} + \frac{1}{n_{10}} + \frac{1}{4n_{02}} + \frac{1}{4n_{00}})}},$$

$$t_{(0.025,\nu=n_{22}+n_{20}+n_{12}+n_{10}+n_{02}+n_{00}-6)},$$

with

$$s^2 = \frac{(n_{22}-1)s_{22}^2 + (n_{20}-1)s_{20}^2 + (n_{12}-1)s_{12}^2}{n_{22}+n_{20}+n_{12}+n_{10}+n_{02}+n_{00}-6} + \frac{(n_{10}-1)s_{10}^2 + (n_{02}-1)s_{02}^2 + (n_{00}-1)s_{00}^2}{n_{22}+n_{20}+n_{12}+n_{10}+n_{02}+n_{00}-6};$$

(8.25)
$$t_{\mathbf{B}a} = \frac{(m_{22}+m_{02}) - (m_{20}+m_{00})}{\sqrt{s^2(\frac{1}{n_{22}} + \frac{1}{n_{20}} + \frac{1}{n_{02}} + \frac{1}{n_{00}})}}, \quad t_{(0.025,\nu=n_{22}+n_{20}+n_{02}+n_{00}-4)},$$

with

$$s^2 = \frac{(n_{22}-1)s_{22}^2 + (n_{20}-1)s_{20}^2 + (n_{02}-1)s_{02}^2 + (n_{00}-1)s_{00}^2}{n_{22}+n_{20}+n_{02}+n_{00}-4};$$

(8.26)
$$t_{\mathbf{B}d} = \frac{(m_{21}+m_{01}) - \frac{1}{2}(m_{22}+m_{20}+m_{02}+m_{00})}{\sqrt{s^2(\frac{1}{4n_{22}} + \frac{1}{n_{21}} + \frac{1}{4n_{20}} + \frac{1}{4n_{02}} + \frac{1}{n_{01}} + \frac{1}{4n_{00}})}},$$

$$t_{(0.025,\nu=n_{22}+n_{21}+n_{20}+n_{02}+n_{01}+n_{00}-6)},$$

with

$$s^2 = \frac{(n_{22}-1)s_{22}^2 + (n_{21}-1)s_{21}^2 + (n_{20}-1)s_{20}^2}{n_{22}+n_{21}+n_{20}+n_{02}+n_{01}+n_{00}-6} + \frac{(n_{02}-1)s_{02}^2 + (n_{01}-1)s_{01}^2 + (n_{00}-1)s_{00}^2}{n_{22}+n_{21}+n_{20}+n_{02}+n_{01}+n_{00}-6};$$

(8.27)
$$t_{aa} = \frac{(m_{22}+m_{00}) - (m_{20}+m_{02})}{\sqrt{s^2(\frac{1}{n_{22}} + \frac{1}{n_{20}} + \frac{1}{n_{02}} + \frac{1}{n_{00}})}}, t_{(0.025,\nu=n_{22}+n_{20}+n_{02}+n_{00}-4)},$$

with

$$s^2 = \frac{(n_{22}-1)s_{22}^2 + (n_{20}-1)s_{20}^2 + (n_{02}-1)s_{02}^2 + (n_{00}-1)s_{00}^2}{n_{22}+n_{20}+n_{02}+n_{00}-4};$$

(8.28)
$$t_{ad} = \frac{[m_{21} - \frac{1}{2}(m_{22}+m_{20})] - [m_{01} - \frac{1}{2}(m_{02}+m_{00})]}{\sqrt{s^2(\frac{1}{4n_{22}} + \frac{1}{n_{21}} + \frac{1}{4n_{20}} + \frac{1}{4n_{02}} + \frac{1}{n_{01}} + \frac{1}{4n_{00}})}},$$

$$t_{(0.025,\nu=n_{22}+n_{21}+n_{20}+n_{02}+n_{01}+n_{00}-6)},$$

with

$$s^2 = \frac{(n_{22}-1)s_{22}^2 + (n_{21}-1)s_{21}^2 + (n_{20}-1)s_{20}^2}{n_{22}+n_{21}+n_{20}+n_{02}+n_{01}+n_{00}-6} + \frac{(n_{02}-1)s_{02}^2 + (n_{01}-1)s_{01}^2 + (n_{00}-1)s_{00}^2}{n_{22}+n_{21}+n_{20}+n_{02}+n_{01}+n_{00}-6};$$

(8.29)
$$t_{da} = \frac{[m_{12} - \frac{1}{2}(m_{22}+m_{02})] - [m_{10} - \frac{1}{2}(m_{20}+m_{00})]}{\sqrt{s^2(\frac{1}{4n_{22}} + \frac{1}{4n_{20}} + \frac{1}{n_{12}} + \frac{1}{n_{10}} + \frac{1}{4n_{02}} + \frac{1}{4n_{00}})}},$$

$$t_{(0.025,\nu=n_{22}+n_{20}+n_{12}+n_{10}+n_{02}+n_{00}-6)},$$

with

$$s^2 = \frac{(n_{22}-1)s_{22}^2 + (n_{20}-1)s_{20}^2 + (n_{12}-1)s_{12}^2}{n_{22}+n_{20}+n_{12}+n_{10}+n_{02}+n_{00}-6} + \frac{(n_{10}-1)s_{10}^2 + (n_{02}-1)s_{02}^2 + (n_{00}-1)s_{00}^2}{n_{22}+n_{20}+n_{12}+n_{10}+n_{02}+n_{00}-6};$$

$$t_{dd} = \frac{[m_{11} - \frac{1}{2}(m_{21}+m_{01})] - \frac{1}{2}\{[m_{12} - \frac{1}{2}(m_{22}+m_{02})] + [m_{10} - \frac{1}{2}(m_{20}+m_{00})]\}}{\sqrt{s^2(\frac{1}{4n_{22}} + \frac{1}{n_{12}} + \frac{1}{4n_{02}} + \frac{1}{4n_{21}} + \frac{1}{n_{11}} + \frac{1}{4n_{01}} + \frac{1}{4n_{20}} + \frac{1}{n_{10}} + \frac{1}{4n_{00}})}},$$

(8.30)
$$t_{(0.025,\nu=n_{22}+n_{12}+n_{02}+n_{21}+n_{11}+n_{01}+n_{20}+n_{10}+n_{00}-9)},$$

with

$$s^2 = \frac{(n_{22}-1)s_{22}^2 + (n_{12}-1)s_{12}^2 + (n_{02}-1)s_{02}^2 + (n_{21}-1)s_{21}^2 + (n_{11}-1)s_{11}^2}{n_{22}+n_{12}+n_{02}+n_{21}+n_{11}+n_{01}+n_{20}+n_{10}+n_{00}-9} + \frac{(n_{01}-1)s_{01}^2 + (n_{20}-1)s_{20}^2 + (n_{10}-1)s_{10}^2 + (n_{00}-1)s_{00}^2}{n_{22}+n_{12}+n_{02}+n_{21}+n_{11}+n_{01}+n_{20}+n_{10}+n_{00}-9},$$

where $s_{22}^2, \ldots, s_{00}^2$ are the sample variances for nine genotypes at markers **A** and **B**.

Example 8.8. Revisit Example 3.2. An F_2 intercross population derived from the Large (LG/J) and Small (SM/J) inbred strains of mice is used to describe marker interaction effects on adult body weight (Cheverud et al. 1996). A total of 535 F_2 mice were weighed weekly from 1 to 10 weeks of age. The 10-week weight, referred to as adult weight, will be analyzed. Seventy-six microsatellite polymorphisms were scored throughout the 19 mouse autosomes. As an example, our analysis will focus on one marker (D1Mit7) on chromosome 1 and the second marker (D2Mit17) on chromosome 2. Table 8.5 tabulates the estimates of the mean squares, F-values, genetic variance components, and the proportions of each effect contributing to the total phenotypic variance.

Table 8.5. ANOVA results of a two-marker analysis in the F_2 population of mice.

Source of Variation	df	Mean Square	F-value	Pr $> F$	Proportion
Main effect due to marker D1Mit7	2	115.20	7.61	< 0.0001	0.026
Main effect due to marker D2Mit17	2	87.40	5.77	0.0065	0.019
Interaction effect	4	12.25	0.81	0.5195	
Residual error	838	15.14			

Both markers, D1Mit7 and D2Mit17, are significant for their main effects on the adult body weight of mice, which explain 2.6 percent and 1.9 percent of the phenotypic variance, respectively. The interaction effect between the two markers is found to be nonsignificant. Based on equations (8.23) and (8.24), we calculate the t statistics for the additive and dominance effects associated with marker D1Mit7, which are $t_{\mathbf{A}a} = -1.7866$ ($p = 0.0756$) and $t_{\mathbf{A}d} = 1.7232$ ($p = 0.0857$), respectively. These two effects are not significant compared with the corresponding critical values $t_{(0.025,\nu=186)} = 1.9728$ and $t_{(0.025,\nu\,=\,378)} = 1.9663$. The t statistics for the additive and dominance effects at marker D2Mit17 are calculated as $t_{\mathbf{B}a} = -3.1693$ ($p = 0.0018$) and $t_{\mathbf{B}d} = -0.2349$ ($p = 0.8144$) with equations (8.25) and (8.26), suggesting that the additive effect is significant based on $t_{(0.025,\nu\,=\,186)} = 1.9728$ and that the dominance effect is not significant based on $t_{(0.025,\nu\,=\,408)} = 1.9658$.

8.6.2 Genetic Analysis

The main effect of a marker reflects only the additive effect at the marker level for the backcross, but both the additive and dominance effects at the marker level for the F_2. The interaction effect between two different markers virtually represents the additive $\times$ additive genetic effect at the marker level for the backcross and can be partitioned into four components, additive $\times$ additive, additive $\times$ dominance, dominance $\times$ additive, and dominance $\times$ dominance genetic effects, at the marker level for the F_2. In practice, it is of great importance to estimate each of these interaction components (Cheverud and Routman 1995).

As shown for a single-marker case, marker analysis will confound the QTL location and effect. This is also true for two-marker analysis, but it is interesting to explore how the QTL locations and different genetic factors are confounded.

Consider two unlinked markers, **A** (with alleles A and a) and **B** (with alleles B and b). Assume each marker is linked with a different QTL, **P** (with alleles P and p) or **Q** (with alleles Q and q), with the recombination fractions of r_1 and r_2, respectively. Let $\omega_{j_1 j_2|k_1 k_2}$ be the conditional probability of two-QTL genotype $j_1 j_2$, conditional upon two-marker genotype k_1k_2. If these two markers are located on different chromosomes, conditional probabilities $\omega_{j_1j_2|k_1k_2}$ can be expressed as follows:

Marker	QTL **P**			Marker	QTL **Q**	
M	$Pp(0)$	$pp(0)$	$\otimes$	genotype	$Qq(1)$	$qq(0)$
$Aa(1)$	$1-r_1$	r_1		$Bb(1)$	$1-r_2$	r_2
$aa(0)$	r_1	$1-r_1$		$bb(0)$	r_2	$1-r_2$

for the backcross and

Marker	QTL **P**				Marker	QTL **Q**		
M	$PP(2)$	$Pp(1)$	$pp(0)$		genotype	$QQ(2)$	$Qq(1)$	$qq(0)$
$AA(2)$	$(1-r_1)^2$	$2r_1(1-r_1)$	r_1^2	$\otimes$	$BB(2)$	$(1-r_2)^2$	$2r_2(1-r_2)$	r_2^2
$Aa(1)$	$r_1(1-r_1)$	$(1-r_1)^2+r_1^2$	$r_1(1-r_1)$		$Bb(1)$	$r_2(1-r_2)$	$(1-r_2)^2+r_2^2$	$r_2(1-r_2)$
$aa(0)$	r_1^2	$2r_1(1-r_1)$	$(1-r_1)^2$		$bb(0)$	r_2^2	$2r_2(1-r_2)$	$(1-r_2)^2$

Let $\mu_{j_1j_2}$ ($j_1, j_2 = 1, 0$ for the backcross and 2, 1, 0 for the F_2) and $m_{k_1k_2}$ ($k_1, k_2 = 1, 0$ for the backcross and 2, 1, 0 for the F_2) be the genotypic values at two QTLs and at two markers, respectively. The QTL genotypic values can be partitioned into different components: i.e., the overall mean (μ), the additive effects at QTLs **P** (a_1) and **Q** (a_2) and the additive $\times$ additive effect (i_{aa}) for the backcross, or the overall mean (μ), the additive (a_1) and dominance effects (d_1) at QTL **P**, the additive (a_2) and dominance effects (d_2) at QTL **Q**, as well as the additive $\times$ additive (i_{aa}), additive $\times$ dominance (i_{ad}), dominance $\times$ additive (i_{da}) and dominance $\times$ dominance (i_{dd}), for the F_2 (see equation (1.9) and matrix (9.9)).

We use a general formula to express the marker genotypic values in terms of the QTL genotypic values:

(8.31) $$m_{k_1k_2} = \sum_{j_1=0}^{2}\sum_{j_2=0}^{2} \omega_{j_1j_2|k_1k_2}\mu_{j_1j_2}.$$

More specifically, based on equation (8.31), we derive the marker genotypic values in terms of individual QTL effects for the backcross as

$$\begin{aligned}
m_{11} &= \mu + (1-r_1)a_1 + (1-r_2)a_2 + (1-r_1)(1-r_2)i_{aa},\\
m_{10} &= \mu + (1-r_1)a_1 + r_2a_2 + (1-r_1)r_2i_{aa},\\
m_{01} &= \mu + r_1a_1 + (1-r_2)a_2 + r_1(1-r_2)i_{aa},\\
m_{00} &= \mu + r_1a_1 + r_2a_2 + r_1r_2i_{aa}.
\end{aligned}$$

The main effects of markers **A** and **B**, respectively, are expressed as

(8.32) $$M_{\mathbf{A}} = \tfrac{1}{2}(m_{11}+m_{10}) - \tfrac{1}{2}(m_{01}+m_{00}) = (1-2r_1)a_1 + \tfrac{1}{2}(1-2r_1)i_{aa},$$

(8.33) $$M_{\mathbf{B}} = \tfrac{1}{2}(m_{11}+m_{01}) - \tfrac{1}{2}(m_{10}+m_{00}) = (1-2r_2)a_2 + \tfrac{1}{2}(1-2r_2)i_{aa},$$

whereas the interaction effect between the two markers is expressed as

$$
\begin{aligned}
I_{\mathbf{AB}} &= (m_{11} + m_{00}) - (m_{10} + m_{01}) \\
&= (1 - 2r_1)(1 - 2r_2)i_{aa}.
\end{aligned}
\tag{8.34}
$$

It is interesting to find that for the two-marker analysis the estimate of a marker main effect is not only confounded by the position of a QTL that is linked with the marker but also contaminated by the epistatic effect between this QTL and any other QTL located at a different position. The estimate of the interaction effect based on marker information is confounded by the additive $\times$ additive epistatic effect between the two putative QTLs and their respective recombination fractions with markers.

The main and interaction effects of two different QTLs in the F_2 are shown by equation (1.9) and tabulated in matrix (9.9), which allows for the characterization of additive, dominance, and all four possible epistatic effects. Using equation (8.31), we derive the marker genotypic values in terms of QTL effects for the F_2 as follows:

$$
\begin{aligned}
m_{22} &= \mu + (1 - 2r_1)a_1 + 2r_1(1 - r_1)d_1 + (1 - 2r_2)a_2 + 2r_2(1 - r_2)d_2 \\
&\quad +(1 - 2r_1)(1 - 2r_2)i_{aa} + 2r_2(1 - r_2)(1 - 2r_1)i_{ad} \\
&\quad +2r_1(1 - 2r_2)(1 - r_1)i_{da} + 4r_1(1 - r_1)r_2(1 - r_2)i_{dd}, \\
m_{21} &= \mu + (1 - 2r_1)a_1 + 2r_1(1 - r_1)d_1 + (1 - 2r_2 + 2r_2^2)d_2 \\
&\quad +(1 - 2r_2 + 2r_2^2)(1 - 2r_1)i_{ad} + 2r_1(1 - r_1)(1 - 2r_2 + 2r_2^2)i_{dd}, \\
m_{20} &= \mu + (1 - 2r_1)a_1 + 2r_1(1 - r_1)d_1 - (1 - 2r_2)a_2 + 2r_2(1 - r_2)d_2 \\
&\quad -(1 - 2r_1)(1 - 2r_2)i_{aa} + 2r_2(1 - r_2)(1 - 2r_1)i_{ad} \\
&\quad -2r_1(1 - 2r_2)(1 - r_1)i_{da} + 4r_1(1 - r_1)r_2(1 - r_2)i_{dd}, \\
m_{12} &= \mu + (1 - 2r_1 + 2r_1^2)d_1 + (1 - 2r_2)a_2 + 2r_2(1 - r_2)d_2 \\
&\quad +(1 - 2r_1 + 2r_1^2)(1 - 2r_2)i_{da} + 2(1 - 2r_1 + 2r_1^2)r_2(1 - r_2)i_{dd}, \\
m_{11} &= \mu + (1 - 2r_1 + 2r_1^2)d_1 + (1 - 2r_2 + 2r_2^2)d_2 \\
&\quad +(1 - 2r_1 + 2r_1^2)(1 - 2r_2 + 2r_2^2)i_{dd}, \\
m_{10} &= \mu + (1 - 2r_1 + 2r_1^2)d_1 - (1 - 2r_2)a_2 + 2r_2(1 - r_2)d_2 \\
&\quad -(1 - 2r_1 + 2r_1^2)(1 - 2r_2)i_{da} + 2(1 - 2r_1 + 2r_1^2)r_2(1 - r_2)i_{dd}, \\
m_{02} &= \mu - (1 - 2r_1)a_1 + 2r_1(1 - r_1)d_1 + (1 - 2r_2)a_2 + 2r_2(1 - r_2)d_2 \\
&\quad -(1 - 2r_1)(1 - 2r_2)i_{aa} - 2r_2(1 - r_2)(1 - 2r_1)i_{ad} \\
&\quad +2r_1(1 - 2r_2)(1 - r_1)i_{da} + 4r_1(1 - r_1)r_2(1 - r_2)i_{dd}, \\
m_{01} &= \mu - (1 - 2r_1)a_1 + 2r_1(1 - r_1)d_1 + (1 - 2r_2 + 2r_2^2)d_2 \\
&\quad -(1 - 2r_2 + 2r_2^2)(1 - 2r_1)i_{ad} + 2r_1(1 - r_1)(1 - 2r_2 + 2r_2^2)i_{dd}, \\
m_{00} &= \mu - (1 - 2r_1)a_1 + 2r_1(1 - r_1)d_1 - (1 - 2r_2)a_2 + 2r_2(1 - r_2)d_2 \\
&\quad +(1 - 2r_1)(1 - 2r_2)i_{aa} - 2r_2(1 - r_2)(1 - 2r_1)i_{ad} \\
&\quad -2r_1(1 - 2r_2)(1 - r_1)i_{da} + 4r_1(1 - r_1)r_2(1 - r_2)i_{dd}.
\end{aligned}
$$

The marker additive and dominant effects for **A** and **B** are derived, respectively, as

$$
\begin{aligned}
A_{\mathbf{A}} &= \tfrac{1}{4}(m_{22}+m_{20}) - \tfrac{1}{4}(m_{02}+m_{00}) \\
&= (1-2r_1)a_1 + 2(1-2r_1)r_2(1-r_2)i_{ad}, \\
D_{\mathbf{A}} &= \tfrac{1}{2}(m_{12}+m_{10}) - \tfrac{1}{4}(m_{22}+m_{20}+m_{02}+m_{00}) \\
&= (1-2r_1)^2 d_1 + 2(1-2r_1)^2 r_2(1-r_2)i_{dd}, \\
A_{\mathbf{B}} &= \tfrac{1}{4}(m_{22}+m_{02}) - \tfrac{1}{4}(m_{20}+m_{00}) \\
&= (1-2r_2)a_2 + 2r_1(1-r_1)(1-2r_2)i_{da}, \\
D_{\mathbf{B}} &= \tfrac{1}{2}(m_{21}+m_{01}) - \tfrac{1}{4}(m_{22}+m_{20}+m_{02}+m_{00}) \\
&= (1-2r_1)d_2 + 2r_1(1-r_1)(1-2r_2)^2 i_{dd}.
\end{aligned}
$$

The additive × additive (I_{aa}), additive × dominance (I_{ad}), dominance × additive (I_{da}), and dominance × dominance (I_{dd}) epistatic effects at the marker level can be expressed as

$$
\begin{aligned}
I_{aa} &= \tfrac{1}{4}(m_{22}+m_{00}) - \tfrac{1}{4}(m_{20}+m_{02}) \\
&= (1-2r_2)(1-2r_1)i_{aa}, \\
I_{ad} &= \tfrac{1}{2}[m_{21} - \tfrac{1}{2}(m_{22}+m_{20})] - \tfrac{1}{2}[m_{01} - \tfrac{1}{2}(m_{02}+m_{00})] \\
&= (1-2r_2)^2(1-2r_1)i_{ad}, \\
I_{da} &= \tfrac{1}{2}[m_{12} - \tfrac{1}{2}(m_{22}+m_{02})] - \tfrac{1}{2}[m_{10} - \tfrac{1}{2}(m_{20}+m_{00})] \\
&= (1-2r_1)^2(1-2r_2)i_{da}, \\
I_{dd} &= [m_{11} - \tfrac{1}{2}(m_{21}+m_{01})] - \tfrac{1}{2}\{[m_{12} - \tfrac{1}{2}(m_{22}+m_{02})] + [m_{10} - \tfrac{1}{2}(m_{20}+m_{00})]\} \\
&= [m_{11} - \tfrac{1}{2}(m_{12}+m_{10})] - \tfrac{1}{2}\{[m_{21} - \tfrac{1}{2}(m_{22}+m_{20})] + [m_{01} - \tfrac{1}{2}(m_{02}+m_{00})]\} \\
&= (1-2r_1)^2(1-2r_2)^2 i_{aa}.
\end{aligned}
$$

From the analysis above, we can see how much the estimates of each marker effect are contaminated by the QTL locations and QTL–QTL interaction effects.

8.7 Whole-Genome Marker Analysis

The purpose of constructing a genetic map is the identification of the QTLs that affect a quantitative trait, their number, genomic distribution, genetic effects, and sensitivity to various environmental signals. Separate analyses of individual markers are not adequate for precisely capturing all the information about QTL because their interacting network cannot be systematically identified. Xu (2003) proposed a method for simultaneously estimating marker effects of the entire genome. This method assumes that the marker density is relatively high, and thus marker effects approximately represent the QTL effects associated with markers.

Consider a backcross in which a quantitative trait is measured for individual i, denoted as y_i. The statistical model for y_i at a total of m markers, each with two genotypes 1 and 0, in the entire genome is written as

$$
y_i = b_0 + \sum_{k=1}^{m} x_{ik} b_k + e_i, \tag{8.35}
$$

where b_0 is the overall mean, x_{ik} is a dummy variable indicating the genotype of the kth marker for individual i, b_k is the effect of the kth marker, and e_i is the residual error following $N(0, \sigma^2)$. The dummy variable is defined as $x_{ik} = 1, 0$, depending on marker genotype. Equation (8.35) is the multiple regression model, with the partial regression coefficient b_k being the effect of marker k associated with the trait.

How to estimate partial regression coefficients in equation (8.35) is statistically a challenging issue because the number of markers in the linear model may be larger than the number of individuals and because the effects of many markers actually are close to zero. Xu (2003) implemented a Bayesian approach for parameter estimation. Within the Bayesian context, everything, including the parameters, is treated as a random variable with a particular distribution. For example, each b_k is assumed to be sampled from a normal distribution with mean zero and variance σ_k^2. All variables are sorted into observables and unobservables. The observables are phenotypic values, $\mathbf{y} = \{y_i\}$, and marker data, whereas the unobservables include $\mathbf{b} = \{b_k\}$ and $\mathbf{v} = \{\sigma_e^2, \sigma_k^2\}$ $(k = 1, \ldots, m)$. The Bayesian framework is composed of three elements, the prior distribution, likelihood, and posterior distribution. The prior distribution is the distribution of the unobservables. The likelihood is the distribution of the observables, expressed as a function of the unobservables. The posterior distribution, which needs to be inferred from Bayesian analysis, is the conditional distribution of the parameters given the observed data.

The Markov chain Monte Carlo (MCMC) method makes Bayesian analysis tractable. The MCMC-implemented Bayesian analysis does not need an explicit form of the posterior distribution; rather, it draws a sample of the unobservables from the joint posterior distribution. From the joint posterior sample, the desired Bayesian estimates, such as the posterior means and posterior variances, can be obtained. The subsequent description of Bayesian analysis will be based on the idea of Xu (2003), who chose the following prior distributions:

$$P(b_0) \propto 1,$$

$$\begin{aligned} P(\sigma_e^2) &\propto 1/\sigma_e^2, \\ P(b_k) &= N(0, \sigma_k^2), \\ P(\sigma_k^2) &\propto 1/\sigma_k^2. \end{aligned}$$

The joint prior distribution of the unobservables $P(\mathbf{b}, \mathbf{v})$ takes the product of the priors of individual parameters. The likelihood is

$$\begin{aligned} P(\mathbf{y}|\mathbf{b}, \mathbf{v}) &= \prod_{i=1}^{n} P(y_i|\mathbf{b}, \sigma_e^2) \\ &\propto [\sigma_e^2]^{-n/2} \exp\left[-\frac{1}{2\sigma_e^2} \sum_{i=1}^{n} \left(y_i - b_0 - \sum_{k=1}^{m} x_{ik} b_k\right)^2\right]. \end{aligned} \tag{8.36}$$

The joint posterior distribution has a form of

$$P(\mathbf{b}, \mathbf{v}|\mathbf{y}) \propto P(\mathbf{y}|\mathbf{b}, \mathbf{v})P(\mathbf{b}, \mathbf{v}). \tag{8.37}$$

For the MCMC-implemented Bayesian analysis, the unobservables are sampled from the joint posterior distribution above. The sampling is performed in the following steps.

(1) Determine initial values for all unobservables denoted as

$$(b_0^{(0)}, b_1^{(0)}, \ldots, b_k^{(0)}, \sigma_e^{2(0)}, \sigma_1^{2(0)}, \ldots, \sigma_k^{2(0)}).$$

Parameters $\mathbf{b}$ are all initialized with zero value and the scale parameters $\mathbf{v}$ initialized with a positive number.

(2) Update the population mean b_0. The conditional posterior distribution of b_0 is normal with mean

$$\bar{b}_0 = \frac{1}{n}\sum_{i=1}^{n}\left(y_i - \sum_{k=1}^{m} x_{ik}b_k^{(0)}\right)$$

and variance

$$\bar{s}_0^2 = \sigma_c^{2(0)},$$

from which we sample a new b_0, denoted as $b_0^{(1)}$, which replaces $b_0^{(0)}$ in all subsequent sampling processes.

(3) Update partial regression coefficients b_k for $k = 1, \ldots, m$. The conditional posterior distribution for b_k is normal with mean

$$\bar{b}_k = \left[\sum_{i=1}^{n} x_{ik}^2 + \frac{\sigma_e^{2(0)}}{\sigma_k^{2(0)}}\right]^{-1} \sum_{i=1}^{n} x_{ik}\left(y_i - b_0^{(0)} - \sum_{l\neq k}^{m} x_{il}b_l^{(0)}\right)$$

and variance

$$\bar{s}_k^2 = \left[\sum_{i=1}^{n} x_{ik}^2 + \frac{\sigma_e^{2(0)}}{\sigma_k^{2(0)}}\right]^{-1} \sigma_e^{2(0)},$$

which are used to sample b_k to generate $b_k^{(1)}$ which replaces $b_k^{(0)}$ in all subsequent sampling processes.

(4) Update the residual variance $\sigma^{2(0)}$. The residual variance is sampled from a scaled inverted χ^2-square distribution,

$$\sigma_e^{2(1)} = \frac{1}{\chi_n^2}\sum_{i=1}^{n} x_{ik}\left(y_i - b_0^{(0)} - \sum_{l\neq k}^{m} x_{il}b_l^{(0)}\right),$$

where χ_n^2 is a random number sampled from a χ^2–distribution with n degrees of freedom. The variance $\sigma_e^{2(0)}$ is immediately updated by $\sigma_e^{2(1)}$.

(5) Update the variances of partial regression coefficients σ_k^2 for $k = 1, \ldots, m$. They can be sampled from a scaled inverted χ^2–distribution,

$$\sigma_k^{2(1)} = \frac{b_k^{2(0)}}{\chi_1^2},$$

where χ_1^2 is a random number sampled from a χ^2 distribution with one degree of freedom.

(6) Repeat steps 2–5. At this point, one sweep of the MCMC is complete, and sampling for the next round can be continued. The sampled parameters will follow the joint posterior distribution when the chain converges to the stationary distribution.

Example 8.9. Xu (2003) used the Bayesian approach to detect marker effects throughout the entire genome in barley (Tinker et al. 1996). The mapping population is composed of $n = 145$ DH lines for which the model developed for the backcross can be used directly. A total of $m = 127$ molecular markers constructed seven linkage groups, covering 1500 cM of the barley genome. The study material was planted in a range of environments. Several agronomic traits were measured for each plant and analyzed on the basis of across-environment means, but only results for kernel weight are shown here.

Figure 8.5a illustrates the marker effects across the seven ligated linkage groups. Four candidate regions were detected to show significant associations between markers and kernel weight, which were consistent with those by single-marker analysis (Fig. 8.5b). The significant difference between the two approaches is that the marker effects detected by multimarker analysis are more sharply contrasting than those by single-marker analysis. This may suggest that multimarker analysis has identified fewer spurious significant associations, compared with single-marker analysis.

Xu (2003) also did an interesting test for the distribution of marker effects over the genome. The marker effects detected by multimarker and single-marker analyses can be fit by the Gamma distribution with the estimated scale (0.0579 vs. 0.1134) and shape parameters (0.2233 vs. 1.1396) dramatically different between the two approaches. Thus, multimarker analysis generated an L-shaped distribution (Fig. 8.6a), whereas single-marker analysis generated a bell-shaped distribution (Fig. 8.6b). Multiple analyses implemented within the Bayesian context provide a powerful tool for studying the overall genetic architecture of a quantitative trait.

8.8 Exercises

8.1 Referring to the data in Table 8.1, for marker **B** test whether the difference in body weight between genotypes 1 and 0 is statistically significant.

8.2 Referring to Example 8.5, construct the ANOVA table for marker **B** and perform any subsequent t–tests. What can you conclude about this marker?

8.3 For the backcross model (8.9), note that if genotype M_1M_1 is observed, the density of y is given by

$$\frac{(1-r)}{\sqrt{2\pi\sigma^2}} \exp\left\{\frac{(y-\mu_1)^2}{2\sigma^2}\right\} + \frac{r}{\sqrt{2\pi\sigma^2}} \exp\left\{\frac{(y-\mu_{12})^2}{2\sigma^2}\right\}.$$

Use this fact to verify the means and variances in equation (8.10)

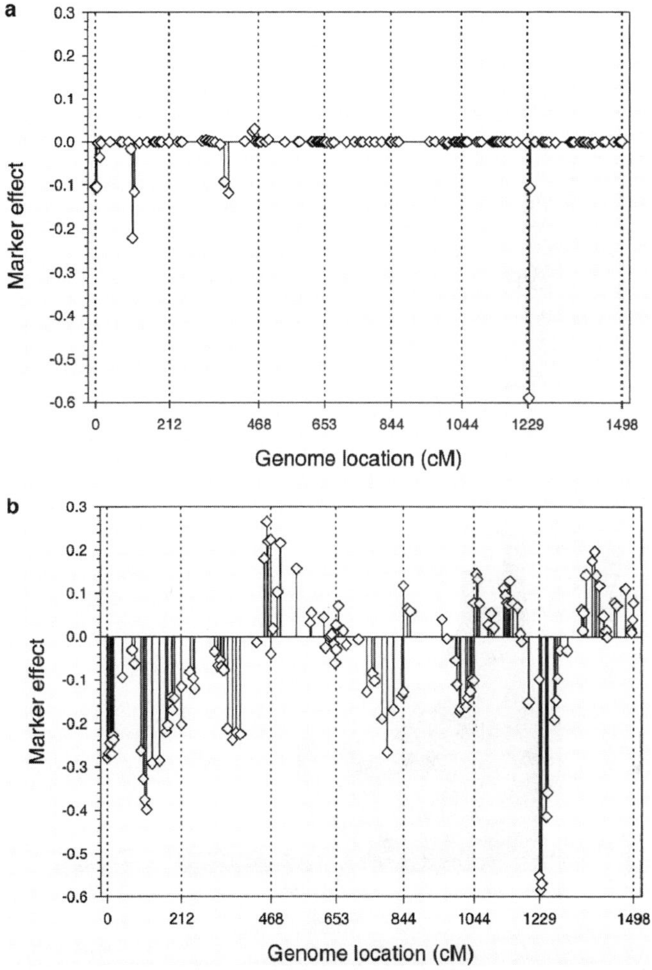

Fig. 8.5. Marker effects of kernel weight in barley across different markers on the genome. (a) Multimarker Bayesian analysis; (b) single-marker regression analysis. The seven linkage groups are separated by dotted vertical lines. Adapted from Xu (2003).

8.4 If f is the normal density $f(y) = \frac{(1)}{\sqrt{2\pi\sigma^2}} \exp\left\{\frac{(y-\mu)^2}{2\sigma^2}\right\}$, show that:

(a) $\frac{\partial}{\partial \mu} f = (y-\mu) f$.

(b) $\frac{\partial}{\partial \sigma^2} f = \left(\frac{(y-\mu)^2-\sigma^2}{2\sigma^{3/2}}\right) f$.

8.5 . Referring to (8.13), verify the expressions for the likelihood and the MLEs.

8.6 Verify equations (8.17) and (8.18).

8.7 For the log-likelihood in part (3) of the iteration in Section 8.4.1, show that

$$\frac{\partial}{\partial r} \log L(\hat{\mu}_1, \hat{\mu}_{12}, \hat{\sigma}^2, r|\mathbf{y}) = \sum_{i=1}^{n_1} \left(\frac{1-P_1(y_i)}{r} - \frac{P_1(y_i)}{1-r}\right) + \sum_{i=n_1+1}^{n} \left(\frac{P_2(y_i)}{r} - \frac{1-P_2(y_i)}{1-r}\right)$$

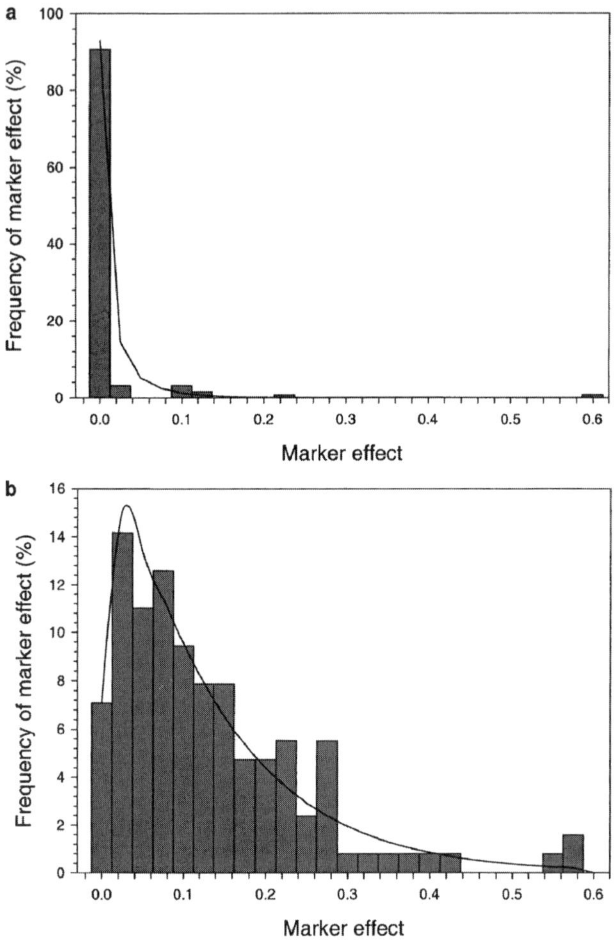

Fig. 8.6. Gamma approximation of marker effects for kernel weight in barley. (a) Multi-marker Bayesian analysis; (b) single-marker regression analysis. Adapted from Xu (2003).

and setting this equal to zero yields

$$r = \frac{1}{n}\left(\sum_{i=1}^{n_1} 1 - P_1(y_i) + \sum_{i=n_1+1}^{n} P_2(y_i)\right).$$

8.8 In Section 8.6.2, we assumed that two markers are independent from each other. But if the markers are linked on the same genomic region, do the following things separately for the backcross and F_2:

(a) Derive the conditional probabilities of QTL genotypes given marker genotypes.
(b) Express the marker genotypic values in terms of the QTL genotypic values.
(c) Show how the interaction effects at the marker level are underestimated when they are used to represent QTL interaction effects.

8.9 Doebley et al. (1995) genotyped two markers as the candidate QTL for the average length of vegetative internodes in the primary lateral branch in an F_2 population of 183 corns derived from inbred lines, *Teosinte-M1L* × *Teosinte-M3L*. The observations and trait values for each genotype at these two QTLs are given below (see Kao and Zeng 2002)

	Q			
P	2	1	0	Total
	Observations			
2	8	22	3	33
1	20	42	24	86
0	11	21	10	42
Total	39	85	37	161
	Phenotypic values			Mean
2	101.60	66.50	61.11	74.52
1	83.62	47.55	40.94	54.09
0	47.80	54.57	17.98	44.08
Total	77.21	54.19	36.37	55.67

(a) Test whether these two candidate QTLs are linked.
(b) Estimate the additive effects of each candidate QTL.
(c) Estimate the dominant effects of each candidate QTL.
(d) Estimate the epistatic genetic effects between the two QTL.
(e) If these two candidate QTLs are viewed as markers, discuss how the estimates of the marker additive, dominance, and epistatic genetic effects are confounded by the QTL positions and epistatic effects.

8.10 Missing marker data in whole-genome marker analysis.
It is common for genotypic data to be missing at some markers. Genotypes of missing markers can be generated randomly on the basis of the probability inferred jointly from the nearest nonmissing flanking markers and the phenotype (Xu 2003). The probability from the markers is treated as the prior probability. After incorporation of the marker effects through the phenotype, the probability becomes the posterior probability, which is used to generate the missing marker genotype. Show the algorithm for sampling missing data within the MCMC framework as shown in Section 8.7.

9

The Structure of QTL Mapping

9.1 Introduction

In the previous chapter, statistical approaches were described for evaluating associations between markers and phenotypes. Such an association analysis can provide evidence for the genetic control of trait variation but is not very precise because the genetic effects associated with marker genotypes are confounded by the position of a functional QTL and its actual effect. Of course, if markers are so highly dense that they are generated at QTL positions, a simple marker-phenotype association analysis may be useful. The generation of such high-density maps is not possible for a majority of species in practice. Powerful analytical techniques are needed to separate the effects of a QTL from its location.

Unlike molecular markers, the genomic locations of QTLs are unknown and should be inferred on the basis of the association analysis of the phenotypes and markers. The role of statistical methods is in the identification, mapping and estimation of functional QTLs using location-known, neutral markers. Starting in this chapter, we will systematically introduce the genetic principles behind QTL mapping strategies and statistical methods proposed to estimate the locations and effects of QTLs based on genetic linkage maps. All of these issues form the core of this book.

One of the most important statistical foundations for QTL mapping is laid out in the mixture model (McLachlan and Peel 2000), in which each observation is assumed to have arisen from one of g unobservable QTL genotype groups, each group being suitably modelled by a density from some parametric family. This model provides a framework by which observations may be clustered together into genotype groups for discrimination. QTL mapping methods based on the mixture model include three tasks:

(1) Model the full mixture for possible phenotype-genotype correspondence.
(2) Characterize the types of QTL effects, random or fixed.
(3) Estimate the unknown QTL parameters using statistical methods.

In this chapter, we will introduce several fundamental genetic and statistical issues related to the characterization of mixture models for QTL mapping. The exploration of these issues helps us to better interpret the results and develop novel statistical models for new problems.

9.2 The Mixture Model

9.2.1 Formulation

The genetic foundation of statistical mapping methods is that a QTL being identified is segregating to form two or more different genotypes in a mapping progeny population. Thus, QTL mapping can be performed using any segregating population. The statistical models for studying QTL effects can be specified most precisely by assuming that the underlying QTL is known using equation (8.1) for the backcross or equation (8.2) for the F_2. Table 9.1 gives the examples for the data structure of a backcross design that corresponds to this so-called QTL regression model.

In practice, the QTL genotypes for a trait are unobserved, although they actually determine the genetic variation of the trait. For a mapping population, the pattern of QTL segregation can be predicted; for example, a 1:1 ratio for the backcross (BC) and 1:2:1 ratio for the F_2. Randomly selecting a mouse, i, from Table 9.1, the probability that this mouse will carry QTL genotype Qq (1) or qq (0) is 1/2 and 1/2, respectively. Similarly, we have such probabilities 1/4 for QQ (2), 1/2 for Qq (1), and 1/4 for qq (0) in the F_2 design (Table 9.2). Thus, each mouse can be assumed to arise from one and only one of these QTL genotypes with a probability specified above for different designs.

Table 9.1. Mouse body weight (y) and a putative QTL in a backcross design.

Sample	QTL genotype (z)	Body weight (y)
1	Qq (1)	30
2	Qq (1)	32
3	Qq (1)	28
4	Qq (1)	29
5	Qq (1)	29
6	qq (0)	22
7	qq (0)	20
8	qq (0)	21
9	qq (0)	20
10	qq (0)	21

For mouse i, a likelihood function of its phenotype, y_i, can be formulated in the mixture model context, expressed as

$$(9.1)\ y_i \sim p(\phi_j, \eta | y_i) = \begin{cases} \frac{1}{2} f_1(y_i; \phi_1, \eta) + \frac{1}{2} f_0(y_i; \phi_0, \eta) & \text{for BC} \\ \frac{1}{4} f_2(y_i; \phi_2, \eta) + \frac{1}{2} f_1(y_i; \phi_1, \eta) + \frac{1}{4} f_0(y_i; \phi_0, \eta) & \text{for } \mathrm{F}_2, \end{cases}$$

where $f_j(y_i; \phi_j, \eta)$ is a type of probability density function with ϕ_j specific to each component (i.e., genotype) and η common to all components. The mixing proportions of each component in mixture model (9.1) are the genotype frequencies in the entire mapping population.

Table 9.2. Mouse body weight affected by a putative QTL segregating with different prior probabilities in a backcross or F_2 design.

	Body	Backcross		F_2		
Sample	Weight	Qq (1)	qq (0)	QQ (2)	Qq (1)	qq (0)
1	30	$\frac{1}{2}$	$\frac{1}{2}$	$\frac{1}{4}$	$\frac{1}{2}$	$\frac{1}{4}$
2	32	$\frac{1}{2}$	$\frac{1}{2}$	$\frac{1}{4}$	$\frac{1}{2}$	$\frac{1}{4}$
3	28	$\frac{1}{2}$	$\frac{1}{2}$	$\frac{1}{4}$	$\frac{1}{2}$	$\frac{1}{4}$
4	29	$\frac{1}{2}$	$\frac{1}{2}$	$\frac{1}{4}$	$\frac{1}{2}$	$\frac{1}{4}$
5	29	$\frac{1}{2}$	$\frac{1}{2}$	$\frac{1}{4}$	$\frac{1}{2}$	$\frac{1}{4}$
6	22	$\frac{1}{2}$	$\frac{1}{2}$	$\frac{1}{4}$	$\frac{1}{2}$	$\frac{1}{4}$
7	20	$\frac{1}{2}$	$\frac{1}{2}$	$\frac{1}{4}$	$\frac{1}{2}$	$\frac{1}{4}$
8	21	$\frac{1}{2}$	$\frac{1}{2}$	$\frac{1}{4}$	$\frac{1}{2}$	$\frac{1}{4}$
9	20	$\frac{1}{2}$	$\frac{1}{2}$	$\frac{1}{4}$	$\frac{1}{2}$	$\frac{1}{4}$
10	21	$\frac{1}{2}$	$\frac{1}{2}$	$\frac{1}{4}$	$\frac{1}{2}$	$\frac{1}{4}$

Model (9.1) can be used to diagnose the existence of a segregating QTL in a mapping population but is unable to localize the chromosomal location of the QTL. The QTL location can be estimated by inferring the QTL genotypes from the segregation pattern of markers that are linked with the QTL. The raw data for QTL mapping consists of the ordered marker information and measured phenotypes of different individuals (Tables 8.1 and 8.2). With additional marker information (**M**), the likelihood function of mouse i for the backcross, as described by equation (9.1), can be rewritten as

$$(9.2) \qquad y_i \sim p(\phi_j, \eta | y_i, \mathbf{M}) = \omega_{1|i} f_1(y_i; \phi_1, \eta) + \omega_{0|i} f_0(y_i; \phi_0, \eta),$$

where $\omega_{j|i}$ is the conditional probability of QTL genotype j given marker genotypes for mouse i.

Similarly, the likelihood of a mouse for the F_2 can be written as

$$
\begin{aligned}
y_i &\sim p(\phi_j, \eta | y_i, \mathbf{M}) \\
&= \omega_{2|i} f_2(y_i; \phi_2, \eta) + \omega_{1|i} f_1(y_i; \phi_1, \eta) + \omega_{0|i} f_0(y_i; \phi_0, \eta).
\end{aligned} \tag{9.3}
$$

The conditional probability matrices for the backcross and F_2 are provided in Chapter 10.

9.2.2 Structure, Setting, and Estimation

Finite mixtures of distributions can be used to model a wide variety of random phenomena (see McLachlan and Peel 2000) and can provide a sound statistical approach for distinguishing unknown QTL genotypes. With an approach based on a normal mixture model for separating QTL genotypes for a general mapping population, it is assumed that each observation, which includes the phenotype (y_i) and marker information ($\mathbf{M}$), is from a mixture of a specified number (J) of probability densities expressed as

$$
p(\boldsymbol{\Omega} | y_i, \mathbf{M}) = \omega_{1|i} f_1(y_i; \mu_1, \sigma^2) + \cdots + \omega_{J|i} f_J(y_i; \mu_J, \sigma^2), \tag{9.4}
$$

where $\boldsymbol{\Omega} = (\omega_{1|i}, \cdots, \omega_{J|i}, \mu_1, \cdots, \mu_J, \sigma^2)$ is the vector of unknown parameters, the $\omega_{j|i}$'s are the mixing proportions, summing to 1, which are the conditional probabilities of the QTL genotypes given the marker genotypes, and $f_j(y_i; \mu_j, \sigma^2)$ is generally assumed to be a normal density for a particular QTL genotype j with mean μ_j and variance σ^2, expressed as

$$
f_j(y_i; \mu_j, \sigma^2) = \frac{1}{\sqrt{2\pi}\sigma} \exp\left[-\frac{(y_i - \mu_j)^2}{2\sigma^2}\right]. \tag{9.5}
$$

All the component distributions are assumed to be from the same parametric family.

With basic knowledge about a mixture model, the development of a statistical method for QTL mapping contains three major tasks:

(1) Derive the *structures* of the mixture model including the mixture proportions (denoted as the frequencies of QTL genotypes) and the density functions (expressed in terms of QTL effects and residual variance).
(2) Optimize the *setting* of QTL mapping from an experimental design perspective by specifying QTL effects and other model effects as fixed or random.
(3) Develop a statistical technique for the *estimation* of the unknown parameters defined in the mixture model.

The first task needs the formulation of thorough genetic models based on different mating designs, different marker types, different population structures, and different genetic effects. The second task is relevant to the modeling and analysis of a mapping experiment according to particular sampling strategies and the manner in which results from QTL mapping are explained and utilized. The third task depends upon statistical and computational algorithms with the aid of computer techniques. All three tasks are coupled under the mixture model framework to efficiently solve for the unknown parameters of interest.

In practice, some statistical mapping methods focus on *the structure of the mixture model* by deriving genetic models that better reveal the genetic architecture of a complex trait in a particular population. Some methods attempt to examine *the setting of the mixture model* based on different experimental designs so as to make better use of mapping results. Others may focus on *the estimation of the mixture model* by implementing a more efficient computational algorithm for a complicated pedigree. We will describe these three aspects of QTL mapping from genetic, experimental, and statistical perspectives.

9.3 Population Genetic Structure of the Mixture Model

As described above, a given normal mixture model contains two different structures, mixture proportions $\omega = (\omega_{1|i}, \cdots, \omega_{J|i})$ and normal distributions determined by model parameters $\mathbf{\Omega}_m = (\mu_1, \cdots, \mu_J, \sigma^2)$. The mixture proportions are actually the frequencies of QTL genotypes at a putative QTL in a population. For an entire mapping population initiated with two inbred lines, the frequencies of QTL genotypes can be predicted on the basis of the first Mendelian law. For example, the frequencies of QTL genotypes are 1/2 and 1/2 for the backcross or 1/4, 1/2, and 1/4 for the F_2. However, when marker information is associated with a putatively linked QTL, the frequencies of QTL genotypes given a particular marker genotype (i.e., conditional probabilities) will not obey the Mendelian law but rather depend on the recombination fraction or linkage disequilibrium between the marker and QTL. The conditional probabilities of QTL genotypes upon the marker genotypes are the "bridge" associating the known marker information with unknown QTL information. Below, we will give the procedure for deriving these conditional probabilities in different mapping populations.

9.3.1 Backcross/F_2

Two different inbred lines are crossed to generate the heterozygous F_1. When the F_1 is backcrossed to the original parents, or selfed or sibling-mated, different genes will be cosegregating, which leads to nonparental, recombinant types whose proportion depends on the degree of linkage between different genes. By observing the number of recombinant types, we can then estimate the linkage. However, the recombinants between a marker and QTL are not observable because the QTL position is unknown, and we need to derive the conditional probabilities of QTL genotypes given known marker genotypes in terms of their recombination fractions. The advantage of the backcross or F_2 as a mapping population lies in the clear linkage phase between all genes, from which the parental origin of alleles can be precisely determined. In Sections 10.3 and 10.4, we provide the procedure for deriving these conditional probabilities for the backcross (Table 10.3) and F_2 populations (Table 10.6).

9.3.2 Outbred Crosses

Many species, such as forest trees, cannot generate inbred lines because of their complicated biological features, but they can be crossed to generate a segregating progeny population, called an outbred cross. It is possible to use outcross progeny as an outbred population because crossover events occur during meioses. In fact, outbred lines as parents can be homozygous at some loci but are heterozygous at many loci, and thus their controlled crosses can be backcross-like for some loci, F_2-like for other loci, or present new cross types. The QTL genotypes in outbred crosses can be inferred by their conditional probabilities given marker genotypes expressed as a function of the recombination fraction. A procedure is necessary to determine a correct linkage phase prior to the estimation of linkage. A detailed description of QTL mapping in outbred populations is given in the last chapter.

9.3.3 Recombinant Inbred Lines

Recombinant inbred lines (RILs) are powerful material for genetic mapping. They can be derived either by repeated selfing or by repeated brother–sister mating of the progeny from an F_1 cross between two inbred lines. RILs can serve as a permanent mapping population for multiple uses because they are fixed and homozygous for two alternative alleles at all genes. Some lines are the same as parental (nonrecombinant) types, whereas the others are recombinant types. The conditional probabilities of QTL genotypes given marker genotypes are derived in terms of the proportion of recombinant zygotes (see Table 10.7).

9.3.4 Natural Populations

For some species in which crosses are not possible, the cosegregation between the marker and QTL can be specified by linkage disequilibrium (LD). The LD represents nonrandom associations between different loci in a population and can be used to analyze an unstructured population. Linkage analysis has been widely used for genetic mapping by detecting the degree of LD between the marker and QTL. Unlike controlled crosses, the conditional probabilities of QTL genotypes given marker genotypes are expressed in terms of LD values (Lou et al. 2003).

9.4 Quantitative Genetic Structure of the Mixture Model

It is generally assumed that a quantitative trait (y) of interest at a putative QTL is normally distributed with expected mean μ_j and variance σ^2. The genetic contributions of the QTL to the trait phenotype are reflected in the mean or variance of the normal distribution. For a given QTL genotype j, we can partition its expected mean (μ_j) into the different components: overall mean (μ), additive effect (a), dominance effect (d), and additive $\times$ additive (i_{aa}), additive $\times$ dominance (i_{ad}), dominance $\times$ additive (i_{da}), and dominance $\times$ dominance (i_{dd}) epistatic effects. If we assume that these effects are fixed, they can be directly estimated in the mixture model. As the variances of the fixed effects are viewed as zero, the variance contains only the residual variance σ^2 within a particular QTL genotype. The major task of a *fixed-model*–based mapping approach is to specify the genetic components of μ_j based on different genetic problems or mapping purposes.

9.4.1 Additive-Dominance Model

In many cases, it is reasonable to assume that there are no nonallelic interactions (epistasis) between different QTLs. Consider a QTL of three possible genotypes whose values and residual variances are defined as

(9.6)

Genotype	QQ (2)	Qq (1)	qq (0)
Overall mean	μ	μ	μ
Effect	a	d	$-a$
Genotypic value	$\mu_2 = \mu + a$	$\mu_1 = \mu + d$	$\mu_0 = \mu - a$
Residual	σ^2	σ^2	σ^2

In most situations, the residual variance is assumed to be identical among the three groups of QTL genotypes for the sake of computational simplicity. Statistical techniques are also available when the residual variances are genotype-specific (*heteroscedasticity*).

When two or more QTLs are fit, we will have more QTL genotypes included in the mixture model. For an F_2 population, the genotypic values, $\mu_{j_1j_2}$ $(j_1, j_2 = 0, 1, 2)$, of nine QTL genotypes for two given QTLs, **P** and **Q**, under the additive-dominance model can be defined as

$$\mu_{j_1j_2} = \begin{array}{c} \mathbf{P} \\ 2 \\ 1 \\ 0 \end{array} \overset{\begin{array}{ccc} & \mathbf{Q} & \\ 2 & 1 & 0 \end{array}}{\begin{bmatrix} \mu_{22} = \mu + a_1 + a_2 & \mu_{21} = \mu + a_1 + d_2 & \mu_{20} = \mu + a_1 - a_2 \\ \mu_{12} = \mu + d_1 + a_2 & \mu_{11} = \mu + d_1 + d_2 & \mu_{10} = \mu + d_1 - a_2 \\ \mu_{02} = \mu - a_1 + a_2 & \mu_{01} = \mu - a_1 + d_2 & \mu_{00} = \mu - a_1 - a_2 \end{bmatrix}}, \quad (9.7)$$

where a_1 and a_2 are the additive effects and d_1 and d_2 are the dominance effects of these two QTLs, respectively.

Example 9.1. In mapping body weight for an F_2 progeny of mouse, joint genotypic values at two QTLs, **P** and **Q**, are estimated. Genotypic values at individual QTLs are then calculated as marginal means. The estimates of joint or marginal genotypic means are given in Table 9.3. The additive and dominance effects of individual QTL can be estimated on the basis of marginal means using equation (9.6) or joint genotypic values using equation (9.7).

Table 9.3. Joint genetic values and marginal means at two QTLs for body weight in the F_2 progeny.

P	**Q**			
	2	1	0	Marginal
2	$\mu_{22} = 30$	$\mu_{21} = 32$	$\mu_{20} = 28$	$\mu_{2\cdot} = 30$
1	$\mu_{12} = 29$	$\mu_{11} = 29$	$\mu_{10} = 21$	$\mu_{1\cdot} = 25$
0	$\mu_{02} = 21$	$\mu_{01} = 20$	$\mu_{00} = 21$	$\mu_{0\cdot} = 20.67$
Marginal	$\mu_{\cdot 2} = 26.67$	$\mu_{\cdot 1} = 27$	$\mu_{\cdot 0} = 22.75$	

Let $\mathbf{m}$ be the genotypic mean vector, $\mathbf{a}$ be the effect vector, and $\mathbf{D}$ be the design matrix that relates $\mathbf{m}$ and $\mathbf{a}$. The relationship $\mathbf{m} = \mathbf{Da}$ leads to

$$\mathbf{a} = \mathbf{D}^{-1}\mathbf{m}. \tag{9.8}$$

For the marginal means, we have $\mathbf{m} = (\mu_2, \mu_1, \mu_0)^{\mathrm{T}}$, $\mathbf{a} = (\mu, a, d)^{\mathrm{T}}$, and

$$\mathbf{D} = \begin{bmatrix} 1 & 1 & 0 \\ 1 & 0 & 1 \\ 1 & -1 & 0 \end{bmatrix}.$$

For joint genotypic values, we have $\mathbf{m} = (\mu_{22}, \mu_{21}, \mu_{20}, \mu_{12}, \mu_{11}, \mu_{10}, \mu_{02}, \mu_{01}, \mu_{00})^{\mathrm{T}}$, $\mathbf{a} = (\mu, a_1, d_1, a_2, d_2)^{\mathrm{T}}$, and

$$\mathbf{D} = \begin{bmatrix} 1 & 1 & 0 & 1 & 0 \\ 1 & 1 & 0 & 0 & 1 \\ 1 & 1 & 0 & -1 & 0 \\ 1 & 0 & 1 & 1 & 0 \\ 1 & 0 & 1 & 0 & 1 \\ 1 & 0 & 1 & -1 & 0 \\ 1 & -1 & 0 & 1 & 0 \\ 1 & -1 & 0 & 0 & 1 \\ 1 & -1 & 0 & -1 & 0 \end{bmatrix}.$$

Using the data in Table 9.3, the genetic effects are estimated as $\mu = 25.335$ and 24.71, $a = 4.665$ and 1.96, and $d = -0.335$ and 2.29 for **P** and **Q**, respectively, from the marginal means, and as $\mu = 24.56$, $a_1 = 4.67$, $d_1 = 0.357$, $a_2 = 1.99$, and $d_2 = 2.32$ from the joint genotypic values.

9.4.2 Additive-Dominance-Epistasis Model

If two or more QTLs interact to affect a quantitative trait, their epistatic effects should be modeled in the QTL genotypic values within the mixture model. We use Mather and Jinks' (1982) notations to describe the epistasis under which the genotypic values are expressed, such as when two QTLs are assumed,

$$
\begin{array}{cc} & \\ & \mathbf{P} \\ \mu_{j_1j_2} = & \begin{array}{c} 2 \\ 1 \\ 0 \end{array} \end{array}
\begin{array}{c} \mathbf{Q} \\ \begin{array}{ccc} 2 & 1 & 0 \end{array} \\
\left[\begin{array}{ccc}
\mu + a_1 + a_2 + i_{aa} & \mu + a_1 + d_2 + i_{ad} & \mu + a_1 - a_2 - i_{aa} \\
\mu + d_1 + a_2 + i_{da} & \mu + d_1 + d_2 + i_{dd} & \mu + d_1 - a_2 - i_{da} \\
\mu - a_1 + a_2 - i_{aa} & \mu - a_1 + d_2 - i_{ad} & \mu - a_1 - a_2 + i_{aa}
\end{array}\right] \end{array},
\tag{9.9}
$$

where i_{aa}, i_{ad}, i_{da}, and i_{dd} are the additive × additive, additive × dominance, dominance × additive, and dominance × dominance epistatic effects, respectively.

Example 9.2. Revisit Example 9.1. Based on Table 9.3, we can also estimate the four kinds of epistatic effects. At this time, $\mathbf{a} = (\mu, a_1, d_1, a_2, d_2, i_{aa}, i_{ad}, i_{da}, a_{dd})^{\mathrm{T}}$ and

$$
\mathbf{D} = \begin{bmatrix}
1 & 1 & 0 & 1 & 0 & 1 & 0 & 0 & 0 \\
1 & 1 & 0 & 0 & 1 & 0 & 1 & 0 & 0 \\
1 & 1 & 0 & -1 & 0 & -1 & 0 & 0 & 0 \\
1 & 0 & 1 & 1 & 0 & 0 & 0 & 1 & 0 \\
1 & 0 & 1 & 0 & 1 & 0 & 0 & 0 & 1 \\
1 & 0 & 1 & -1 & 0 & 0 & 0 & -1 & 0 \\
1 & -1 & 0 & 1 & 0 & -1 & 0 & 0 & 0 \\
1 & -1 & 0 & 0 & 1 & 0 & -1 & 0 & 0 \\
1 & -1 & 0 & -1 & 0 & 1 & 0 & 0 & 0
\end{bmatrix}.
$$

Using equation (9.8), we estimate the genetic effects as $\mu = 25$, $a_1 = 4$, $d_1 = 0$, $a_2 = 0.5$, $d_2 = 1$, $i_{aa} = 0.5$, $i_{ad} = 2$, $i_{da} = 3.5$, and $i_{dd} = 3$.

9.4.3 Multiplicative-Epistatic Model

The physiological basis of epistasis has been studied by modeling the relationship between genes and their products in many plant and animal experiments. Minvielle (1987) showed, from a theoretical perspective, that multiplicative interaction between a pair of loci may be an important form of epistasis for controlling complex traits, as anticipated by Arunachlum (1977). The multiplicative-epistasis model assumes that genotypes at a pair of loci have genotypic values equal to the product of genotypic values at the two different loci (Schnell and Cockerham 1992; Li and Wu 1996; Otto and Feldman 1997). For example, if genotypic values are μ_2 for **P** genotype 2 and μ_2' for **Q** genotype 2, then the value of joint QTL genotype 22 is $\mu_{22} = \mu_2\mu_2'$. Under the multiplicative-epistatic model, the genotypic values of two QTL can be modelled by

$$
\begin{array}{cc} & \\ & \mathbf{P} \\ \mu_{j_1j_2} = & \begin{array}{c} 2 \\ 1 \\ 0 \end{array} \end{array}
\begin{array}{c} \mathbf{Q} \\ \begin{array}{ccc} 2 & 1 & 0 \end{array} \\
\left[\begin{array}{ccc}
\mu_{22} = \mu_2\mu_2' & \mu_{21} = \mu_2\mu_1' & \mu_{20} = \mu_2\mu_0' \\
\mu_{12} = \mu_1\mu_2' & \mu_{11} = \mu_1\mu_1' & \mu_{10} = \mu_1\mu_0' \\
\mu_{02} = \mu_0\mu_2' & \mu_{01} = \mu_0\mu_1' & \mu_{00} = \mu_0\mu_0'
\end{array}\right] \end{array}.
\tag{9.10}
$$

Although multiplicative interactions between a pair of QTL are considered, two special cases, completely multiplicative action (both between and within loci) and pure additive action (without dominance), can also be manipulated by setting restrictions.

Example 9.3. Revisit Example 9.1. Here, we attempt to estimate the genetic-effect parameters using the data in Table 9.3 under the multiplicative-epistasis model. We use a nonlinear least squares approach to estimate three genotypic values for each QTL based on matrix (9.10). These estimates are $\mu_2 = 5.6954$, $\mu_1 = 5.0308$, $\mu_0 = 3.9065$, $\mu'_2 = 5.4624$, $\mu'_1 = 5.5649$, and $\mu'_0 = 4.755$.

Schnell and Cockerham (1992) proposed an analytical model for estimating additive, dominance, and epistatic effects of various kinds due to the two multiplicative–interacting QTLs. The estimators of these effects are expressed as

$$\begin{aligned} &\mu = uu', a_1 = \alpha u', d_1 = \delta u', a_2 = u\alpha', d_2 = u\delta', \\ &i_{aa} = \alpha\alpha', i_{ad} = \alpha\delta', i_{da} = \delta\alpha', i_{dd} = \delta\delta', \end{aligned} \tag{9.11}$$

where $u = (\frac{1}{4}\mu_2 + \frac{1}{2}\mu_1 + \frac{1}{4}\mu_0), u' = (\frac{1}{4}\mu'_2 + \frac{1}{2}\mu'_1 + \frac{1}{4}\mu'_0), \alpha = \frac{1}{2}(\mu_2 - \mu_1) + \frac{1}{2}(\mu_1 - \mu_0), \alpha' = \frac{1}{2}(\mu'_2 - \mu'_1) + \frac{1}{2}(\mu'_1 - \mu'_0), \delta = \mu_2 - 2\mu_1 + \mu_0$, and $\delta' = \mu'_2 - 2\mu'_1 + \mu'_0$.

We further estimate the genetic effects as $\mu = 26.2349$, $a_1 = 4.7736$, $d_1 = -2.45331$, $a_2 = 1.73856$, $d_2 = -4.48501$, $i_{aa} = 0.31634$, $i_{ad} = -0.816074$, $i_{da} = -0.162578$, and $i_{dd} = 0.419407$.

9.4.4 Mechanistic Model

The genotypic values of QTLs contained in the mixture model can also be modeled on the basis of the mechanistic principles causing them. For example, the genotypic value of a QTL genotype QQ for a quantitative trait can be expressed as a function of other independent variables, time, or environmental factor,

$$y = \begin{cases} f(x) & \text{for allometric laws} \\ g(t) & \text{for growth models} \\ h(z) & \text{for reaction norms} \end{cases} \tag{9.12}$$

where y is the biological trait of interest, x is the body size, t is the age, and z is an environmental variable such as temperature, nutrition, or light intensity. The forms of mathematical functions $f(x)$, $g(t)$, and $h(z)$, which can be linear or nonlinear, are generally different, depending on specific questions of interest. Generally, the establishment of appropriate mathematical functions is based on the goodness of fit to observational data (Niklas 1994). Alternatively, these mathematical functions are derived from an optimality perspective. For example, West et al. (1997, 1999) proposed a fractal-like network system for the absorption and internal distribution of metabolites to explain quarter-power scaling laws pervasive in the living world. In addition, West et al. (1997, 1999, 2001) explained why the growth of an organism follows a sigmoid curve based on fundamental principles for the allocation of metabolic energy between the maintenance of existing tissue and their production of new biomass.

So far, we have discussed two key structures in the mixture model: mixing proportions and genetic-effect models. As can be seen, the conditional probabilities are related to allele frequencies and the linkage or association between the markers and QTL (the theme of Mendelian or population genetics), whereas the normal distributions concern the quantitative effects of QTL on traits (the theme of quantitative genetics). Depending on the nature of the question considered, the complexities of these two structures will be different. Some methods stress the derivations of the conditional probabilities, such as linkage disequilibrium-based mapping or polyploid mapping. Yet, other methods stress genetic-effect models that can be used to approximate a biological phenomenon. These two structures determine the originality of a new statistical method for QTL mapping.

9.5 Experimental Setting of the Mixture Model

In the mixture model (9.4), we assume that different normal components are characterized by a known or unknown number of QTL genotypes. Genetic effects of putative QTLs on the phenotype, which are embedded within normal distributions, can be directly estimated by incorporating the *fixed model* approach. The fixed model approach is useful if the underlying genetic effects can be readily specified, as in the case of controlled crosses derived from homozygous inbred lines (Lander and Botstein 1989; Haley and Knott 1992; Zeng 1994) or outbred lines (Lin et al. 2003).

The progeny from a controlled cross should be grown in a regular experimental design for phenotypic data collection for quantitative traits of interest. The mixture model allows the estimation of the experimental effects due to any covariates such as sex or discrete environments and the interaction effects between QTL and treatments. Two experimental designs are used to estimate QTL $\times$ environment interactions. In design 1, the same set of genotypes recorded for markers are grown in different environments. Such a design is possible for species that can generate genetically identical individuals, such as clones and RILs. In design 2, only different sets of individuals from a mapping population are reared in different environments, but for these sets the same marker systems are genotyped. This design is used for many species in which genotypes cannot be duplicated.

The same genotype may perform differently across a range of environments. Genetic variation that underlies such *phenotypic plasticity* provides the organism with the capacity to buffer against environmental fluctuations (Schlichting 1986). This role is thought to be affected by allelic sensitivity and gene regulation (Via et al. 1995). The concept of allelic sensitivity proposes that plasticity arises from differential effects of loci directly contributing to variation in plastic traits. The gene regulation hypothesis states that specific loci influence trait changes between environments without altering the means within a given environment. These hypotheses are not mutually exclusive, but their difference lies within the effect of the environment on the expression of the genes underlying the trait: either directly for allelic sensitivity or indirectly for regulatory loci (Schlichting and Pigliucci 1998).

These two general hypotheses can be tested within the mixture model context for QTL mapping (Wu 1998; Leips and Mackay 2000). Allelic sensitivity says that alleles have varying effects on the phenotype in different environments, which implies that the QTL affecting the sensitivity of a trait to environmental changes should map to the same regions as those that explain the genetic variation in the trait within an environment. Differential expression of alleles in these QTL regions across environments would explain the covariation between trait performance and the environment. Gene regulation states that special regulatory genes respond to the environment by turning on or adjusting the expression of structural genes that directly influence the trait. Thus, according to this hypothesis, the QTL regions that explain the genetic variation in phenotypic sensitivity will be distinct from those that contribute to the variation within a given environment.

9.6 Estimation in the Mixture Model

The mixture model (9.4) lays a statistical foundation for mapping QTLs in a segregating population. The estimation of unknown parameters contained in the mixture model is an important next step for a mapping project. There are many statistical methods that can provide estimates of unknown parameters under the mixture-model framework designed to map QTLs. In this section, we detail the method of maximum likelihood; a discussion of other methods is given in Section 9.7.

The maximum likelihood approach to parameter estimation in mixture models obtains point estimates $(\hat{\omega}_{1|i}, \hat{\mu}_i, \cdots, \hat{\omega}_{J|i}, \hat{\mu}_J, \hat{\sigma}^2)$ of the parameters

$$\mathbf{\Omega} = (\omega_{1|i}, \mu_1, \cdots, \omega_{J|i}, \mu_J, \sigma^2)$$

by maximizing the likelihood. The unknown parameters $\mathbf{\Omega}$ contain the mixture proportions $\omega = (\omega_{1|i}, \cdots, \omega_{J|i})$ specifying the QTL locations and model parameters $\mathbf{\Omega}_m = (\mu_1, \cdots, \mu_J, \sigma^2)$ specifying the QTL effects and residual variance.

An observed data set in a mapping project (see, for example, Tables 8.1 and 8.2) includes the phenotypes of a quantitative trait (y) and marker genotype ($\mathbf{M}$) for all genotyped individuals. A complete data set is composed of these observations, along with the QTL genotypes (g) and the locations of the QTL (d), which are missing data.

The likelihood of the unknown model parameters θ and the unknown mixture proportions ω (and therefore the unknown QTL locations d) for individual i can be expressed as

$$p(d, \theta | y_i, \mathbf{M}_i) = \sum_{j=1}^{J} \omega_{j|i}(g_i | d, \mathbf{M}_i) f_j(y_i | g_i, \theta), \tag{9.13}$$

with the sum over all possible QTL genotypes for individual i, and we explicitly show the dependence of the weights on g_i, d, and $\mathbf{M}_i$.

Suppose there are n independent offspring in a mapping population. The likelihood function of the mixture model (9.13) for the parameter vector $\mathbf{\Omega}$ can be formed by taking the products of the log mixture densities at each y_i to give

$$L(d,\theta|y,\mathbf{M}) = \prod_{i=1}^{n} p(d,\theta|y_i,\mathbf{M}_i).$$

It is almost always easier to work with the log-likelihood

$$\begin{aligned}\log L(d,\theta|y,\mathbf{M}) &= \log \prod_{i=1}^{n} p(d,\theta|y_i,\mathbf{M}_i) \\ &= \sum_{i=1}^{n} \log p(d,\theta|y_i,\mathbf{M}_i) \qquad (9.14) \\ &= \sum_{i=1}^{n} \log \sum_{j=1}^{J} \left[\omega_{j|i}(g_i|d,\mathbf{M}_i) f_j(y_i|g_i,\mathbf{\Omega}_m)\right].\end{aligned}$$

The maximum likelihood estimate of $\mathbf{\Omega}$ is obtained as an appropriate root of the log-likelihood equation

$$\frac{\partial}{\partial \mathbf{\Omega}} \log L(d,\theta|y_i,\mathbf{M}_i) = \sum_{i=1}^{n} \log \frac{\partial}{\partial \mathbf{\Omega}} p(d,\theta|y_i,\mathbf{M}_i) = 0. \qquad (9.15)$$

To ease notation, we write $\omega_{j|i}(g_i) = \omega_{j|i}(g_i|d,\mathbf{M}_i)$ and $f_j(y_i) = f_j(y_i|g_i,\mathbf{\Omega}_m)$ and note that

$$\begin{aligned}\log \frac{\partial}{\partial \mathbf{\Omega}} p(d,\theta|y_i,\mathbf{M}_i) &= \frac{1}{p(d,\theta|y_i,\mathbf{M}_i)} \frac{\partial}{\partial \mathbf{\Omega}} \sum_{j=1}^{J} \omega_{j|i}(g_i) f_j(y_i) \\ &= \sum_{j=1}^{J} \frac{\omega_{j|i}(g_i) f_j(y_i)}{p(d,\theta|y_i,\mathbf{M}_i)} \frac{\partial}{\partial \mathbf{\Omega}} \log[\omega_{j|i}(g_i) f_j(y_i)] \\ &= \sum_{j=1}^{J} \Pi_{j|i} \frac{\partial}{\partial \omega} \log \omega_{j|i}(g_i) + \sum_{j=1}^{J} \Pi_{j|i} \frac{\partial}{\partial \theta} \log f_j(y_i), \qquad (9.16)\end{aligned}$$

where we define

$$\mathbf{\Pi}_{j|i} = \frac{\omega_{j|i}(g_i) f_j(y_i)}{p(d,\theta|y_i,\mathbf{M}_i)}, \qquad (9.17)$$

which could be thought of as a posterior probability that individual i has a QTL genotype j. We then implement the EM algorithm with the expanded parameter set $\{\mathbf{\Omega},\mathbf{\Pi}\}$, where $\mathbf{\Pi} = \{\Pi_{j|i}; j=1,\cdots,J, i=1,\cdots,n\}$.

To solve equations (9.15), we iterate equations between (9.16) and (9.17). Assuming that we know $\mathbf{\Pi}_{j|i}$, we solve for the zero of equation (9.16). Using those values of

ω and θ, we update $\mathbf{\Pi}_{j|i}$ and continue the iterations until convergence. This scheme can be thought of as an implementation of the EM algorithm (Dempster et al. 1977; Meng and Rubin 1993; see also Section 9.7.1).

Once the mixture model has been fit, a probabilistic clustering of the data into g clusters can be obtained in terms of the fitted posterior probabilities of component membership for the data, $\Pi_{j|i}(y_i|\hat{\mathbf{\Omega}})$, where $\hat{\mathbf{\Omega}}$ denotes the maximum likelihood estimate of $\mathbf{\Omega}$. An outright assignment of the data into J clusters is achieved by assigning each data point to the component to which it has the highest estimated posterior probability of belonging.

9.7 Computational Algorithms for the Mixture Model

In this section, we review some of the options in solving the likelihood equations of Section 9.6, and also discuss some other aspects of fitting mixture models.

9.7.1 EM Algorithm

Each iteration consists of two steps. First, in the E step, the complete-data log-likelihood is averaged over the conditional distribution of the indicator variables given the observed data using the current estimate of the parameter vector. Since the complete-data log-likelihood is linear in these indicator variables (a result of the normality assumption), the E step of the EM algorithm simply involves replacing them by the current values of their conditional expectations; that is, the posterior probabilities of component membership expressed in equation (9.17). Next, in the M step, conditional on $\mathbf{\Pi}$, we solve for the zeros of equation (9.15) (likelihood equations) to get our estimates of $\mathbf{\Omega}$. The likelihood equation can be split into the two terms of equation (9.16). The first term refers to the genetic linkage between loci as specified in a controlled cross or the genetic association as specified in a natural population, and the second term refers to the phenotype–genotype relationship. The estimates are then used to update $\mathbf{\Pi}$ in the E step, and the process is repeated until convergence. The values at convergence are the maximum likelihood estimates.

The EM algorithm has reliable convergence in that, regardless of the starting point, the likelihood (9.15) is increased after each EM iteration and that convergence is to a local maximum (or a stationary point).

9.7.2 Monte Carlo EM

In each cycle of the EM algorithm, the likelihood equation can be estimated using a number (N) of Monte Carlo realizations

$$\frac{\partial}{\partial \mathbf{\Omega}} \log L(d, \theta | y, \mathbf{M}) \hat{=} \frac{1}{N} \sum_{k=1}^{N} \frac{\partial}{\partial \mathbf{\Omega}} \sum_{i=1}^{n} \log \pi_j^{(k)}(g_i) + \frac{1}{N} \sum_{k=1}^{N} \frac{\partial}{\partial \mathbf{\Omega}} \log \sum_{i=1}^{n} f_j^{(k)}(y_i),$$

where in the k Monte Carlo samples, QTL genotypes j are generated given y, $\mathbf{M}$, and the current parameter estimate. For sufficiently large genotype samples, Monte Carlo EM will inherit the properties of exact EM. The Monte Carlo samples can also be used for likelihood–ratio estimation in the final EM step.

9.7.3 Stochastic EM

In each cycle of the EM algorithm, the likelihood equation can be estimated by using a single Monte Carlo realization

$$\frac{\partial}{\partial\Omega}\log L(d,\theta|y,\mathbf{M})\hat{=}\frac{\partial}{\partial\Omega}\log\sum_{i=1}^{n}\omega_{j|i}^{(k)}(g_i)+\frac{\partial}{\partial\Omega}\sum_{i=1}^{n}\log f_j^{(k)}(y_i),$$

where in the kth EM cycle a single QTL genotype j is generated given y and m and the current parameter estimates by using the distribution $\mathbf{\Pi}^{(k)}$. This expression can be treated as a standard likelihood equation. The posterior distribution of parameter estimates obtained over many EM cycles and after a suitable burn-in period is approximately centered at the maximum likelihood estimate, and the mean of the distribution can be used as an ML estimate (Celeux and Diebolt 1985). This posterior distribution can also be plotted for parameters of interest. Also, a preliminary short stage of stochastic EM can be run to get starting values for Monte Carlo EM.

9.7.4 An EM Algorithm/Newton-Raphson Hybrid

For some QTL parameters being estimated using the EM algorithm, the maximum likelihood equations obtained by setting the scores to zero may not be easily solvable because of complex nonlinear structures. These parameters can be estimated by incorporating other numerical methods. The most direct method is simply to evaluate the likelihood for a range of values of the parameters to be estimated and choose the value that gives the largest likelihood. This *grid search* procedure can be made quite sophisticated–the range of parameter values could be divided into tenths, and then the tenth that appears to contain the maximum likelihood estimate itself could be divided into ten parts. As many rounds of the search could be performed as significant digits are required in the solution. It is helpful to plot the likelihood to show how it is responding to changes in parameter values and to guard against a local maximum being confused with a global maximum.

An alternative approach is given by Newton-Raphson iterations, the incorporation of which into the EM algorithm forms a *hybrid* method for estimating those parameters expressed in nonlinear log-likelihood equations. Briefly, some initial value is chosen for the estimate and then this value is modified by using the score function. The modified value is modified in turn, and the process continues until successive iterates differ by less than some specified amount (see Weir 1996 for an excellent description).

For a parameter φ, denote the required MLE as $\hat{\varphi}$ and the initial value, or guess, as φ'. The score S_φ, the derivative of the log-likelihood with respect to φ, is to be zero at $\hat{\varphi}$, and that score can be expanded by Taylor's theorem as

$$S_{\hat{\varphi}} = 0 = S_{\varphi'} + (\hat{\varphi} - \varphi') \left[\frac{\partial S_{\varphi}}{\partial \varphi} \right]_{\varphi = \varphi'},$$

where higher-order terms in $(\hat{\varphi} - \varphi')$ are ignored. Rearranging this expression provides an approximation value φ'' for $\hat{\varphi}$,

$$\varphi'' = \varphi' - S_{\varphi'} / \left[\frac{\partial S_{\varphi}}{\partial \varphi} \right]_{\varphi = \varphi'} = \varphi' + S_{\varphi'}/I(\varphi'),$$

where $I(\varphi) = \left[\frac{\partial S_{\varphi}}{\partial \varphi} \right]_{\varphi = \varphi'}$ is the *information matrix*.

The initial value φ' is modified by adding to it the score divided by the information, with both evaluated at the initial value. The new value then serves as an initial value for a further modification,

$$\varphi''' = \varphi'' + S(\varphi'')/I(\varphi''),$$

and the iteration continues until convergence.

Since the information $I(\varphi)$ is typically a matrix, the iteration procedure requires matrix inversion:

$$\varphi'' = \varphi' + I^{-1}(\varphi')S(\varphi').$$

Obviously, the method breaks down if the information is zero, or the information matrix is singular.

9.7.5 Some Cautions

There can sometimes be problems when the maximum likelihood method is used for QTL mixture models. First, for some choices of parametric families, the likelihood can be unbounded. Second, in complex situations, the likelihood function can have many local maxima, each of which may give different (and possibly reasonable) plug-in estimates for quantities of interest. In these cases, it could be difficult to choose one of these point estimates of the parameters above the others.

Third and most importantly, when a QTL mapping strategy is incorporated by population genetic properties of genes, we will encounter an increased dimensional space for the unknown parameters. Although the mere existence of a high-dimensional parameter space is not necessarily detrimental, extra care must be taken in searching for the ML estimator. An extra complication (not only of ML) is that the uncertainty about the actual number of QTLs for a quantitative trait results in extra difficulty in model fitting and selection.

9.7.6 Bayesian Methods

The use of a Bayesian approach can avoid some of the problems related to the maximum likelihood approach. In maximum likelihood, the unknown parameters are treated as unknown variables (unobservables) and the likelihood function is maximized in these variables. In the Bayesian paradigm, each unobservable parameter is given a prior distribution, and we then infer the posterior distribution of each unobservable conditional on the data (the observables). The summary statistics of the posterior distribution (e.g., the mean, the mode, or the median) can be regarded as Bayesian estimates of unobservables (Carlin and Louis 1998). An interval estimate can be obtained by examining the posterior distribution.

Using the same notation as in the maximum likelihood analysis, denote the observables by y (phenotype data) and $\mathbf{M}$ (marker data) and the unobservables by θ (model parameters), d (QTL locations), g (QTL genotypes) and $\mathbf{\Omega}_m = (\mu_1, \cdots, \mu_J, \sigma^2)$, specifying the QTL effects and residual variance. The posterior distribution of the unobservables is given by

$$\begin{aligned}\pi(d, \mathbf{\Omega}_m, g|y, \mathbf{M}) &= \frac{\pi(y, \mathbf{M}, d, \mathbf{\Omega}_m, g)}{\pi(y, \mathbf{M})} \\ &= \frac{p(y, \mathbf{M}|g, \mathbf{\Omega}_m)\pi(g|d)\pi(\mathbf{d}, \mathbf{\Omega}_m)}{\pi(y)} \\ &\propto \pi(y, \mathbf{M}|g, \mathbf{\Omega}_m)\pi(g|d)\pi(d, \mathbf{\Omega}_m),\end{aligned} \tag{9.18}$$

where $\pi(\cdot)$ is a generic expression for a probability density, $p(y, \mathbf{M}|g, \mathbf{\Omega}_m)$ is the data probability mass given the QTL genotypes, $\pi(g|d)$ is the probability mass of the QTL genotypes of all the observations given their locations d, and $\pi(d, \mathbf{\Omega}_m)$ is the prior probability distribution of the unobservables. Because the denominator is not a function of the parameters, it can be ignored.

We assume prior independence of the parameters:

$$\pi(d, \mathbf{\Omega}_m) = \pi(d)\pi(\mu)\pi(\sigma^2)\prod_{k=1}^{\dot{n}}[\pi(a_k)\pi(d_k)].$$

Note that k denotes the kth QTL and $\dot{n}$ denotes the number of QTLs rather than the number of QTL genotypes. The extent to which the choice of the prior distribution over the parameter space affects the final inference is a measure of robustness and requires checking in each application. The prior distribution could be chosen based on related studies or information from the literature. In general, when no information regarding the locations is available, a natural choice for the prior of d is the uniform distribution. Specifying a conjugate prior for μ, a_k, d_k, and σ^2 makes its form simple, while increasing diffuseness decreases the influence of the prior.

In the Bayesian approach, we infer the genetic parameters based on their marginal posterior distribution, which can be obtained from the joint posterior (9.18) by integrating over the other unknowns. We partition the vector $\mathbf{\Omega} = (d, g, \mathbf{\Omega}_m)$ into $\mathbf{\Omega} = [\mathbf{\Omega}_l \ \mathbf{\Omega}_{-l}]$, where $\mathbf{\Omega}_l$ is a single element of the unobservables and $\mathbf{\Omega}_{-l}$ is the rest of the unobservables that exclude $\mathbf{\Omega}_l$. The marginal posterior distribution of $\mathbf{\Omega}_l$ is expressed by

$$\Pi(\mathbf{\Omega}_l|y) = \int \pi(\mathbf{\Omega}_l, \mathbf{\Omega}_{-l}|y)d\mathbf{\Omega}_{-l} \propto \int \pi(y|\mathbf{\Omega}_l, \mathbf{\Omega}_{-l})\pi(\mathbf{\Omega}_l, \mathbf{\Omega}_{-l})d\mathbf{\Omega}_{-l}. \tag{9.19}$$

The mean of this marginal posterior distribution is a candidate Bayesian estimator of $\mathbf{\Omega}_l$. This marginal distribution rarely has an explicit form, and numerical integration is often prohibitive because the dimensionality of $\mathbf{\Omega}_{-l}$ may be high. But the distribution (9.19) can be approximated by constructing it using a Markov chain Monte Carlo algorithm (such as a Gibbs sampler) over the product space of the parameters.

The Markov chain is a random sequence of states

$$\{(d^0, g^0, \mathbf{\Omega}_m^0), (d^1, g^1, \mathbf{\Omega}_m^1), \ldots, d^N, g^N, \mathbf{\Omega}_m^N)\}$$

started at an arbitrary point $(d^0, g^0, \mathbf{\Omega}_m^0)$ having positive posterior density and proceeding by simple rules that modify the three unknowns d, g, and $\mathbf{\Omega}_m$. Each step in this chain is a cycle of three smaller steps, first updating d, then g, followed by $\mathbf{\Omega}_m$. The step updating d is typically a Metropolis-Hastings step (Hastings 1970), while the steps updating g and $\mathbf{\Omega}_m$ are Gibbs sampler steps (Geman and Geman 1984). See Robert and Casella (2004) for a full treatment of these Markov chain Monte Carlo (MCMC) algorithms.

The MCMC algorithm is justified by the fact that if h is any function of the unknowns that is square-integrable with respect to the equilibrium distribution π, then

$$\begin{aligned} \overline{h}_N = \frac{1}{N}\sum_{t=1}^{N} h(d^t, g^t, \mathbf{\Omega}_m^t) \to E_\pi[h(d, g, \mathbf{\Omega}_m)|y], \\ \text{almost surely as } N \to \infty, \end{aligned} \tag{9.20}$$

where $(d^t, g^t, \mathbf{\Omega}_m^t)$ are samples from the Markov chain. The empirical means of the MCMC samples from equation (9.20) can be used to obtain a Bayes estimator of the unknown parameters $(d, g, \mathbf{\Omega}_m)$. The marginal posterior densities of the parameters can also be obtained from the sample values.

9.7.7 Estimating the Number of Components in a Mixture Model

In many practical situations, the number of QTL genotypes (and therefore the number of QTLs) in a mixture model is unknown. The estimation of the number of QTLs included in a quantitative trait is an important issue in quantitative genetic research. Statistically, testing the number of QTLs is equivalent to testing the number of components in a mixture distribution. This difficult problem has not been well solved yet in the statistical literature (see the review by Lo et al. 2001).

From the viewpoint of a likelihood analysis, the determination of the number of components in a mixture model can be based on a formal test of the null hypothesis that a random sample is from a J_0-component mixture versus the alternative hypothesis that the sample is from a J_1-component mixture, where $J_1 > J_0$. It is tempting to use the likelihood ratio test with an asymptotic chi-squared null reference distribution. However, the classic asymptotic chi-squared theory does not hold for the likelihood ratio test in the context of mixtures. Lo et al. (2001) suggested the use of a likelihood ratio test procedure based on the Kullback-Leibler information criterion with a theorem proposed by Vuong (1989). These authors showed that, under certain regularity conditions, the limiting distribution of the likelihood ratio statistic is a weighted sum of independent χ_1^2 random variables when the competing models are nested or overlapping and a normal distribution when the competing models are nonnested.

An alternative approach for detecting the number of QTLs is based on Bayesian model selection criteria. This approach uses the samples from Markov chains to calculate Bayes factor (Jeffreys 1961; Kass and Raftery 1995).

Let us compare two models, M_1 and M_2, that are postulated to fit two different numbers of QTLs within a mixture model context. The posterior odds in favor of M_1 over M_2 for data y can be expressed, using Bayes' theorem, as

$$\frac{\Pr(M_1|y)}{\Pr(M_2|y)} = \frac{\Pr(M_1)}{\Pr(M_2)} \frac{\Pr(y|M_1)}{\Pr(y|M_2)},$$

where the first factor is the prior odds and

$$B = \frac{\Pr(M_1|y)}{\Pr(M_2|y)}$$

is the ratio of marginal probabilities of y given the two models and is called the *Bayes factor*. Newton and Raftery (1994) and Kass and Raftery (1995) discuss approaches to estimate the marginal probability of the data under any two given models.

Green (1995) and Richardson and Green (1997) proposed a reversible jump MCMC method for simulating the posterior distribution of the number of components. This method has been applied to estimate the number of QTL in a mapping experiment (Stephens and Fisch 1998). Reversible jump methods allow the construction of an ergodic Markov chain with the joint posterior distribution of the parameters and the model as its stationary distribution. Moves between models are achieved by periodically proposing a move to a different model and rejecting it with appropriate probability to ensure that the chain possesses the required stationary distribution. Ideally, these proposed moves are designed to have a high probability of acceptance, so that the algorithm explores the different models adequately, though this is not always easy to achieve in practice.

9.8 Exercises

9.1 What are three major tasks for mixture-model–based QTL mapping?

9.2 Referring to Example 9.1, use equation (9.8) and the data in Table 9.3 to verify the effect estimates given in the example. Use both **D** matrices, doing the calculations for both the marginal means and the joint means.

9.3 Similar to Exercise 9.2, verify the values of the effect estimates given in Example 9.2.

9.4 How can you determine the number of QTLs for a complex trait in the mixture model context?

10

Interval Mapping with Regression Analysis

10.1 Introduction

The genetic analysis of quantitative traits includes two major tasks: (1) identifying the location of QTLs affecting a quantitative trait using a genetic linkage map constructed from molecular markers, and (2) estimating the genetic effects of the QTLs on the phenotype. If the genotypes of a putative QTL were known for all individuals, its genomic location could be readily determined using a marker linkage analysis. Furthermore, the genetic effects of the QTL could be precisely estimated and tested by simple t tests or ANOVA. However, it is not possible for the genotypes of QTLs to be directly observed; instead they should be inferred from observed marker and phenotypic information. As was seen in Chapter 8, a marker analysis cannot unambiguously separate the genetic effects of a QTL from the recombination fraction between the markers and QTL.

To rule out the genetic effect and position of a QTL, a more advanced statistical analysis should be adopted. The central idea of individually estimating the QTL effect and position is to formulate a statistical model for observed marker and phenotypic data in terms of the underlying QTL that is located between two flanking markers. This so-called *interval mapping* approach can overcome the confounding problem of the marker–QTL recombination fraction and QTL effects through the conditional probabilities of unknown QTL genotypes given observed marker genotypes. The QTL and the two markers that bracket it allow the application of three-point analysis to derive the conditional probabilities for a particular segregating population, such as the backcross, RIL, or F_2.

Models for interval mapping can be formulated within the context of regression or maximum likelihood (ML) theories. Through analytical and numerical approaches, Kao (2000) compared the differences between regression- and ML-based interval mapping models in terms of the mean squared errors of parameter estimates and the power of QTL detection. These differences increase when the following are truce: (1) there is a large proportion of the variance explained by the QTL, (2) the QTL is located in the middle of the marker interval, (3) the marker interval is wide, (4) there is epistasis between different QTLs, (5) the QTLs differ in their effects, and (6) different QTLs are linked. In practice, the regression method is computationally more efficient than the ML method, especially when the number of QTLs considered is large.

In this chapter, we will introduce the principles of the regression-based interval mapping methods, their statistical inferences about parameter estimation and hypothesis testing, and the procedure for their practical application. Examples will be used to demonstrate the usefulness of QTL mapping by a regression analysis. In Chapter 11, the approaches for QTL interval mapping based on maximum likelihood will be described.

10.2 Linear Regression Model

A QTL linear model conditional on marker interval genotypes for the backcross can be expressed as

$$y_i = \mu + x_{j|i}a + e_i, \tag{10.1}$$

where a is the true effect of the QTL, and $x_{j|i}$ is the indicator variable that is defined as the conditional probability of QTL genotype j given the marker genotype of progeny i. According to equation (10.1), as long as the indicator variable is determined, the QTL effect (a) can be estimated.

The conditional regression model for the F_2 should be formulated as

$$y_i = \mu + x_{1j|i}a + x_{2j|i}d + e_i, \tag{10.2}$$

where $x_{1j|i}$ and $x_{2j|i}$ are the indicator variables that specify conditional probabilities of QTL additive and dominance genetic effects given the marker interval genotype of progeny i.

The regression approach for mapping QTLs is to regress the phenotypic values of a quantitative trait on the conditional expected genotypic values and estimate the unknown parameters by using a classic least squares approach. The conditional expected genotypic values associated with each marker genotype are calculated from the conditional probabilities of the QTL genotypes given a marker genotype and from the genotypic values of different QTL genotypes.

10.3 Interval Mapping in the Backcross

10.3.1 Conditional Probabilities

While Table 9.1 is a QTL table and Table 8.1 is a marker table for a backcross population with ten mice, Table 10.1 integrates information from the two tables. Suppose there is a putative QTL that is bracketed by two linked markers **M** and **N**. For an observable marker genotype, there are two possibilities to carry a QTL genotype, Qq (1) or qq (0). Table 10.1 also provides possible marker-QTL-marker

Table 10.1. QTL genotypes, marker interval genotypes, joint marker-QTL genotypes and mouse body weight in a backcross design.

Mouse	QTL (z_i)	Interval **M-N** (x_i^*)	**M**-QTL-**N** Genotype	Prob(QTL|interval) ($x_{j\|i}$)	Body Weight (y_i)
1	1	11	111	$\omega_{1\|11}$	30
			101		
2	1	11	111	$\omega_{1\|11}$	32
			101		
3	1	11	111	$\omega_{1\|11}$	28
			101		
4	1	11	111	$\omega_{1\|11}$	29
			101		
5	1	10	110	$\omega_{1\|10}$	29
			100		
6	0	01	001		22
			011	$\omega_{0\|01}$	
7	0	00	000		20
			010	$\omega_{0\|00}$	
8	0	00	000		21
			010	$\omega_{0\|00}$	
9	0	00	000		20
			010	$\omega_{0\|00}$	
10	0	00	000		21
			010	$\omega_{0\|00}$	

Note: The first column stands for QTL genotypes, the second column stands for interval genotypes for two markers **M** and **N** that bracket the QTL, and the third column stands for three-point (marker-QTL-marker) genotypes.

genotypes. Let $\omega_{1|i}$ and $\omega_{0|i}$ be the conditional probabilities of two QTL genotypes given a two-marker genotype for mouse i.

The values of $\omega_{1|i}$ and $\omega_{0|i}$ depend on the recombination fractions between the two markers (r), between marker **M** and QTL (r_1), and between QTL and marker **N** (r_2). A triply heterozygous F_1 backcrossed to a parent will generate eight three-point (marker-QTL-marker) genotypes: 111, 101, 110, 100, 011, 001, 010, and 000. As shown by three-point analysis in Chapter 4, these genotype frequencies are derived under the assumption of no double crossover and are expressed in Table 10.2. Thus, the conditional probabilities of the QTL genotypes given the marker genotypes of the interval in the backcross can be derived according to Bayes' theorem, and are expressed in Table 10.3.

If the two markers are highly linked, we make the simplifying assumption that

$$r_1 + r_2 \approx r, \tag{10.3}$$

which follows from the fact that $r = r_1 + r_2 - 2r_1r_2$ and the assumption that $r_1r_2 \approx 0$ and Table 10.3 can be approximated by Table 10.4. For three loci in the order **M**-QTL-**N**, we use g_{11} (see Section 3.6) to denote the frequency of double recombinations (a recombination in each of the two intervals **M**-QTL and QTL-**N**). As shown by equation (3.19), $g_{11} = \frac{1}{2}(r_1 + r_2 - r)$. Thus, if there is no double recombination, we have equation (10.3). In other words, for a highly dense map, we can assume that no double recombinations occur between the adjacent intervals.

Table 10.2. Joint marker-QTL-marker genotype frequencies in the backcross.

Marker Interval		QTL Genotype	
Genotype	Frequency	1	0
11	$\frac{1}{2}(1-r)$	$\frac{1}{2}(1-r_1)(1-r_2)$	$\frac{1}{2}r_1r_2$
10	$\frac{1}{2}r$	$\frac{1}{2}(1-r_1)r_2$	$\frac{1}{2}r_1(1-r_2)$
01	$\frac{1}{2}r$	$\frac{1}{2}r_1(1-r_2)$	$\frac{1}{2}(1-r_1)r_2$
00	$\frac{1}{2}(1-r)$	$\frac{1}{2}r_1r_2$	$\frac{1}{2}(1-r_1)(1-r_2)$

10.3.2 Conditional Regression Model

By substituting the conditional probabilities of Table 10.3 or Table 10.4 into Table 10.1, we can construct the dependent variable $x_{j|i}$ for conditional regression model (10.1). In general, for the backcross of size n, we write equation (10.1) in matrix notation as

$$\mathbf{y} = \mathbf{X}\mathbf{b} + \mathbf{e}, \tag{10.4}$$

Table 10.3. Conditional probabilities of QTL genotypes given the marker interval genotype in the backcross.

Marker Interval	QTL Genotype	
Genotype	1	0
11	$\frac{(1-r_1)(1-r_2)}{1-r}$	$\frac{r_1 r_2}{1-r}$
10	$\frac{(1-r_1)r_2}{r}$	$\frac{r_1(1-r_2)}{r}$
01	$\frac{r_1(1-r_2)}{r}$	$\frac{(1-r_1)r_2}{r}$
00	$\frac{r_1 r_2}{1-r}$	$\frac{(1-r_1)(1-r_2)}{1-r}$

Table 10.4. Approximate conditional probabilities of QTL genotypes given the marker interval genotype in the backcross, assuming no double recombination.

Marker Interval	QTL Genotype	
Genotype	1	0
11	1	0
10	$1-\theta$	θ
01	θ	$1-\theta$
00	0	1

Note: The ratio $\theta = r_1/r$ and $1-\theta = r_2/r$.

where $\mathbf{y} = (y_i)_{n\times 1}$, $\mathbf{b} = (\mu,\ a)^{\mathrm{T}}$, $\mathbf{e} = (e_i)_{n\times 1}$, and

$$\mathbf{X} = \begin{pmatrix} \mathbf{X}_1 \\ \mathbf{X}_2 \\ \mathbf{X}_3 \\ \mathbf{X}_4 \end{pmatrix}_{n\times 2},$$

with, when matrix (10.3) is considered,

$$\begin{aligned}
\mathbf{X}_1 &= \begin{pmatrix} 1 & \frac{(1-r_1)(1-r_2)}{1-r} \end{pmatrix}_{n_1\times 1}, \\
\mathbf{X}_2 &= \begin{pmatrix} 1 & \frac{(1-r_1)r_2}{r} \end{pmatrix}_{n_2\times 2}, \\
\mathbf{X}_3 &= \begin{pmatrix} 1 & \frac{r_1(1-r_2)}{r} \end{pmatrix}_{n_3\times 1}, \\
\mathbf{X}_4 &= \begin{pmatrix} 1 & \frac{r_1 r_2}{1-r} \end{pmatrix}_{n_4\times 1},
\end{aligned}$$

where n_1, n_2, n_3, and n_4 are the sample sizes for four different marker genotypes, 11, 10, 01, and 00, respectively. In the specific example in Table 10.1, these observations are 4, 1, 1, and 4, respectively.

10.3.3 Estimation and Test

If r_1 or r_2 is known, it would be possible to substitute these values into equation (10.3) and then solve it as a simple linear regression with μ as the y-intercept and a as the slope. In fact, even if r_1 or r_2 is unknown, we can compute the design matrix $\mathbf{X}$ by assuming the position of a QTL at several positions (e.g., every 1 or 2 cM) between the two markers.

Note that when a QTL is scanned at every 1 or 2 cM from marker $\mathbf{M}$ to $\mathbf{N}$, we need to use a map function to convert the map distance to the recombination fraction. Given a point x, we have

$$r_1(x) = \frac{1}{2}(1 - e^{-2d(x)}), \; r = \frac{1}{2}(1 - e^{-2d}),$$

and

$$\theta = \frac{r_1(x)}{r},$$

where the Haldane map function is assumed and $d(x)$ and d are the map distances (in Morgans) between the left marker $\mathbf{M}$ and QTL and between the two markers, respectively (Fig. 10.1).

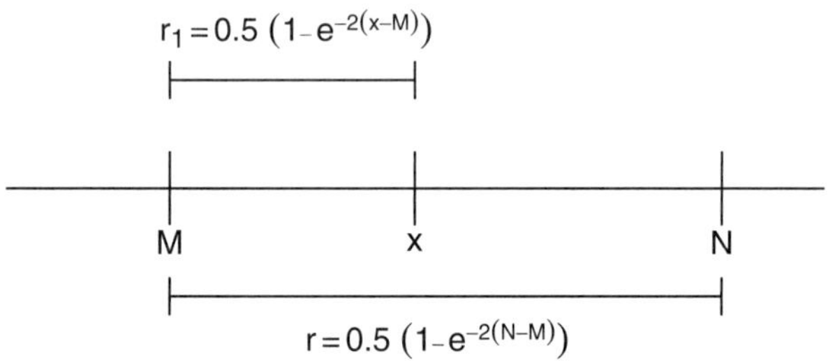

Fig. 10.1. Illustration of QTL interval mapping based on two flanking markers $\mathbf{M}$ and $\mathbf{N}$.

Linear regression is then used to fit μ and a for each assumed QTL position. This provides the least squares estimates of the vector $\mathbf{b}$ and the residual variance with

$$\hat{\mathbf{b}} = (\mathbf{X}'\mathbf{X})^{-1}(\mathbf{X}'\mathbf{y}), \tag{10.5}$$

$$\hat{\sigma}^2 = \frac{1}{n}(\mathbf{y} - \mathbf{X}\mathbf{b})^{\mathrm{T}}(\mathbf{y} - \mathbf{X}\mathbf{b}), \tag{10.6}$$

as well as giving regression and residual sums of squares and mean squares to allow the calculation of the regression variance F ratio and thus a test for a. The position that gives the best-fitting model (i.e., produces the smallest residual mean square) gives the most likely position of a QTL and the best estimate of its effect.

More specifically, the hypothesis test for the existence of a QTL at a given position within a marker interval (i.e., the significance of $\hat{a}$) can be formulated by

$$H_0 : a = 0 \text{ vs. } H_1 : a \neq 0.$$

The model under H_1 is a full one with a QTL, expressed as

$$\hat{y}_i = \hat{\mu} + x_{j|i}\hat{a},$$

whereas the model under H_0 is a reduced one without a QTL, expressed as

$$\tilde{y}_i = \tilde{\mu}.$$

The total sum of squares (SST) is the sum of $(y_i - \tilde{\mu})^2$, whereas the residual sum of squares (SSE) is the sum of $(y_i - \hat{\mu} - x_{j|i}\hat{a})^2$. To test H_0, we can construct an F test statistic as

$$F = \frac{(\text{SST} - \text{SSE})/(2-1)}{\text{SSE}/(n-2)}. \tag{10.7}$$

By comparing the F value with $F_{0.05,(1,n-2)}$, the existence of a QTL can be tested.

Example 10.1. Revisit Example 3.1. A DH population that is equivalent to a backcross was founded using two inbred lines, semi-dwarf IR64 and tall Azucena (Huang et al. 1997). This DH population contains 123 lines, each genotyped for 135 RFLP and 40 isozyme and RAPD markers and phenotyped for various phenotypic traits. In this example, the phenotype chosen for QTL mapping is plant height measured at 10 weeks after rice was transplanted to the field. Arranging the marker and phenotypic data in the form shown by Table 8.1, we wish to use LS-based least squares regression approaches to map a QTL for plant height using a genetic linkage map constructed from these genotyped markers (Fig. 3.3).

Figure 10.2 illustrates the profile of the F-values, calculated by equation (10.7), at different positions across a linkage group of chromosome 1 constructed by 18 markers. There is a clear peak at 199 cM in the marker interval [RZ730–RZ801], whose F-value, 88.02, is largely beyond the critical $F_{0.05,(1,85)} = 3.95$ (87 rice left due to missing data in these two markers). This suggests the existence of a significant QTL around that position.

The LS estimates of the model parameters, μ, a, and σ^2, are obtained at the peak of the F-value profile. They are $\hat{\mu} = 86.9$, $\hat{a} = 42.9$, and $\hat{\sigma}^2 = 268.6$. The genetic variance of plant height due to this detected QTL is calculated as $\sigma_g^2 = 271.8$ with equation (8.4). This allows the calculation of the proportion of the phenotypic variance explained by this QTL ($R^2 = 0.50$) with equation (8.5).

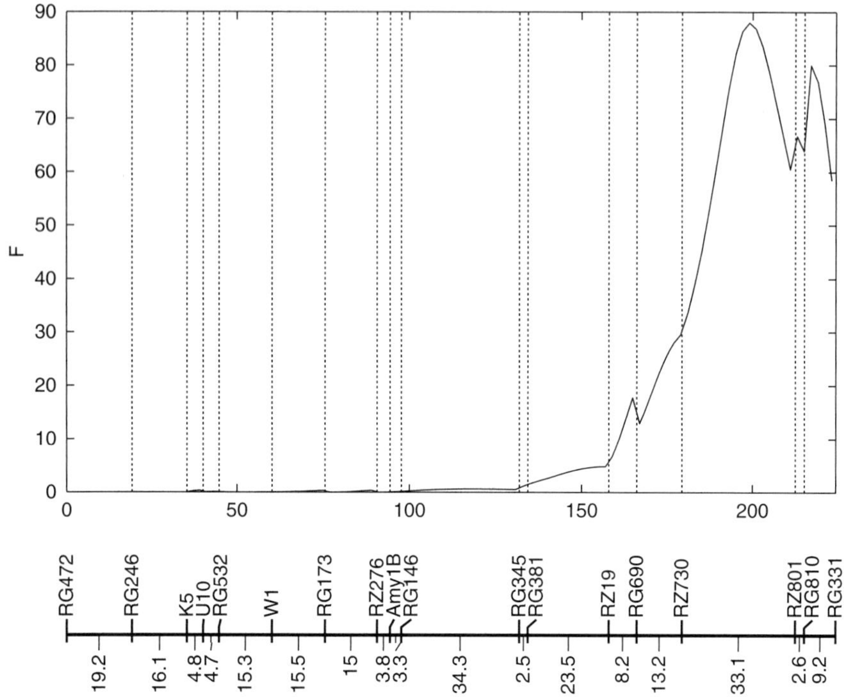

Fig. 10.2. Profile of the F-value across chromosome 1 for the test of QTL that controls plant height at age 10 weeks in a rice DH population (Huang et al. 1997). The marker names and distances are given below the profile.

10.4 Interval Mapping in an F_2

A similar strategy can also be used to map a QTL segregating in an F_2 population based on equation (10.2). As shown for the backcross, we need to derive the conditional probabilities of QTL genotypes given marker intervals. Table 10.2 gives the frequency matrix for three-point (marker-QTL-marker) genotypes for the backcross, which is equivalent to one for three-point gamete genotypes generated by the double heterozygote F_1. Since the F_2 is derived from the combination of F_1 gametes, the frequency matrix for three-point genotypes in the F_2 is the Krockner product of two matrices of (10.2), each for a different parent with the same genotypes collapsed to become a (9×3) matrix. The same genotypes may be produced by different gamete combinations. For example, marker-QTL-marker genotype 12/12/12 in the F_2 can be produced by four different combinations:

$$12/12/12 = \begin{cases} [111] \times [222] \\ [121] \times [212] \\ [112] \times [221] \\ [211] \times [122] \end{cases}.$$

The joint probability of this triply heterozygous genotype is thus the sum of the probabilities of these four combinations. A collapsed (9×3) matrix is expressed in Table 10.5.

Table 10.5. Joint marker-QTL-marker genotype frequencies in the F_2.

Marker Interval		QTL Genotype		
Geno-type	Frequency	2	1	0
22	$\frac{1}{4}(1-r)^2$	$\frac{1}{4}(1-r_1)^2(1-r_2)^2$	$\frac{1}{2}r_1r_2(1-r_1)(1-r_2)$	$\frac{1}{4}r_1^2r_2^2$
21	$\frac{1}{2}r(1-r)$	$\frac{1}{4}r_2(1-r_1)^2(1-r_2)$	$\frac{1}{2}r_1(1-r_1)(1-2r_2+2r_2^2)$	$\frac{1}{4}r_1^2r_2(1-r_2)$
20	$\frac{1}{4}r^2$	$\frac{1}{4}(1-r_1)^2r_2^2$	$\frac{1}{2}r_1r_2(1-r_1)(1-r_2)$	$\frac{1}{4}r_1^2(1-r_2)^2$
12	$\frac{1}{2}r(1-r)$	$\frac{1}{4}r_1(1-r_1)(1-r_2)^2$	$\frac{1}{2}r_2(1-r_2)(1-2r_1+2r_1)^2$	$\frac{1}{4}r_1(1-r_1)r_2^2$
11	$\frac{1-2r+2r^2}{2}$	$\frac{1}{4}r_1r_2(1-r_1)(1-r_2)$	$\frac{1}{2}(1-2r_1+2r_1^2)$ $(1-2r_2+2r_2^2)$	$\frac{1}{4}r_1r_2(1-r_1)(1-r_2)$
10	$\frac{1}{2}r(1-r)$	$\frac{1}{4}r_1(1-r_1)r_2^2$	$\frac{1}{2}r_2(1-r_2)(1-2r_-2r_1^2)$	$\frac{1}{4}r_1(1-r_1)(1-r_2)^2$
02	$\frac{1}{4}r^2$	$\frac{1}{4}r_1^2(1-r_2)^2$	$\frac{1}{2}r_1r_2(1-r_1)(1-r_2)$	$\frac{1}{4}(1-r_1)^2r_2^2$
01	$\frac{1}{2}r(1-r)$	$\frac{1}{4}r_1^2r_2(1-r_2)$	$\frac{1}{2}r_1(1-r_1)(1-2r_2+2r_2^2)$	$\frac{1}{4}r_2(1-r_1)^2(1-r_2)$
00	$\frac{1}{4}(1-r)^2$	$\frac{1}{4}r_1^2r_2^2$	$\frac{1}{2}r_1r_2(1-r_1)(1-r_2)$	$\frac{1}{4}(1-r_1)^2(1-r_2)^2$

From Table 10.5, the coefficients associated with the additive effect (a) and dominant effect (d) of the QTL in the F_2 (Table 10.6), which are the explanatory variables, can be derived in terms of the genomic position of a QTL described by the recombination fraction between the QTL and markers. Because the QTL position is unknown, a grid search strategy, in which a QTL is assumed at every position between two markers, is used to regress the trait value on the explanatory variable to calculate μ, a, and d and the F–statistic for the significance of the regression and identify the position with the largest F-value for the regression as the most likely position for a QTL.

Example 10.2. Revisit Example 3.2. Cheverud et al. (1996) constructed a linkage map using 75 microsatellite markers in a population of 535 F_2 progeny derived from two

Table 10.6. Coefficients for the additive (a) and dominant effects (d) of a QTL for all possible flanking marker genotypes in the F_2.

Marker Genotype	Coefficients of a	Coefficients of d
22	$\frac{(1-r_1)^2(1-r_2)^2 - r_1^2 r_2^2}{(1-r)^2}$	$\frac{2r_1(1-r_1)r_2(1-r_2)}{(1-r)^2}$
21	$\frac{(1-r_1)^2 r_2(1-r_2) - r_1^2 r_2(1-r_2)}{r(1-r)}$	$\frac{r_1(1-r_1)(1-r_2)^2 + r_1(1-r_1)r_2^2}{r(1-r)}$
20	$\frac{(1-r_1)^2 r_2^2 - r_1^2(1-r_2)^2}{r^2}$	$\frac{2r_1(1-r_1)r_2(1-r_2)}{r^2}$
12	$\frac{r_1(1-r_1)(1-r_2)^2 - r_1(1-r_1)r_2^2}{r(1-r)}$	$\frac{(1-r_1)^2 r_2(1-r_2) + r_1^2 r_2(1-r_2)}{r(1-r)}$
11	0	$\frac{(1-2r_1+r_1^2)(1-2r_2+r_2^2)}{r^2+(1-r)^2}$
10	$\frac{r_1(1-r_1)r_2^2 - r_1(1-r_1)(1-r_2)^2}{r(1-r)}$	$\frac{(1-r_1)^2 r_2(1-r_2) + r_1^2 r_2(1-r_2)}{r(1-r)}$
02	$\frac{r_1^2(1-r_2)^2 - (1-r_1)^2 r_2^2}{r^2}$	$\frac{2r_1(1-r_1)r_2(1-r_2)}{r^2}$
01	$\frac{r_1^2 r_2(1-r_2) - (1-r_1)^2 r_2(1-r_2)}{r(1-r)}$	$\frac{r_1(1-r_1)(1-r_2)^2 + r_1(1-r_1)r_2^2}{r(1-r)}$
00	$\frac{r_1^2 r_2^2 - (1-r_1)^2(1-r_2)^2}{(1-r)^2}$	$\frac{2r_1(1-r_1)r_2(1-r_2)}{(1-r)^2}$

strains, the Large (LG/J) and Small (SM/J). The F_2 progeny were measured for body mass at 10 weekly intervals starting at age 7 days. The raw weights were corrected for the effects of each covariate due to dam, litter size at birth and parity but not for the effect due to sex. We use chromosome 1 composed of nine markers to map a QTL affecting body weight at age 10 weeks with LS-based regression approaches.

Figure 10.3 illustrates the profile of the F-values calculated by equation (10.7) at different positions across a linkage group of chromosome 1 constructed by 18 markers. There is a clear peak at 46 cM in the marker interval [D2MIT389–D2MIT17], whose F-value, 16.434, is largely beyond the critical $F_{0.05,(1,451)} = 3.862$. This suggests the existence of a significant QTL around that position.

The LS estimates of the model parameters, μ, a, d, and σ^2, are obtained at the peak of the F-value profile. They are $\hat{\mu} = 23.0$, $\hat{a} = 1.6$, $\hat{d} = 1.4$, and $\hat{\sigma}^2 = 21.7$. The genetic variance of plant height due to this detected QTL is calculated as $\sigma_g^2 = 1.5$ with equation (8.4). This allows the calculation of the proportion of the phenotypic variance explained by this QTL ($R^2 = 0.06$) with equation (8.5).

Compared with the backcross, the F_2 population is more informative because it allows the significance test of both the additive and dominance effects of a detected QTL. The full model with the additive effect is written as

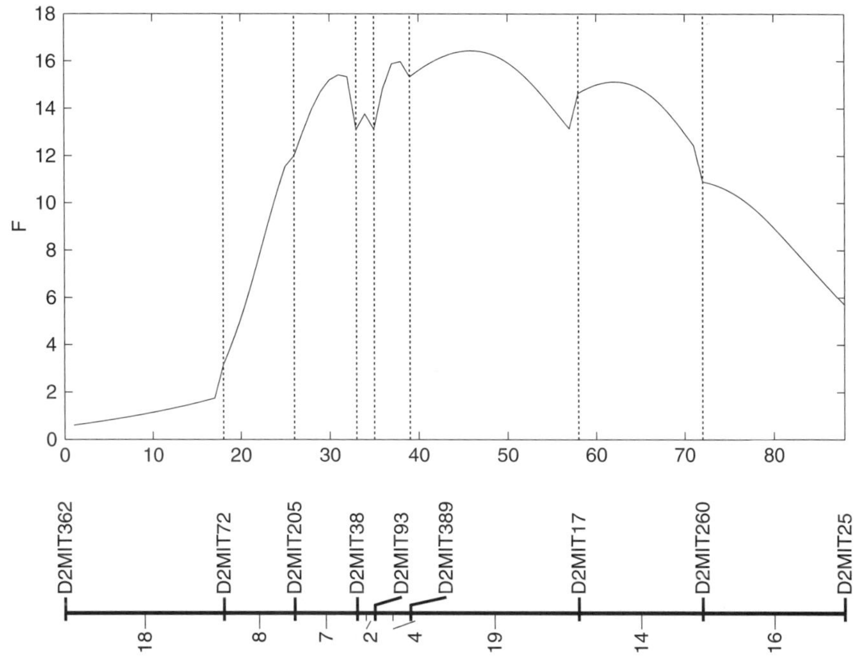

Fig. 10.3. Profile of the F-value across chromosome 1 for the test of a QTL that controls body weight at age 10 weeks in a mouse F_2 population (Cheverud et al. 1996). The marker names and distances are given below the profile.

$$\hat{y}_i = \hat{\mu} + x_{1j|i}\hat{a} + x_{2j|i}\hat{d},$$

whereas the reduced model with no additive effect is written as

$$\tilde{y}_i = \tilde{\mu} + x_{2j|i}\tilde{d}.$$

With these two alternative models, we calculate the total sum of squares (SST) and residual sum of squares (SSE) and ultimately the F–test statistic (see equation (10.7)). The reduced model with no dominant effect is formulated as

$$\tilde{y}_i = \tilde{\mu} + x_{1j|i}\tilde{a},$$

from which the test statistics can be similarly calculated.

10.5 Remarks

The simple regression method of mapping a QTL has been investigated in comparison with the mixture-model–based maximum likelihood method to be described in the next section. No significant difference between the two methods is detected in terms of errors of parameter estimation and statistical power. Although the test statistic profiles show some difference between these two methods, the difference is only detectable at the micro level (Haley and Knott 1992). A main advantage of the regression method is its tremendous computational simplicity and ease of implementation using many available computer statistical packages.

Xu (1995) noted that the estimation of residual variance provided by the regression method is confounded with part of the QTL variance. He later proposed an alternative method, referred to as iteratively reweighted least squares (IRLS), to correct the deficiency of parameter confounding in the regression method (Xu 1998). The IRLS approach retains the properties of simplicity and rapidity of the ordinary regression method. Like the existing regression method, this method can be useful in QTL mapping in conjunction with the permutation tests and construction of confidence intervals by bootstrapping.

10.6 Exercises

10.1 After reading Chapters 8 and 10, compare the advantages of interval mapping over a single marker analysis. You may use an example for either rice or mouse mapping as used in the book to demonstrate these advantages.

10.2 Use Bayes' theorem to derive Table 10.3

10.3 **Missing data in QTL mapping**

Missing-data problems are common in genetic mapping. A commonly used treatment for missing data is simply to drop those individuals with either phenotypes or markers missing. But this should not be the most effective approach. If data are missing completely at random, one may wish to replace missing data by their expected values given the available data at other markers or individuals using an algorithm by Lander and Green (1987). Read the relevant statistical and genetic literature for treating missing-data problems, and figure out how these treatments can be implemented for QTL mapping.

10.4 **Interval mapping with recombinant inbred lines (RILs)** RILs can be derived either by repeated selfing or by repeated brother–sister mating of the progeny from an F_1 cross between two inbred lines. RILs are fixed, with homozygous genotypes 2 and 0, for all genes and can serve a permanent mapping population for multiple uses.

Consider two flanking markers **M** and **N** that bracket a putative QTL in an RIL population. Let R_1 and R_2 be the proportions of recombinant zygotes between marker **M** and QTL and between QTL and marker **N**, respectively. Martin and Hospital (2006) derived the conditional probabilities of QTL genotypes given the marker genotypes in terms of R_1 and R_2 in the selfing RIL population, which are tabulated in Table 10.7.

Let r_1 and r_2 be the recombination fractions between marker **M** and QTL and between QTL and marker **N**, respectively. The relationship between R and r for two loci in a selfing RIL population has been derived by Haldane and Waddington (1931), expressed as

$$\begin{aligned} R_1 &= \frac{2r_1}{1+2r_1}, \qquad r_1 = \frac{R_1}{2(1-R_1)}, \\ R_2 &= \frac{2r_2}{1+2r_2}, \qquad r_2 = \frac{R_2}{2(1-R_2)}. \end{aligned} \tag{10.8}$$

Table 10.7. Conditional probabilities of QTL genotypes given marker genotypes in the selfing RIL population.

Marker Interval		QTL Genotype	
Genotype	Size	2	0
22	n_{22}	$1-\dfrac{R_1R_2(3-2R_1-2R_2)}{2(1-R_1)(1-R_2)}$	$\dfrac{R_1R_2(3-2R_1-2R_2)}{2(1-R_1)(1-R_2)}$
20	n_{20}	$1-\dfrac{2R_1-R_1R_2(3+2R_1-2R_2)}{2R_2+R_1(2-6R_2)}$	$\dfrac{2R_1-R_1R_2(3+2R_1-2R_2)}{2R_2+R_1(2-6R_2)}$
02	n_{02}	$\dfrac{2R_1-R_1R_2(3+2R_1-2R_2)}{2R_2+R_1(2-6R_2)}$	$1-\dfrac{2R_1-R_1R_2(3+2R_1-2R_2)}{2R_2+R_1(2-6R_2)}$
00	n_{00}	$\dfrac{R_1R_2(3-2R_1-2R_2)}{2(1-R_1)(1-R_2)}$	$1-\dfrac{R_1R_2(3-2R_1-2R_2)}{2(1-R_1)(1-R_2)}$

(a) Using f_2 and f_0 to denote the densities of two homozygous QTL genotypes 2 and 0, respectively, write down the mixture-model–based likelihood.
(b) Define the posterior probabilities of QTL genotypes based on the prior conditional probabilities in Table 10.7.
(c) Based on the procedures described for the backcross and F_2, show a detailed computational EM algorithm for estimating the genotypic values, μ_2 and μ_0, and residual variance, σ^2.
(d) It is difficult to derive a closed form for estimating the proportions of recombinant zygotes, R_1 and R_2. Use a grid approach for estimating these proportions. Note $0 \leq R_1, R_2 \leq 0.5$.
(e) After R_1 and R_2 are estimated, use equation (10.8) to estimate the MLEs of the recombination fractions and therefore the location of the putative QTL by a map function.
(f) Test whether the QTL is significant by formulating the null hypothesis

$$H_0 : \mu_2 = \mu_0.$$

The critical threshold for declaring the existence of a QTL can be determined from permutation tests.

11

Interval Mapping by Maximum Likelihood Approach

11.1 Introduction

As discussed in Chapter 10, interval mapping is powerful for the separation of QTL effects and QTL marker linkage. Interval mapping based on a regression approach is computationally faster than a maximum likelihood (ML) approach, with comparable results between the two approaches in some particular cases. Kao (2000) discussed analytically and through simulation studies the conditions under which the regression- and ML-based approaches generate different results. Further comparisons by Mayer (2005) between the two approaches were made in terms of power, accuracy of position and effect estimates, and estimation of the residual variance when multiple linked QTLs are involved in the genetic control of a quantitative trait.

It is, however, recognized that the ML method has several attractive statistical properties, such as consistency and asymptotic efficiency, and therefore it has great potential for the precise estimation of QTL parameters. Furthermore, the ML method has better interpretability than the regression model in terms of the genetic model, suggesting its applicability to practical genetic mapping problems.

In situations with multiple QTLs linked on a similar genomic region, ML-based interval mapping generally outperformed regression interval mapping with regard to the power of QTL detection and the precision of parameter estimation (Mayer 2005). This superiority increases with wider marker intervals and larger population sizes. Also, if linked QTLs are in repulsion, the differences between the two approaches are substantial. The ML approach is regarded as a powerful way to simultaneously map multiple QTL as proposed by Kao et al. (1999).

In this chapter, we will describe basic principles of the ML-based interval mapping method, its statistical inferences about parameter estimation and hypothesis testing, and the procedure for its practical application. Examples for ML interval mapping are provided. All of the current literature on interval mapping assumes no double recombination or no meiotic interference in different intervals between markers and QTLs. We will describe the QTL interval mapping that relaxes these two assumptions for both a backcross and F_2 design.

11.2 QTL Interval Mapping in a Backcross

The mixture model is the central theme of maximum-likelihood–based interval mapping proposed by Lander and Botstein (1989). In Chapter 9, we described a general framework for the structure of a mixture mapping model from population and quantitative genetic points of view and further reviewed statistical algorithms used to estimate QTL parameters specified in the mixture model. Under this framework, interval mapping of QTLs in an inbred line cross can be viewed as a special case in which statistical issues about significance tests and parameter estimation have been well explored. We will focus our discussion on two simple genetic designs, backcross and F_2 or recombinant inbred lines (RILs), initiated with two contrasting inbred lines.

Consider a possible QTL with left flanking marker **M** and right flanking marker **N**, with recombination fractions r_1 between **M** and QTL, r_2 between QTL and **N**, and r between **M** and **N**. We assume that we are working in a backcross population and denote the two possible alleles of each marker by 1 and 0. Table 10.2 lists the possible marker genotypes and their probabilities. Again, we do not observe the QTL genotype, which has to be inferred from the marker information.

11.2.1 The Likelihood

To construct a likelihood function, we assume that we have densities f_1 and f_0 corresponding to two QTL genotypes, 1 and 0. Then the densities of the observations at the marker classes are a mixture of the two possible QTL genotypes, given by

	Marker class		Density		
	Genotype	Observation	1		0
1	11	n_1	$1f_1(y)$	+	$0f_0(y)$
2	10	n_2	$(1-\theta)f_1(y)$	+	$\theta f_0(y)$
3	01	n_3	$\theta f_1(y)$	+	$(1-\theta)f_0(y)$
4	00	n_4	$0f_1(y)$	+	$1f_0(y)$
Overall		n	$\frac{1}{2}f_1(y)$	+	$\frac{1}{2}f_0(y)$

where no double recombination is assumed between two marker–QTL intervals. Because phenotypic data are observed for each marker class, we have a likelihood function that is dissolved into four classes:

$$L = \prod_{i=1}^{n} \left[\frac{1}{2} f_1(y_i) + \frac{1}{2} f_0(y_i) \right]$$

$$
\begin{aligned}
&= \prod_{i=1}^{n_1} f_1(y_i) \\
&\times \prod_{i=1}^{n_2} [(1-\theta) f_1(y_i) + \theta f_0(y_i)] \\
&\times \prod_{i=1}^{n_3} [\theta f_1(y_i) + (1-\theta) f_0(y_i)] \\
&\times \prod_{i=1}^{n_4} f_0(y_i)
\end{aligned} \tag{11.1}
$$

The null hypothesis of no QTL is

$$H_0 : f_1(y_i) = f_0(y_i) = f(y_i), \text{ i.e., } \mu_1 = \mu_0 = \mu$$

and to test this at position x, where $\theta = r_1(x)/r$, we map the likelihood ratio statistic

$$\lambda(x) = \frac{\max_{H_0} L(f(y_i))}{\max_{H_1} L(f_1(y_i), f_0(y_i), \theta)}, \tag{11.2}$$

according to Fig. 10.1. For values of x between markers **M** and **N**, typically taken in 2 cM steps, we calculate the value of equation (11.2), or $-2\log\lambda$, and hence "map" the likelihood. We then find the maximum of the test statistic over the **M**–**N** interval.

Calculation of the significance level of $\max_x \lambda(x)$ can be a problem. Although often the likelihood ratio test statistic is approximately χ^2-distributed, the *maximum* likelihood ratio test statistic is not. There are a number of ways to proceed:

(a) Asymptotics for a dense map: Lander and Botstein (1989) used an elegant probability argument to show that the LOD score, a transformation of λ given by

$$\mathrm{LOD}(x) = -2\log_{10}\lambda(x),$$

converges to a χ_2^2 as the markers get dense (as the distance between the markers goes to zero). In particular,

$$P(\max_x \mathrm{LOD}(x) > T | H_0) \to (C + 2Gt)\chi_2^2(t),$$

where $t = 2\log_{10} T$, C is number of chromosomes, and G is genetic length.

(b) Approximations such as those of Piepho (2001) and Davies (1977, 1987).

(c) Permutation tests: If there are sufficient computational resources, permutation tests, following the work of Churchill and Doerge (1994) can be used. Here we generate N permuted samples, and for each we calculate $\max_x \lambda(x)$.

In this section, we illustrate the use of the permutation test, comparing the statistic for the observed data with this reference distribution. All of these approximations are discussed further in Chapter 12.

11.2.2 Maximizing the Likelihood

We now assume that the distribution of the phenotypic traits can be described with a normal distribution. Specifically,

$$f_1(y_i) = f(y_i|\mu_1, \sigma^2) = N(\mu_1, \sigma^2) \text{ and } f_0(y_i) = f(y_i|\mu_0, \sigma^2) = N(\mu_0, \sigma^2),$$

and the likelihood becomes

$$\begin{aligned} L(\mu_1, \mu_2, \sigma^2, \theta|\mathbf{y}) = &\prod_{i=1}^{n_1} f(y_i|\mu_1, \sigma^2) \\ &\times \prod_{i=1}^{n_2} [(1-\theta) f(y_i|\mu_1, \sigma^2) + \theta f(y_i|\mu_0, \sigma^2)] \\ &\times \prod_{i=1}^{n_3} [\theta f(y_i|\mu_1, \sigma^2) + (1-\theta) f(y_i|\mu_0, \sigma^2)] \\ &\times \prod_{i=1}^{n_4} f(y_i|\mu_0, \sigma^2) \end{aligned} \tag{11.3}$$

and now

$$\lambda(\theta) = \frac{\max_{\mu_1=\mu_0=\mu, \sigma^2} L(\mu, \sigma^2, \theta|y)}{\max_{\mu_1, \mu_0, \sigma^2} L(\mu_1, \mu_0, \sigma^2, \theta|y)}. \tag{11.4}$$

To maximize the numerator, we note that if $\mu_1 = \mu_0 = \mu$, then the likelihood becomes

$$L(\mu, \sigma^2, \theta|y) = L(\mu, \sigma^2|y) = \prod_{i=1}^{n} f(y_i|\mu, \sigma^2), \tag{11.5}$$

which does not depend on θ and has MLEs $\hat{\mu} = \bar{y}$ and $\hat{\sigma}^2 = (1/n)\sum_i^n (y_i - \bar{y})^2$. To maximize the full likelihood (11.3), we proceed as in Section 2.2 and define

$$\begin{aligned} P_{1|i} &= \frac{(1-\theta) f(y_i|\mu_1, \sigma^2)}{(1-\theta) f(y_i|\mu_1, \sigma^2) + \theta f(y_i|\mu_0, \sigma^2)}, \\ P_{0|i} &= \frac{\theta f(y_i|\mu_0, \sigma^2)}{(1-\theta) f(y_i|\mu_1, \sigma^2) + \theta f(y_i|\mu_0, \sigma^2)}, \end{aligned} \tag{11.6}$$

with $P_{1|i} + P_{0|i} = 1$.

Differentiating $\log L$, letting the derivatives equal zero, and solving the log-likelihood equations gives

$$\begin{aligned} \frac{\partial}{\partial \mu_1} \log L = 0 \Rightarrow \hat{\mu}_1 &= \frac{\sum_{i=1}^{n_1} y_i + \sum_{i=1}^{n_2+n_3} P_{1|i} y_i}{n_1 + \sum_{i=1}^{n_2+n_3} P_{1|i}}, \\ \frac{\partial}{\partial \mu_0} \log L = 0 \Rightarrow \hat{\mu}_0 &= \frac{\sum_{i=1}^{n_2+n_3} P_{0|i} y_i + \sum_{i=1}^{n_4} y_i}{\sum_{i=1}^{n_2+n_3} P_{0|i} + n_4}, \end{aligned} \tag{11.7}$$

$$\frac{\partial}{\partial\sigma^2}\log L = 0 \Rightarrow \hat{\sigma}^2 =$$

$$\frac{1}{n}\left(\sum_{i=1}^{n_1}(y_i-\hat{\mu}_1)^2 + \sum_{i=1}^{n_2+n_3}\left[P_{1|i}(y_i-\hat{\mu}_1)^2 + P_{0|i}(y_i-\hat{\mu}_0)^2\right] + \sum_{i=1}^{n_4}(y_i-\hat{\mu}_0)^2\right),$$

$$\frac{\partial}{\partial\theta}\log L = 0 \Rightarrow \hat{\theta} = \frac{\sum_{i=1}^{n_2} P_{0|i} + \sum_{i=1}^{n_3} P_{1|i}}{n_2+n_3}.$$

Equation (11.7) is a system of likelihood equations in which the solutions for the unknown parameters are not in analytical form because each estimate depends on estimates of other parameters. Unlike in other numerical problems, each estimate is also a function of the posterior probability for each individual with an expected QTL genotype. Thus, this iteration is actually a special case of the EM algorithm proposed by Dempster et al. (1977) and Meng and Rubin (1993), which guarantees the convergence of the iterations (see Exercise 11.4).

The EM algorithm is derived to estimate the parameters. In the E step, the posterior probabilities for a backcross progeny i to carry a QTL genotype 1 or 0 are calculated with equation (11.6). In the M step, the calculated posterior probabilities are used to estimate the parameters with equation (11.7). The iteration between equations (11.6) and (11.7) is repeated until the estimates are stable. The stable estimates are regarded as the maximum likelihood estimates (MLEs) that will maximize the likelihood.

Example 11.1. (Procedures for Interval Mapping). Revisit Table 10.1, a hypothesized small example, in which ten mice are genotyped for two markers **M** and **N** and phenotyped for body weight. The observed marker and phenotype data, along with the conditional probabilities of a QTL genotype (1 or 0) given genotypes of the marker interval, are expressed as

Mouse	Interval **M-N**	Body Weight (y_i)	QTL Genotype 1	QTL Genotype 0
1	11	30	1	0
2	11	32	1	0
3	11	28	1	0
4	11	29	1	0
5	10	29	$1-\theta$	θ
6	01	22	θ	$1-\theta$
7	00	20	0	1
8	00	21	0	1
9	00	20	0	1
10	00	21	0	1

To use the EM algorithm to estimate the QTL position (θ), QTL genotypic values (μ_1 and μ_0), and residual variance (σ^2) in interval mapping, we need to provide the initial values for these parameters. Usually, this can be determined from their sampling estimates, $\mu = 25.2$ and $s^2 = 22.84$. Because μ_1 and μ_0 are different, we use two slightly different values, say $\mu_1^{(0)} = 25.2$ and $\mu_0^{(0)} = 23.2$, as their initial values. Meanwhile, we assume $\sigma^{2(0)} = 22.84$ and $\theta^{(0)} = 0.5$. The iterative steps of the EM algorithm are described as follows.

Step 1. Calculate the posterior probabilities ($P_{1|i}$ and $P_{0|i}$) that individual mice (i) carry QTL genotype 1 or 0. In this example, these probabilities are calculated only for mice 5 and 6 using equation (11.6); i.e.,

$$
\begin{aligned}
P_{1|5}^{(1)} &= \frac{(1-\theta^{(0)})\exp\left[-\frac{(y_5-\mu_1^{(0)})^2}{2\sigma^{2(0)}}\right]}{(1-\theta^{(0)})\exp\left[-\frac{(y_5-\mu_1^{(0)})^2}{2\sigma^{2(0)}}\right]+\theta^{(0)}\exp\left[-\frac{(y_5-\mu_0^{(0)})^2}{2\sigma^{2(0)}}\right]} \\
&= \frac{(1-0.5)\exp\left[-\frac{(29-25.2)^2}{2\times 22.84}\right]}{(1-0.5)\exp\left[-\frac{(29-25.2)^2}{2\times 22.84}\right]+0.5\exp\left[-\frac{(29-23.2)^2}{2\times 22.84}\right]} \\
&= 0.6035,
\end{aligned}
$$

$$
\begin{aligned}
P_{0|5}^{(1)} &= \frac{\theta^{(0)}\exp\left[-\frac{(y_5-\mu_0^{(0)})^2}{2\sigma^{2(0)}}\right]}{(1-\theta^{(0)})\exp\left[-\frac{(y_5-\mu_1^{(0)})^2}{2\sigma^{2(0)}}\right]+\theta^{(0)}\exp\left[-\frac{(y_5-\mu_0^{(0)})^2}{2\sigma^{2(0)}}\right]} \\
&= \frac{0.5\exp\left[-\frac{(29-23.2)^2}{2\times 22.84}\right]}{(1-0.5)\exp\left[-\frac{(29-25.2)^2}{2\times 22.84}\right]+0.5\exp\left[-\frac{(29-23.2)^2}{2\times 22.84}\right]} \\
&= 0.3965,
\end{aligned}
$$

and

$$
\begin{aligned}
P_{1|6}^{(1)} &= \frac{(1-\theta^{(0)})\exp\left[-\frac{(y_6-\mu_1^{(0)})^2}{2\sigma^{2(0)}}\right]}{(1-\theta^{(0)})\exp\left[-\frac{(y_6-\mu_1^{(0)})^2}{2\sigma^{2(0)}}\right]+\theta^{(0)}\exp\left[-\frac{(y_6-\mu_0^{(0)})^2}{2\sigma^{2(0)}}\right]} \\
&= \frac{(1-0.5)\exp\left[-\frac{(22-25.2)^2}{2\times 22.84}\right]}{(1-0.5)\exp\left[-\frac{(22-25.2)^2}{2\times 22.84}\right]+0.5\exp\left[-\frac{(22-23.2)^2}{2\times 22.84}\right]} \\
&= 0.4520,
\end{aligned}
$$

$$P_{0|6}^{(1)} = \frac{\theta^{(0)} \exp\left[-\frac{(y_6-\mu_0^{(0)})^2}{2\sigma^{2(0)}}\right]}{(1-\theta^{(0)}) \exp\left[-\frac{(y_6-\mu_1^{(0)})^2}{2\sigma^{2(0)}}\right] + \theta^{(0)} \exp\left[-\frac{(y_6-\mu_0^{(0)})^2}{2\sigma^{2(0)}}\right]}$$

$$= \frac{0.5 \exp\left[-\frac{(22-23.2)^2}{2\times 22.84}\right]}{(1-0.5) \exp\left[-\frac{(22-25.2)^2}{2\times 22.84}\right] + 0.5 \exp\left[-\frac{(22-23.2)^2}{2\times 22.84}\right]}$$

$$= 0.5480.$$

We then tabulate the posterior probabilities for each mouse as follows:

Mouse	Interval M–N	Body Weight (y_i)	Posterior Probability $P_{1\|i}^{(1)}$	$P_{0\|i}^{(1)}$
1	11	30	1	0
2	11	32	1	0
3	11	28	1	0
4	11	29	1	0
5	10	29	0.6035	0.3965
6	01	22	0.4520	0.5480
7	00	20	0	1
8	00	21	0	1
9	00	20	0	1
10	00	21	0	1

Step 2. Estimate the QTL genotypic values, residual variance, and QTL position using the log-likelihood equation (11.7); i.e.,

$$\hat{\mu}_1^{(1)} = \frac{30+32+28+29+29\times 0.6986+22\times 0.4049}{1+1+1+1+0.6986+0.4049}$$
$$= 28.97,$$

$$\hat{\mu}_0^{(1)} = \frac{29\times 0.3014+22\times 0.5951+20+21+20+21}{0.3014+0.5951+1+1+1+1}$$
$$= 21.35,$$

$$
\begin{aligned}
\hat{\sigma}^{2(1)} &= \frac{1}{10}[(30-29.03)^2+(32-29.02)^2+(28-29.03)^2+(29-29.03)^2 \\
&\quad +(29-29.03)^2\times 0.6986+(22-29.03)^2\times 0.4049 \\
&\quad +(29-21.21)^2\times 0.3014+(22-21.21)^2\times 0.5951 \\
&\quad +(20-21.21)^2+(21-21.21)^2+(20-21.21)^2+(21-21.21)^2] \\
&= 6.05,
\end{aligned}
$$

$$
\hat{\theta}^{(1)} = \frac{0.3965+0.4520}{1+1} = 0.4242.
$$

Based on equation (11.3), we use the estimated parameters to calculate the log-likelihood value as

$$
\begin{aligned}
&\log L^{(1)} = \log\left(\frac{10}{\sqrt{2\pi}\times 2.46}\right) \\
&+\left[-\frac{(30-28.97)^2}{2\times 6.05}\right]+\left[-\frac{(32-28.97)^2}{2\times 6.05}\right]+\left[-\frac{(28-28.97)^2}{2\times 6.05}\right]+\left[-\frac{(29-28.97)^2}{2\times 6.05}\right] \\
&+\log\left\{(1-0.4242)\exp\left[-\frac{(29-28.97)^2}{2\times 6.05}\right]+0.4242\exp\left[-\frac{(29-28.97)^2}{2\times 6.05}\right]\right\} \\
&+\log\left\{0.4242\exp\left[-\frac{(22-21.35)^2}{2\times 6.05}\right]+(1-0.4242)\exp\left[-\frac{(22-21.35)^2}{2\times 6.05}\right]\right\} \\
&+\left[-\frac{(20-21.35)^2}{2\times 6.05}\right]+\left[-\frac{(21-21.35)^2}{2\times 6.05}\right]+\left[-\frac{(20-21.35)^2}{2\times 6.05}\right]+\left[-\frac{(21-21.35)^2}{2\times 6.05}\right] \\
&= -47.74.
\end{aligned}
$$

Step 3. Repeat steps 1 and 2 until the estimates of all the parameters are stable. In this example, we find that the parameter estimates converge at step x. The MLEs of the parameters are obtained as $\hat{\mu}_1 = 29.6$, $\hat{\mu}_0 = 20.8$, $\hat{\sigma}^2 = 1.2$, $\hat{\theta} \approx 0$, and $\log L = -6.53$.

Example 11.2. (**QTL Mapping in Poplar Trees**). To illustrate interval mapping, we estimate QTL for the poplar data (Yin et al. 2002). Table 11.1 shows a portion of the data, which gives marker information on four markers: CA/CGA-580RD, A7-690, AM11-1060, and AT2-850. Cumulative map distances over the markers are 15.6, 30.7, and 43.9 cM, respectively. The trait measured is the height (m) of each of 57 trees.

Using the iterations above for the MLEs, we implement an R program that is available on the website for this book. The results are given in Table 11.2. The recombination fraction estimate $\hat{\theta}$ locates the possible QTL on the interval. To translate the recombination fraction into mapping distance, we use the Haldane mapping function (10.5). Recall that $\theta = r_1(x)/r$ and hence, for an interval **M–N**

$$
\theta = \frac{r_1(x)}{r} = \frac{\frac{1}{2}(1-e^{-2d(x)})}{\frac{1}{2}(1-e^{-2d})},
$$

Table 11.1. A portion of the poplar marker data and heights.

Obs	CA/CGA-580RD	A7-690	AM11-1060	AT2-850	Height(m)
1	2	2	2	2	10.8
2	2	1	1	1	11.5
3	1	1	1	1	13.8
4	1	1	2	2	13.9
⋮	⋮	⋮	⋮	⋮	⋮
54	2	2	2	2	21.0
55	2	2	1	2	22.0
56	2	1	1	1	22.1
57	2	2	2	2	26.0

Table 11.2. Parameter estimates and likelihood ratios for the poplar data.

Interval	$\hat{\mu}_1$	$\hat{\mu}_0$	$\hat{\sigma}$	$\hat{\theta}$	$-2\log\lambda$
1	18.006	16.356	2.632	0.176	3.257
2	17.826	16.891	2.708	0.101	1.476
3	18.020	16.748	2.673	1.000	3.136

implying that

$$d(x) = -\frac{1}{2}\log\left(1 - 2r\theta\right), \quad r = \frac{1}{2}(1 - e^{-2d}).$$

Applying this to the first interval yields

$$r = \frac{1}{2}(1 - e^{-2\times 0.156}) = 0.134, \quad d_{x\mathbf{M}} = -\frac{1}{2}\log\left(1 - 2\times 0.134 \times 0.176\right) = 0.024,$$

and for all intervals Table 11.3 gives the estimated QTL locations.

Table 11.3. Estimated QTL locations for the poplar data.

Interval	$\hat{\theta}$	$d(x)$
$(0, 0.156)$	0.176	2.4 cM
$(0.156, 0.307)$	0.101	16.9 cM
$(0.307, 0.439)$	1.000	43.9 cM

11.3 Hypothesis Testing

After the parameters are estimated, we must assess the significance of these findings. In testing the null hypothesis

$$H_0 : \text{There is no QTL in any interval},$$

we would reject H_0 if any of the likelihood ratios were significant. If we designate LRT_i to be the likelihood ratio statistic over interval ℓ, then we would reject H_0 if

$$\max_{\ell} LRT_{\ell} > t,$$

where t is an appropriately chosen cutoff. The log-likelihood ratio (LR) test statistic under the H_0 ($\mu_1 = \mu_0 \equiv \mu$) and H_1 ($\mu_1 \neq \mu_0$) hypotheses is calculated by

$$\text{LR} = -2\ln\left[\frac{L(\widetilde{\mu}, \widetilde{\sigma}^2)}{L(\hat{\mu}_1, \hat{\mu}_0, \hat{\sigma}^2)}\right], \tag{11.8}$$

where $\hat{\mu}_1$, $\hat{\mu}_0$, and $\hat{\sigma}^2$ are the MLEs of the corresponding parameters under the H_1 that there is a QTL (full model) and $\widetilde{\mu}$ and $\widetilde{\sigma}^2$ are the MLEs under the H_0 that there is no QTL (reduced model).

Lander and Botstein (1989) used the LOD score as a test statistic, expressed as

$$\text{LOD} = -\log_{10}\left[\frac{L(\widetilde{\mu}, \widetilde{\sigma}^2)}{L(\hat{\mu}_1, \hat{\mu}_0, \hat{\sigma}^2)}\right], \tag{11.9}$$

which is actually equivalent to the LR, with the relationship

$$\text{LOD} = \frac{1}{2}(\log_{10} e)\text{LR} = 0.217\text{LR}.$$

Under H_0, the LOD score or LR statistic is asymptotically distributed as a central χ^2 statistic, with the degrees of freedom being the number of parameters fixed in the null hypothesis. But these degrees are allowed to "float" in the alternative hypothesis because the recombination fraction and the estimated QTL effect are correlated. For example, if the recombination fraction between a marker and QTL is fixed at 0.5, then the magnitude of the putative QTL effect is immaterial for a single-marker analysis. Thus, it is not clear how many degrees of freedom should be used for such a marker-linked QTL analysis. Although this problem is somewhat alleviated with interval mapping, it is still not clear whether both the QTL position and QTL effect are fixed in the null hypothesis.

Since the position of the QTL (measured by the ratio θ) is present only under the alternative hypothesis, it is regarded as a nuisance parameter in significance testing for the QTL. Under the null hypothesis, the nuisance parameter (QTL position) is undefined, so that standard techniques are not applicable for deriving the null distribution of the test statistic unconditionally on the nuisance parameter. For these reasons, a profile of the test statistic across the permissible interval for the nuisance parameter is constructed, and we choose the maximum of that profile to perform just one test. This so-called grid search assumes that the QTL at any position is bracketed by two markers, with no need to include the log-likelihood equation for the QTL position in the iterations but instead allowing θ to vary deterministically in order. Because the QTL is assumed to be somewhere on the genomic interval bracketed by two flanking markers, θ will take a value from 0 to 1. At each assumed position, the LOD score or LR value is calculated, which allows a systematic search for a significant QTL throughout a given linkage group or even the entire genome. If the QTL is first assumed at the right marker and then moved at every 1 or 2 cM from this marker, we can plot the LOD score or LR value against the map position of the QTL. If the test statistic at a region exceeds a predefined critical threshold, a significant QTL is indicated at the maximum (peak) of the LOD or LR profile. The θ value corresponding to the maximum of the likelihood ratio test statistic across a linkage group is the optimal estimate of the QTL position.

Because it is difficult to analytically obtain the distribution of the test statistics, the critical threshold can be determined empirically from permutation tests (Churchill and Doerge 1994). For the poplar data of Example 11.2, we permute the phenotypic values, breaking all associations and mimicking a null distribution. For every permutation, we calculate $\max_\ell \text{LRT}_\ell$ and construct a histogram and obtain a cutoff point. Figure 11.1 is a histogram of 5000 permutations of the data and the resulting maximum LR statistic. Using that as a reference, we have an upper 5 percent cutoff of 5.757, showing that we have not found a significant QTL.

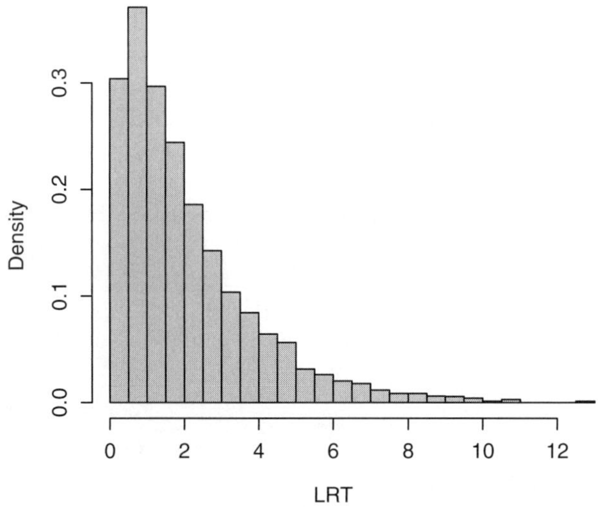

Fig. 11.1. Histogram of the null distribution of $\max -2\log\lambda$ for the data of Example 11.2. The distribution is based on 5000 permutations of the data, with an upper 5 percent cutoff of 5.757.

One can also use the permutation distribution to calculate the p-value. Our observed maximum was 3.257, and of the 5000 permutations, 1016 resulted in larger values of the test statistic, giving a p-value of $1016/5000 = 0.203$.

Example 11.3. Revisit Example 3.1. Two inbred lines, semi-dwarf IR64 and tall Azucena, were crossed to generate a heterozygous F_1. The haploid chromosomes for pollens (gametes) of the F_1 were doubled to produce 123 doubled haploid (DH) plants. These DH plants, equivalent to a backcross progeny, were genotyped for 135 RFLP and 40 isozyme and RAPD markers, from which a linkage map covering the entire genome of 12 chromosomes was constructed (Yan et al. 1998; Fig. 3.3). Each of the DH lines was measured for plant height at each of ten consecutive weeks.

Maximum likelihood is used to scan the existence of a QTL across the linkage map. We use chromosome 1 as an example to describe the procedure. Table 11.4 gives the cumulative and pairwise map distances in centiMorgans for 18 markers on rice chromosome 1, along with the sample sizes, means, and variances of plant height measured at age 10 weeks for each marker interval. Theoretically, these sample estimates should be identical among different marker intervals. They are different because of missing data, which is a common case in QTL mapping.

Using the estimates of these sample parameters as initial values of the unknown parameters, the information in Table 11.4 is incorporated into a chromosome-wide scan of a QTL by assuming the QTL location at every 2 cM within each marker interval. Table 11.5 gives the results for QTL scanning at every 4 cM (to save space), and includes the estimates of two QTL genotype values, residual variance, log-likelihood values under the null and alternative hypotheses, and LR at assumed QTL positions. By plotting the LR values against the length of the linkage group, we can see how the test statistics change over the assumed QTL positions (Fig. 11.2). It is likely that a peak of the LR profile at 217 cM from the first marker within marker interval RG810–RG331 corresponds to the MLE of the QTL location.

To test whether the QTL at the LR peak is statistically significant, we can empirically determine the critical threshold based on permutation tests (see Chapter 12 for details). By reshuffling the phenotypic data, we estimate 1000 maximal LR values, whose 99th percentile is used as the cutoff point (12.64) for testing the chromosome-wide existence of a QTL at the $\alpha = 0.01$ significance level.

A by-product of calculating the test statistic is that the proportion of the phenotypic variance explained by a QTL (R^2) can be calculated as

$$R^2 = \frac{\hat{\sigma}_T^2 - \hat{\sigma}_R^2}{\hat{\sigma}_R^2}, \tag{11.10}$$

where $\hat{\sigma}_T^2$ is the estimate of the total phenotypic variance (i.e., $\tilde{\sigma}^2$ under the null hypothesis) and $\hat{\sigma}_R^2$ is the estimate of the residual variance (i.e., $\hat{\sigma}^2$ under the alternative hypothesis).

We estimated the additive effect of the detected QTL on rice chromosome 1 as $\hat{a} = \hat{\mu}_1 - \hat{\mu}_0 = 30.97$, which is 134.9 percent relative to the standard deviation of the plant height in the DH population. Using equation (11.10), we calculate the proportion of the phenotypic variance explained by the QTL, which is 44.5 percent.

Table 11.4. Map distances, measured in centiMorgans (cM), for 18 markers located on chromosome 1 and sample sizes, means, and variances for each of the 17 marker intervals in a DH population of rice.

Marker	Cumulative Distance	Pairwise Distance	Sample Size	Mean	Variance
RG472	0.0				
		19.2	96	109.63	539.48
RG246	19.2				
		16.1	100	109.85	524.21
K5	35.3				
		4.8	96	109.27	509.49
U10	40.1				
		4.7	78	110.49	506.71
RG532	44.8				
		15.3	100	109.85	537.80
W1	60.1				
		15.5	102	109.57	511.86
RG173	75.6				
		15.0	97	109.68	509.24
RZ276	90.6				
		3.8	104	110.20	529.99
Amy1B	94.4				
		3.3	104	109.85	536.55
RG146	97.7				
		34.3	99	109.44	543.70
RG345	132.0				
		2.5	105	109.71	528.72
RG381	134.5				
		23.5	97	109.37	520.59
RZ19	158.0				
		8.2	91	109.96	510.37
RG690	166.2				
		13.2	104	109.69	531.14
RZ730	179.4				
		33.1	89	110.17	538.40
RZ801	212.5				
		2.6	96	109.90	521.97
RG810	215.1				
		9.2	105	109.93	532.13
RG331	224.3				

Note: Because different markers are missing, the estimates of sample means and variances are different among marker intervals.

Table 11.5. Estimation process of interval QTL mapping for chromosome 1 in a rice DH population.

Interval	Position	μ_1	μ_0	σ^2	LogL_0	LR	n
1	1	109.835	109.798	541.929	-433.818	0	95
1	5	109.789	109.850	541.928	-433.818	0	95
1	9	109.736	109.907	541.922	-433.818	0.001	95
1	13	109.685	109.959	541.910	-433.817	0.002	95
1	17	109.646	109.996	541.899	-433.816	0.004	95
2	21	109.761	108.940	497.122	-425.197	0.029	94
2	25	109.665	109.051	497.197	-425.204	0.014	94
2	29	109.506	109.193	497.266	-425.209	0.003	94
2	33	109.324	109.325	497.290	-425.211	0	94
3	37	108.182	110.686	477.832	-324.262	0.219	72
4	41	109.577	110.522	513.786	-345.029	0.030	76
5	45	108.813	109.880	515.150	-445.035	0.051	98
5	49	109.108	109.652	515.355	-445.055	0.011	98
5	53	109.562	109.383	515.418	-445.060	0.001	98
5	57	110.078	109.149	515.237	-445.045	0.031	98
6	61	108.572	109.336	489.133	-424.434	0.019	94
6	65	107.921	109.519	488.784	-424.408	0.073	94
6	69	107.171	109.688	488.169	-424.361	0.166	94
6	73	106.532	109.779	487.584	-424.303	0.281	94
7	77	109.198	109.789	509.190	-439.930	0.007	97
7	81	107.208	110.150	508.087	-439.882	0.102	97
7	85	104.375	110.519	504.806	-439.756	0.355	97
7	89	104.600	110.332	505.947	-439.705	0.458	97
8	93	111.159	109.952	534.307	-469.621	0.031	103
9	97	107.169	109.662	547.736	-448.043	0.116	98
10	101	105.543	109.882	542.842	-447.655	0.319	98
10	105	104.908	110.177	541.152	-447.601	0.427	98
10	109	104.561	110.451	539.543	-447.552	0.524	98
10	113	104.592	110.641	538.702	-447.519	0.590	98
10	117	104.935	110.721	538.838	-447.506	0.617	98
10	121	105.442	110.710	539.635	-447.508	0.613	98
10	125	105.982	110.641	540.656	-447.519	0.590	98
10	129	106.481	110.545	541.620	-447.534	0.561	98
11	133	105.749	110.980	515.684	-436.028	1.094	96
12	137	104.165	111.730	501.760	-394.034	1.939	87
12	141	103.212	112.374	495.530	-393.686	2.634	87
12	145	102.494	112.917	489.624	-393.330	3.347	87
12	149	102.098	113.277	485.604	-393.014	3.979	87
12	153	102.046	113.419	484.362	-392.777	4.453	87
12	157	102.324	113.349	485.997	-392.638	4.731	87
13	161	99.411	116.583	440.496	-407.358	10.928	91
13	165	98.170	117.625	419.984	-404.359	16.926	91
14	169	97.924	118.561	433.685	-393.583	15.806	88

14	173	96.276	119.960	400.347	-390.697	21.578	88
14	177	96.079	120.437	391.781	-388.902	25.168	88
15	181	94.988	121.898	364.697	-382.823	29.192	87
15	185	91.994	125.186	269.907	-377.980	38.879	87
15	189	91.895	126.999	235.728	-374.168	46.504	87
15	193	92.113	128.047	220.819	-371.591	51.657	87
15	197	92.313	128.583	215.013	-370.005	54.829	87
15	201	92.475	128.694	216.105	-369.339	56.160	87
15	205	92.641	128.359	225.014	-369.701	55.436	87
15	209	92.967	127.307	248.828	-371.535	51.769	87
16	213	94.335	123.975	302.895	-410.949	51.269	96
17	217	93.769	124.741	292.775	-449.026	58.997	105
17	221	93.759	124.256	300.457	-451.654	53.741	105

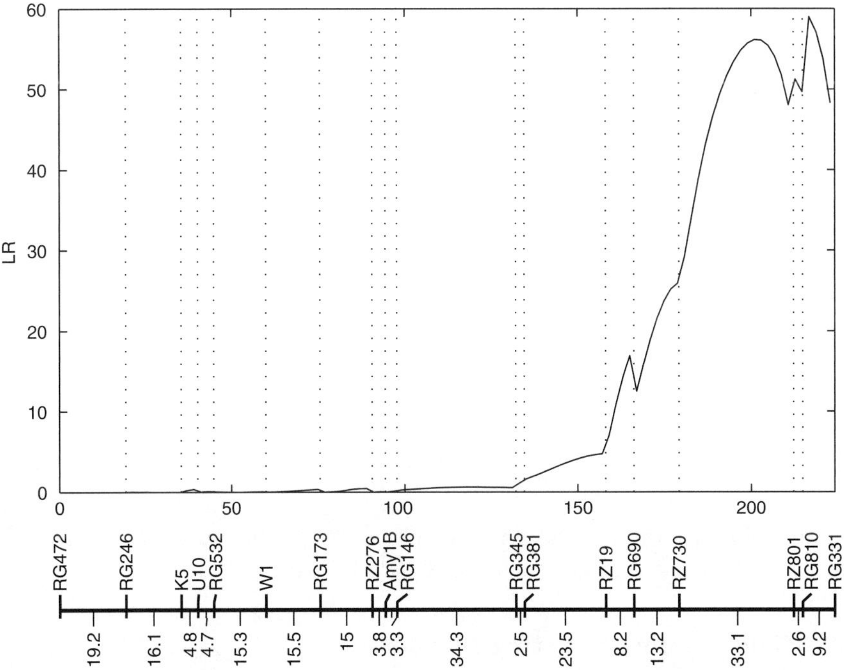

Fig. 11.2. Profile of the LR-value across chromosome 1 for the test of QTL that controls plant height at age 10 weeks in a rice DH population (Huang et al. 1997).

There are many other programs available to produce QTL maps. For example, there is a package `rQTL` available in the statistical package `R`, which is free and available on the Web. There is also a program called `QTL Cartographer`, which is also available as a free download.

11.3.1 Model for Incorporating Double Recombination

The likelihood (11.1) was constructed by assuming no double recombination; i.e., $r = r_1 + r_2$. If double recombination occurs but no interference is assumed, we have $r = r_1 + r_2 - 2r_1r_2$. In this case, the conditional probabilities of QTL genotypes given marker genotypes are expressed in Table 10.3. The EM algorithm can be derived to estimate the r_1 or r_2, QTL genotypic values, and residual variance. Let $\omega_{j|i}$ be the general conditional (prior) probability for a progeny i to have QTL genotype j (1 or 0). In the E step, we define and calculate the corresponding posterior probability for progeny i by

$$P_{j|i} = \frac{\omega_{j|i} f_j(y_i)}{\omega_{1|i} f_1(y_i) + \omega_{0|i} f_0(y_i)}. \tag{11.11}$$

In the M step, the parameters are estimated by

$$\begin{aligned}
\hat{\mu}_j &= \frac{\sum_{i=1}^n P_{j|i} y_i}{\sum_{i=1}^n P_{j|i}}, \\
\hat{\sigma}^2 &= \frac{1}{n}\left[\sum_{i=1}^{n}\sum_{j=0}^{1}(y_i - \hat{\mu}_j)^2 P_{j|i}\right], \\
\hat{r}_1 &= \frac{1}{n}\left[\sum_{i=1}^{n_1+n_2} P_{0|i} + \sum_{i=1}^{n_3+n_4} P_{1|i}\right], \\
\hat{r}_2 &= \frac{1}{n}\left[\sum_{i=1}^{n_1+n_3} P_{0|i} + \sum_{i=1}^{n_2+n_4} P_{1|i}\right].
\end{aligned} \tag{11.12}$$

The E and M steps are iterated until the estimates of the parameters converge. The estimates at the convergence are the MLEs of the parameters.

11.3.2 Model for Incorporating Interference

Consider the marker-QTL order in **M**-QTL-**N**. Let g_{11}, g_{10}, g_{01}, and g_{00} be the probabilities of two crossovers each between **M** and QTL and between QTL and **N**, only one crossover between **M** and QTL, only one crossover between QTL and **N**, and no crossover between the three loci, respectively.

Based on the three-point analysis in Table 4.2, we derive joint marker-QTL-marker genotype frequencies in terms of g_{11}, g_{10}, g_{01}, and g_{00}, which are given in Table 11.9.

Instead of using Table 11.7, we now use the conditional probabilities, generally expressed as $\omega_{j|i}$, derived from Table 11.9 to construct the likelihood similar to equation (11.7). The joint probabilities of marker-QTL-marker genotypes are expressed in terms of these four g's (Table 11.6), from which the conditional probabilities of QTL genotypes given marker interval genotypes ($\omega_{j|i}$) can be derived.

Table 11.6. Joint three-point genotype frequencies in the backcross.

Marker Interval		QTL Genotype	
Genotype	Sample Size	1	0
11	n_1	g_{00}	g_{11}
10	n_2	g_{01}	g_{10}
01	n_3	g_{10}	g_{01}
00	n_4	g_{11}	g_{00}

The EM algorithm is derived to estimate the g probabilities, QTL genotypic values and residual variance. In the E step, define and calculate the posterior probability with a form similar to equation (11.11). In the M step, estimate QTL genotypic values and the residual variance by equation (11.12), and the g probabilities by the following equations:

$$
\begin{aligned}
\hat{g}_{00} &= \frac{1}{n_1+n_4}\left[\sum_{i=1}^{n_1} P_{1|i} + \sum_{i=1}^{n_4} P_{0|i}\right], \\
\hat{g}_{01} &= \frac{1}{n_2+n_3}\left[\sum_{i=1}^{n_2} P_{1|i} + \sum_{i=1}^{n_3} P_{0|i}\right], \\
\hat{g}_{10} &= \frac{1}{n_2+n_3}\left[\sum_{i=1}^{n_3} P_{1|i} + \sum_{i=1}^{n_2} P_{0|i}\right], \\
\hat{g}_{11} &= \frac{1}{n_1+n_4}\left[\sum_{i=1}^{n_4} P_{1|i} + \sum_{i=1}^{n_0} P_{0|i}\right].
\end{aligned}
$$

After the MLEs of g's, the recombination fractions are then estimated as

$$
\begin{aligned}
\hat{r}_1 &= \hat{g}_{10} + \hat{g}_{11}, \\
\hat{r}_2 &= \hat{g}_{01} + \hat{g}_{11}, \\
\hat{r} &= \hat{g}_{01} + \hat{g}_{10}.
\end{aligned} \tag{11.13}
$$

The coefficient of meiotic interference measuring the degree with which the recombinations in two adjacent intervals are affected by one another is calculated by

$$\hat{I} = 1 - \frac{\hat{g}_{11}}{\hat{r}_1 \hat{r}_2}. \tag{11.14}$$

The significance test for the interference can be formulated by calculating the ratio of the two log-likelihoods under the null ($I = 0$) and alternative ($I \neq 0$) hypothesis estimated from Sections 11.3.1 and 11.3.2, respectively. This ratio is asymptotically χ^2-distributed with one degree of freedom.

11.4 QTL Interval Mapping in an F_2

As in the backcross, we will consider three different situations for the F_2: (1) there is no double recombination in the marker interval, (2) the recombinations in different intervals are independent, and (3) meiotic interference occurs between different intervals. In each situation, we will provide the EM algorithm for parameter estimation.

11.4.1 No Double Recombination

In an F_2 design, each marker can have three values, and when taken in pairs to form intervals, this leads to the nine marker classes whose conditional probabilities are given in Table 10.5. The conditional probability of the QTL genotype given a marker genotype is calculated as the ratio of the joint marker-QTL genotype frequency over the marker genotype frequency. The proportions r_1, r_2, and r are the recombination fractions between marker **M** (with two alleles, M and m) and the QTL (with two alleles, Q and q), between the QTL and marker **N** (with two alleles, N and n), and between the two flanking markers, respectively.

Assume that there is no double recombination: we have $r = r_1 + r_2$ or $r_1 r_2 \approx 0$. Then, defining $\theta = r_1/r$ and $\eta = r^2/[(1-r)^2 + r^2]$, we obtain the conditional probabilities of QTL genotypes given different marker classes (Table 11.7).

Likelihood

Similar to the development in Section 11.2, we assume that we have densities f_2, f_1, and f_0 corresponding to the QTL genotypes, and based on Table 11.7 we have the likelihood function

$$\begin{aligned}
L = & \prod_{i=1}^{n_{22}} [1 f_2(y_i) + 0 f_1(y_i) + 0 f_0(y_i)] \\
& \times \prod_{i=1}^{n_{21}} [(1-\theta) f_2(y_i) + \theta f_1(y_i) + 0 f_0(y_i)] \\
& \times \prod_{i=1}^{n_{20}} [(1-\theta)^2 f_2(y_i) + 2\theta(1-\theta) f_1(y_i) + \theta^2 f_0(y_i)] \\
& \times \prod_{i=1}^{n_{12}} [\theta f_2(y_i) + (1-\theta) f_1(y_i) + 0 f_0(y_i)]
\end{aligned}$$

Table 11.7. Conditional probabilities of QTL genotypes given marker genotypes in the F_2.

Marker Interval		QTL Genotype		
Genotype	Sample Size	$QQ(2)$	Qq (1)	qq (0)
$MMNN$ (22)	n_{22}	1	0	0
$MMNn$ (21)	n_{21}	$1-\theta$	θ	0
$MMnn$ (20)	n_{20}	$(1-\theta)^2$	$2\theta(1-\theta)$	θ^2
$MnNN$ (12)	n_{12}	θ	$1-\theta$	0
$MmMn$ (11)	n_{11}	$\eta\theta(1-\theta)$	$1-2\eta\theta(1-\theta)$	$\eta\theta(1-\theta)$
$Mmnn$ (10)	n_{10}	0	$1-\theta$	θ
$mmNN$ (02)	n_{02}	θ^2	$2\theta(1-\theta)$	$(1-\theta)^2$
$mmNn$ (01)	n_{01}	0	θ	$1-\theta$
$mmnn$ (00)	n_{00}	0	0	1
Overall	n	$\frac{1}{4}$	$\frac{1}{2}$	$\frac{1}{4}$

Note: $\theta = r_1/r, \eta = r^2/[(1-r)^2 + r^2]$.

$$
\begin{aligned}
(11.15) \quad & \times \prod_{i=1}^{n_{11}} [\eta\theta(1-\theta)f_2(y_i) + (1-2\eta\theta(1-\theta))f_1(y_i) + \eta\theta(1-\theta)f_0(y_i)] \\
& \times \prod_{i=1}^{n_{10}} [0f_2(y_i) + (1-\theta)f_1(y_i) + \theta f_0(y_i)] \\
& \times \prod_{i=1}^{n_{02}} [\theta^2 f_2(y_i) + 2\theta(1-\theta)f_1(y_i) + (1-\theta)^2 f_0(y_i)] \\
& \times \prod_{i=1}^{n_{01}} [0f_2(y_i) + \theta f_1(y_i) + (1-\theta)f_0(y_i)] \\
& \times \prod_{i=1}^{n_{00}} [0f_2(y_i) + 0f_1(y_i) + 1f_0(y_i)].
\end{aligned}
$$

To test the QTL significance, the null hypothesis of no QTL in the marker is formulated as

$$H_0 : f_2(y_i) = f_1(y_i) = f_0(y_i) \equiv f(y_i),$$

and to test this at position x, where $\theta = r_1(x)/r$, we map the likelihood ratio statistic

$$\lambda(x) = \frac{\max_{H_0} L(f(y_i))}{\max_{H_1} L(f_2(y_i), f_1(y_i), f_0(y_i), \theta)}, \tag{11.16}$$

similar to what was done in Section 11.2.

Maximizing the Likelihood

We now turn to the maximization of the likelihood (11.15) but first make two assumptions:

1. We assume that the densities $f_j(y_i)$ $(j = 2, 1, 0)$ are normal densities,
$$f_j(y_i) = \frac{1}{\sqrt{2\pi}\sigma} \exp\left[-\frac{(y_i - \mu_j)^2}{2\sigma^2}\right],$$
having means $\mu_2 = \mu + a$, $\mu_1 = \mu + d$, and $\mu_0 = \mu - a$ and variance σ^2.
2. We introduce the notation $\omega_{j|i}$ to denote the QTL genotype weights assigned to individual i. For example, if individual i has marker genotype 12, then
$$\omega_{2|i} = \theta, \quad \omega_{1|i} = 1 - \theta, \quad \omega_{0|i} = 0.$$

Using this notation, we can write the log-likelihood (11.15) given observations y and marker interval **M**–**N** as

$$\log L(\mathbf{\Omega}|y, \mathbf{M}\text{-}\mathbf{N}) = \sum_{i=1}^{n} \log \sum_{j=0}^{2} [\omega_{j|i} f_j(y_i)], \tag{11.17}$$

where
$$\mathbf{\Omega} = (\mu_2, \mu_1, \mu_0, \sigma^2, \theta) \equiv (\mathbf{\Omega}_q, \theta).$$

Differentiating the log of the likelihood with respect to any unknown parameter contained in the vector above gives

$$\begin{aligned} \frac{\partial}{\partial \mathbf{\Omega}} \log L(\mathbf{\Omega}|y, \mathbf{M}\text{-}\mathbf{N}) &= \sum_{i=1}^{n} \left(\frac{\omega_{j|i} f_j(y_i)}{\sum_{j'=0}^{2} [\omega_{j'|i} f_{j'}(y_i)]}\right) \left(\frac{\partial}{\partial \mathbf{\Omega}_q} \log f_j(y_i) + \frac{1}{\omega_{j|i}} \frac{\partial}{\partial \theta} \omega_{j|i}\right) \\ &= \sum_{i=1}^{n} P_{j|i} \left(\frac{\partial}{\partial \mathbf{\Omega}_q} \log f_j(y_i) + \frac{1}{\omega_{j|i}} \frac{\partial}{\partial \theta} \omega_{j|i}\right), \end{aligned} \tag{11.18}$$

where

$$P_{j|i} = \frac{\omega_{j|i} f_j(y_i)}{\sum_{j'=0}^{2} [\omega_{j'|i} f_{j'}(y_i)]} \tag{11.19}$$

can be viewed as the posterior probabilities of the QTL genotypes for individual i given a marker genotype, with
$$\sum_{j=0}^{2} P_{j|i} = 1.$$

In the posterior probabilities, $\omega_{j|i}$'s are the prior probabilities for individual i that bear on QTL genotype j, whereas $f_j(y_i)$'s are the likelihoods of the individual. The posterior probability that an individual will carry a particular QTL genotype can be used to infer the individual's identification.

Setting the partial derivative of equation (11.18) with respect to each unknown equal to zero, we derive the MLEs of the genotypic value μ_j for QTL genotype j as

$$\hat{\mu}_j = \frac{\sum_{i=1}^{n} P_{j|i} y_i}{\sum_{i=1}^{n} P_{j|i}}, \quad j = 2, 1, 0, \tag{11.20}$$

and the MLE of σ^2 in a mapping population as

$$\hat{\sigma}^2 = \frac{1}{n} \sum_{i=1}^{n} \sum_{j=0}^{2} [P_{j|i}(y_i - \hat{\mu}_j)^2]. \tag{11.21}$$

The additive (a) and dominant effects (d) of the QTL can be estimated from the MLEs of μ_j's as

$$\hat{a} = \frac{1}{2}(\hat{\mu}_2 - \hat{\mu}_0),$$
$$\hat{d} = \hat{\mu}_1 - \frac{1}{2}(\hat{\mu}_2 + \hat{\mu}_0).$$

Similarly, the MLE of the QTL location (measured by the ratio θ) can be obtained with a scanning approach in which a putative QTL is assumed at every 2 cM in each marker interval. The likelihood ratio statistic ($\lambda(x)$) calculated by equation (11.16) is then plotted against the length (x) of the linkage group. The peak of the $\lambda(x)$ curve corresponds to the optimal estimate of the QTL location.

Example 11.4. (**F_2 Mouse Data**). Cheverud et al. (1996) constructed a linkage map using 75 microsatellite markers in a population of 535 F_2 progeny derived from two strains, the Large (LG/J) and Small (SM/J). The F_2 progeny were measured for body mass at ten weekly intervals starting at age 7 days. The raw weights were corrected for the effects of each covariate due to dam, litter size at birth, and parity but not for the effect due to sex. Chromosome 1 composed of nine markers was used as an example to map the QTL affecting body weight at age 10 weeks (Table 11.8) with the ML-based interval mapping method.

The map of the likelihood is shown in Fig. 11.3. The maximum of the likelihood is at marker interval [D2MIT93–D2MIT389], at 37 cM from the first marker on the left of the chromosome, suggesting there might be a QTL there. The LR at the peak is 33.81, greater than the critical threshold, 14.58, calculated from permutation tests, indicating that the QTL detected is significant.

11.4.2 Independence

In Section 11.4.1, the conditional probabilities were derived by assuming no double recombinations in different intervals. This assumption may be true when the map density is high, in which case the product of the recombination fractions between marker **M** and QTL (r_1) and between QTL and marker **N** (r_2) approaches zero.

Table 11.8. A portion of the mouse data from Cheverud et al. (1996). The first six markers, D1Mit3, D1Mit20, D1Mit7, D1Mit11, D1Mit14, and D1Mit17, on chromosome 1 at 0, 6.3, 41.6, 52.5, 77.6, and 119.2 cM, respectively, are shown. The phenotypic trait is body weight at age 10 weeks.

ID	D1Mit3	D1Mit20	D1Mit7	D1Mit11	D1Mit14	D1Mit17	WtGain
44	1	3	2	3	2	3	29.804
45	3	2	2	2	2	2	20.622
46	2	2	1	2	2	2	25.959
47	1	3	1	1	1	2	27.484
51	3	2	3	2	2	1	24.338
54	2	2	2	2	2	3	34.980
⋮	⋮	⋮	⋮	⋮	⋮	⋮	⋮
634	2	2	3	2	2	2	30.199
636	1	1	1	1	1	2	30.303
637	2	2	2	2	2	2	32.670
638	2	2	2	2	1	2	36.849
640	3	2	3	2	2	2	25.291
641	2	2	3	2	3	1	37.824
642	1	1	3	2	3	2	16.441

Here, we describe a model that relaxes the assumption of no double recombinations but assumes the independence of recombination occurrences.

Table 10.5 tabulates the joint frequencies of marker interval (**M**–**N**) and QTL genotypes in the F_2 under the assumption of recombination independence but allowing for double recombination. As before, let $\omega_{j|i}$ be the conditional probability of a QTL genotype j ($j = 2, 1, 0$), conditional upon the marker interval genotype of progeny i. Similar to likelihood (11.15) for no double recombination, the likelihood incorporated by conditional probabilities derived from Table 10.5 can be constructed. By maximizing the likelihood, the EM algorithm can be implemented to provide the MLEs of parameters.

In the E step, the posterior probability of QTL genotype j for progeny i given its marker interval genotype and phenotypic observation can be calculated with equation (11.19). In the M step, the QTL genotypic values and residual variance are estimated from the posterior probabilities using equations (11.20) and (11.21), while the marker-QTL recombination fractions are estimated by

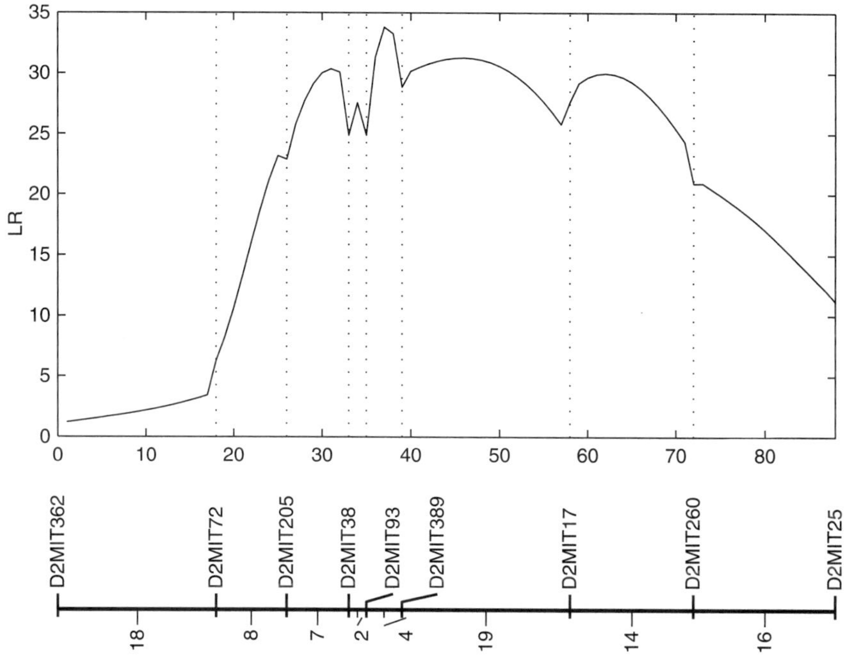

Fig. 11.3. Profile of the LR–value across chromosome 1 for the test of QTL that controls body weight at age 10 weeks in the mouse F_2 population of Example 11.4 (Cheverud et al. 1996).

$$\hat{r}_1 = \frac{1}{n}\left[\sum_{i=1}^{n_{22}+n_{21}+n_{20}} (P_{1|i} + 2P_{0|i}) + \sum_{i=1}^{n_{12}+n_{11}+n_{10}} (P_{2|i} + 2\phi_1 P_{1|i} + P_{0|i}) + \sum_{i=1}^{n_{02}+n_{01}+n_{00}} (2P_{2|i} + P_{1|i})\right],$$

$$\hat{r}_2 = \frac{1}{n}\left[\sum_{i=1}^{n_{22}+n_{12}+n_{02}} (P_{1|i} + 2P_{0|i}) + \sum_{i=1}^{n_{21}+n_{11}+n_{01}} (P_{2|i} + 2\phi_2 P_{1|i} + P_{0|i}) + \sum_{i=1}^{n_{20}+n_{10}+n_{00}} (2P_{2|i} + P_{1|i})\right],$$

where

$$\phi_1 = \frac{r_1^2}{(1-r_1)^2 + r_1^2},$$
$$\phi_2 = \frac{r_2^2}{(1-r_2)^2 + r_2^2}.$$

The E and M steps are iterated until convergence. The estimates at convergence are regarded as the MLEs of parameters.

11.4.3 Interference

It is possible that recombinations are interfered within different intervals so that the coefficient of interference should be incorporated into the mapping model. As in the backcross, define g_{11}, g_{10}, g_{01}, and g_{00} as the probabilities of recombination occurrences in two intervals constructed by order **M**–QTL–**N**. Based on the three-point analysis in Table 4.2, we derive joint marker-QTL-marker genotype frequencies in terms of g_{11}, g_{10}, g_{01}, and g_{00}, which are given in Table (11.9). Instead of using Table 11.7, we now use the conditional probabilities, generally expressed as $\omega_{j|i}$, derived from Table (11.9) to construct the likelihood similar to equation (11.7).

Table 11.9. Joint three-point genotype frequencies in the F_2.

Marker Interval		QTL Genotype		
Genotype	Sample Size	$QQ(2)$	Qq (1)	qq (0)
$MMNN$ (22)	n_{22}	g_{00}^2	$2g_{00}g_{11}$	g_{11}^2
$MMNn$ (21)	n_{21}	$2g_{00}g_{01}$	$2(g_{00}g_{10} + g_{01}g_{11})$	$2g_{11}g_{10}$
$MMnn$ (20)	n_{20}	g_{01}^2	$2g_{01}g_{10}$	g_{10}^2
$MmNN$ (12)	n_{12}	$2g_{00}g_{10}$	$2(g_{00}g_{11} + g_{01}g_{10})$	$2g_{01}g_{11}$
$MmNn$ (11)	n_{11}	$2(g_{00}g_{01} + g_{10}g_{11})$	$2(g_{00}^2 + g_{01}^2 + g_{10}^2 + g_{11}^2)$	$2(g_{00}g_{01} + g_{10}g_{11})$
$Mmnn$ (10)	n_{10}	$2g_{01}g_{11}$	$2(g_{00}g_{11} + g_{01}g_{10})$	$2g_{00}g_{10}$
$mmNN$ (02)	n_{02}	g_{10}^2	$2g_{10}g_{01}$	g_{01}^2
$mmNn$ (01)	n_{01}	$2g_{11}g_{10}$	$2(g_{00}g_{10} + g_{01}g_{11})$	$2g_{00}g_{01}$
$mmnn$ (00)	n_{00}	g_{11}^2	$2g_{00}g_{11}$	g_{00}^2

By defining and calculating the posterior probabilities based on equation (11.19), we derived the closed-form MLEs for th probabilities of occurrence of crossovers. They are given as

$$
\begin{aligned}
\hat{g}_{00} = \frac{1}{2n} \Bigg[& \sum_{i=1}^{n_{22}} (2P_{2|i} + P_{1|i}) + \sum_{i=1}^{n_{21}} (P_{2|i} + \phi_1 P_{1|i}) + \sum_{i=1}^{n_{12}} (P_{2|i} + \phi_2 P_{1|i}) \\
& + \sum_{i=1}^{n_{11}} (\phi_3 P_{2|i} + 2\phi_4 P_{1|i} + \phi_3 P_{0|i}) \\
& + \sum_{i=1}^{n_{10}} (\phi_2 P_{1|i} + P_{0|i}) + \sum_{i=1}^{n_{01}} (\phi_1 P_{1|i} + P_{0|i}) + \sum_{i=1}^{n_{00}} (P_{1|i} + 2P_{0|i}) \Bigg],
\end{aligned}
\tag{11.22}
$$

$$
\begin{aligned}
\hat{g}_{01} = \frac{1}{2n} \Bigg\{ & \sum_{i=1}^{n_{21}} [P_{2|i} + (1-\phi_1)P_{1|i}] + \sum_{i=1}^{n_{20}} (2P_{2|i} + P_{1|i}) + \sum_{i=1}^{n_{12}} [(1-\phi_2)P_{1|i} + P_{0|i}] \\
& + \sum_{i=1}^{n_{11}} (\phi_3 P_{2|i} + 2\phi_4' P_{1|i} + \phi_3 P_{0|i}) \\
& + \sum_{i=1}^{n_{10}} [P_{2|i} + (1-\phi_2)P_{1|i}] + \sum_{i=1}^{n_{02}} (P_{1|i} + 2P_{0|i}) + \sum_{i=1}^{n_{01}} [(1-\phi_1)P_{1|i} + P_{0|i}] \Bigg\},
\end{aligned}
\tag{11.23}
$$

$$
\begin{aligned}
\hat{g}_{10} = \frac{1}{2n} \Bigg\{ & \sum_{i=1}^{n_{21}} (\phi_1 P_{1|i} + P_{0|i}) + \sum_{i=1}^{n_{20}} (P_{1|i} + 2P_{0|i}) + \sum_{i=1}^{n_{12}} [P_{2|i} + (1-\phi_2)P_{1|i}] \\
& + \sum_{i=1}^{n_{11}} [(1-\phi_3)P_{2|i} + 2\phi''{}_4 P_{1|i} + (1-\phi_3)P_{0|i}] \\
& + \sum_{i=1}^{n_{10}} [(1-\phi_2)P_{1|i} + P_{0|i}] + \sum_{i=1}^{n_{02}} (2P_{2|i} + P_{1|i}) + \sum_{i=1}^{n_{01}} (P_{2|i} + \phi_1 P_{1|i}) \Bigg\},
\end{aligned}
\tag{11.24}
$$

$$
\begin{aligned}
\hat{g}_{11} = \frac{1}{2n} \Bigg[& \sum_{i-1}^{n_{22}} (P_{1|i} + 2P_{0|i}) + \sum_{i-1}^{n_{21}} (\phi_1 P_{1|i} + P_{0|i}) + \sum_{i-1}^{n_{12}} (\phi_1 P_{1|i} + P_{0|i}) \\
& + \sum_{i=1}^{n_{11}} [(1-\phi_3)P_{2|i} + 2(1-\phi_4-\phi_4'-\phi_4'')P_{1|i} + (1-\phi_3)P_{0|i}] \\
& + \sum_{i=1}^{n_{10}} (P_{2|i} + \phi_2 P_{1|i}) + \sum_{i=1}^{n_{01}} [P_{0|i} + (1-\phi_1)P_{1|i}] + \sum_{i=1}^{n_{00}} (2P_{2|i} + P_{1|i}) \Bigg],
\end{aligned}
\tag{11.25}
$$

where

$$
\begin{aligned}
\phi_1 &= \frac{g_{00}g_{10}}{g_{00}g_{10} + g_{01}g_{11}}, \\
\phi_2 &= \frac{g_{00}g_{11}}{g_{00}g_{11} + g_{01}g_{10}}, \\
\phi_3 &= \frac{g_{00}g_{01}}{g_{00}g_{01} + g_{10}g_{11}}, \\
\phi_4 &= \frac{g_{00}^2}{g_{00}^2 + g_{01}^2 + g_{10}^2 + g_{11}^2}, \\
\phi_4' &= \frac{g_{01}^2}{g_{00}^2 + g_{01}^2 + g_{10}^2 + g_{11}^2}, \\
\phi_4'' &= \frac{g_{10}^2}{g_{00}^2 + g_{01}^2 + g_{10}^2 + g_{11}^2}.
\end{aligned}
$$

The MLEs of g's can be used to estimate r_1 and r_2 by equations (11.13), and the degree of interference is then estimated by equation (11.14). The significance of =interference can be tested by calculating the likelihood ratio under the null ($I = 0$) and alternative ($I \neq 0$) hypotheses.

11.4.4 Testing Hypotheses

The existence of a segregating QTL linked with known markers can be tested by a likelihood ratio test. In a mapping population, the hypotheses for testing the putative QTL in the F_2 can be formulated as

$$\begin{aligned} H_0 &: \ \mu_2 = \mu_1 = \mu_0 \equiv \mu, \\ H_1 &: \ \text{At least one of the equalities does not hold.} \end{aligned} \tag{11.26}$$

This null hypothesis is equivalent to $a = d = 0$ for the F_2. As in the backcross, the LR value for hypothesis (11.26) can be calculated and compared with the critical threshold determined from permutation tests. But the procedure for the significance tests in the F_2 should be to find:

(1) the existence of a significant QTL,
(2) the additive effect (a), and
(3) the dominance effect (d) of the QTL.

The null hypotheses for these three tests are formulated by posing the constraints

(1) $H_0 : \mu_2 = \mu_1 = \mu_0$,
(2) $H_0 : \frac{1}{2}(\mu_2 - \mu_0) = 0$, and
(3) $H_0 : \mu_1 = \frac{1}{2}(\mu_2 + \mu_0)$,

respectively. Corresponding to each of these tests, the test statistic is calculated as above with equations (11.8) and (11.9).

Example 11.5. Revisit Example 11.4, in which a significant QTL was detected for 10-week body weight on chromosome 1 (Fig. 11.3) in the F_2 mouse population. The F_2 allows the tests of both the additive (a) and dominance effects (d) exerted by the QTL. The MLEs of a and d are 1.59 and 1.71, respectively. The parameters under the null hypothesis of $a = 0$ or $d = 0$ are estimated by maximizing the likelihoods, leading to $L(\widetilde{d}, \widetilde{\sigma}^2)$ and $L(\widetilde{a}, \widetilde{\sigma}^2)$, respectively.

The log-likelihood ratios between the null and alternative hypotheses for the significance tests of a and d are calculated as 12.59 and 10.55, respectively. Because no parameter is nonidentifiable in the null hypothesis for separate tests of a and d, the likelihood ratios of each of these two tests can be thought to asymptotically follow a χ^2 distribution with one degree of freedom. Thus, for a large sample size, the critical threshold can be obtained from the χ^2 distribution table. However, if the sample size used is not adequately large, the threshold can be empirically determined by simulation studies with the data simulated under the null hypothesis. In this example with a good sample size, we obtained the threshold from the table, suggesting that both the additive and dominant effects of the detected QTL are significant.

11.5 Factors That Affect QTL Detection

The statistical power to detect segregating QTLs depends on many factors. They include the size of samples measured for molecular markers and quantitative traits, the magnitude of the type I error allowed, the contribution of the segregating QTL to observed phenotypic variances, the recombination distances between the QTL and the genetic markers, the specific experimental design employed, and the method of statistical analysis. The influences of these factors on the power of QTL detection and parameter estimates can be investigated analytically or through computer simulation. The results from simulation studies are broadly consistent with those obtained analytically for large samples. In most cases of QTL mapping, approaches based on Monte Carlo simulation are widely used because analytical approaches are often difficult to derive.

Numerous misconceptions with respect to the power of QTL detection and experiment design optimization are prevalent. In most cases, the power to detect a segregating QTL of a magnitude likely to be segregating in the population will require genotyping at least 500 individuals and often many more. Most experiments have been too small to find effects of the magnitude that could be reasonably expected. Unless the phenotyping costs are very high relative to genotyping costs, experimental designs with very wide marker spacing are optimum, and decreasing marker intervals below 20 cM will have virtually no effect for most experimental designs. Power per individual genotype can be dramatically increased by replicate progeny, selective genotyping, sample pooling, and sequential sampling, and the effect of these techniques is cumulative. Except for replicate progeny, these other techniques are trait-specific and are therefore most appropriate for experiments that consider only a few traits.

11.6 Procedures for QTL Mapping

The problem of mapping QTLs for observed phenotypic traits can be divided into two stages: (1) initial detection for the existence of segregating QTLs and their enumeration in a mapping population, and (2) genomic localization or fine mapping of the QTLs. The first stage depends on the test and modeling of the number of components within the mixture-model context, with no requirement of marker information, whereas the second stage capitalizes on marker information to locate the positions of QTLs based on marker-QTL cosegregation.

11.6.1 The Number of QTLs

The existence of segregating QTLs in a mapping population is a prerequisite for QTL mapping. Many poorly designed molecular studies fail to detect significant QTLs because of the use of a wrong mapping population in which no such QTLs exist. Also, prior knowledge about the existence of QTLs helps to confirm results from molecular mapping. The pattern of QTL segregation can be well predicted for inbred line crosses based on Mendelian inheritance. In this chapter, maximum likelihood based on a mixture model was implemented to estimate the genetic effects of QTLs. This procedure is extended here to detect any number of QTLs in the mapping population.

Consider a mapping population of size n in which there are k segregating QTLs for a phenotypic trait y that form J genotypes. At each QTL, there are two QTL genotypes, symbolized by 1 and 0 for the backcross, and three QTL genotypes, symbolized by 2, 1 and 0 for the F_2, respectively. Thus, J is 2^k for the backcross or 3^k for the F_2. Considering a backcross, a mixture model that specifies the segregation of all k putative QTLs can be written as

$$L(\boldsymbol{\Omega}|y) = \prod_{i=1}^{n} \left[\left(\tfrac{1}{2}\right)^k f_{11\cdots1}(y_i) + \left(\tfrac{1}{2}\right)^k f_{11\cdots0}(y_i) + \cdots + \left(\tfrac{1}{2}\right)^k f_{00\cdots0}(y_i) \right], \tag{11.27}$$

where $\boldsymbol{\Omega} = (\mu_{11\cdots1}, \mu_{11\cdots0}, \cdots, \mu_{00\cdots0}, \sigma^2)$ and $(\frac{1}{2})^k$ is the prior probability of a QTL genotype in the mapping population. The MLEs of $\boldsymbol{\Omega}$ can be estimated by

$$\begin{aligned}
\hat{\mu}_{11\cdots1} &= \frac{\sum_{i=1}^n P_{11\cdots1|i} y_i}{\sum_{i=1}^n P_{11\cdots1|i}}, \\
\hat{\mu}_{11\cdots0} &= \frac{\sum_{i=1}^n P_{11\cdots0|i} y_i}{\sum_{i=1}^n P_{11\cdots0|i}}, \\
&\cdots \\
\hat{\mu}_{00\cdots0} &= \frac{\sum_{i=1}^n P_{00\cdots0|i} y_i}{\sum_{i=1}^n P_{00\cdots0|i}}, \\
\hat{\sigma}^2 &= \frac{1}{n}[(y_i - \mu_{11\cdots1})^2 + \cdots + (y_i - \mu_{00\cdots0})^2],
\end{aligned} \tag{11.28}$$

where

$$\begin{aligned}
P_{11\cdots1|i} &= \frac{\left(\frac{1}{2}\right)^k f_{11\cdots1}(y_i)}{\left(\frac{1}{2}\right)^k f_{11\cdots1}(y_i) + \left(\frac{1}{2}\right)^k f_{11\cdots0}(y_i) + \cdots + \left(\frac{1}{2}\right)^k f_{00\cdots0}(y_i)}, \\
P_{11\cdots0|i} &= \frac{\left(\frac{1}{2}\right)^k f_{11\cdots0}(y_i)}{\left(\frac{1}{2}\right)^k f_{11\cdots1}(y_i) + \left(\frac{1}{2}\right)^k f_{11\cdots0}(y_i) + \cdots + \left(\frac{1}{2}\right)^k f_{00\cdots0}(y_i)}, \\
&\vdots \\
P_{00\cdots0|i} &= \frac{\left(\frac{1}{2}\right)^k f_{00\cdots0}(y_i)}{\left(\frac{1}{2}\right)^k f_{11\cdots1}(y_i) + \left(\frac{1}{2}\right)^k f_{11\cdots0}(y_i) + \cdots + \left(\frac{1}{2}\right)^k f_{00\cdots0}(y_i)},
\end{aligned} \tag{11.29}$$

are the posterior probabilities of a particular QTL genotype for an individual i given its phenotypic value. The ML estimates are computed as follows:

(1) Provide initial values $\boldsymbol{\Omega}^{(0)} = (\mu_{11\cdots1}^{(0)}, \mu_{11\cdots0}^{(0)}, \cdots, \mu_{00\cdots0}^{(0)}, \sigma^{2(0)})$ for unknown parameters.
(2) Calculate the posterior probabilities $P_{11\cdots1|i}^{(0)}$, $P_{11\cdots0|i}^{(0)}$, $\cdots$. $P_{00\cdots0|i}^{(0)}$ using equation (11.29).
(3) Obtain a new estimate of the unknown parameters $\boldsymbol{\Omega}^{(1)} = (\mu_{11\cdots1}^{(1)}, \mu_{11\cdots0}^{(1)}, \cdots, \mu_{00\cdots0}^{(1)}, \sigma^{2(1)})$ using equation (11.28).

(4) Repeat steps (2) and (3) until the estimate of $\mathbf{\Omega}$ converges at a stable value.

The stable values are regarded as the MLEs of the unknown parameters. These MLEs can be substituted to calculate the plug-in values of the likelihood function (11.27).

The analyses of the mixture model (11.27) above are performed to estimate the number of QTLs involved in the trait. In statistics, the estimation of the QTL number is equivalent to the determination of the number of mixing components that can best explain the phenotypic data. Let k_1 and k_2 be two alternative QTL numbers, which generate the number of QTL genotypes 2^{k_1} and 2^{k_2} for the backcross, respectively. Lo et al. (2001) proposed a statistical model for the characterization of the number of mixture components based on the likelihood ratio statistic calculated from the Kullback-Leibler information criterion. Under a theorem proposed by Vuong (1989), the likelihood ratio between the null hypothesis that the mapping population contains k_1 QTLs and the alternative hypothesis that the mapping population contains k_2 QTLs is asymptotically distributed as a weighted sum of independent χ^2 random variables with one degree of freedom under general regularity conditions. By calculating the likelihood ratio under two alternative hypotheses, Lo et al.'s model can be used to determine the number of QTL that are segregating in the mapping population.

Alternatively, the estimation of the number of segregating QTLs for a quantitative trait in a population can be based on model selection criteria, AIC (Akaike 1997) or BIC (Schwarz 1978). The AIC or BIC, depending the likelihood, the number of parameters being estimated and the sample size used, is calculated. The model that gives the smallest AIC or BIC values is considered to best explain the mapping data analyzed.

A similar procedure can be formulated for the F_2, for which the likelihood function assuming k QTLs is expressed as

$$L(y) = \prod_{i=1}^{n} \left[\left(\tfrac{1}{3}\right)^k f_{22\cdots 2}(y_i) + \left(\tfrac{1}{3}\right)^k f_{22\cdots 0}(y_i) + \cdots + \left(\tfrac{1}{3}\right)^k f_{00\cdots 0}(y_i) \right],$$

where the prior probability of a QTL genotype is 3^k. The calculation of the likelihood ratio in two different cases each for a different number of QTLs, k_1 and k_2, is used to determine the optimal number of actual QTLs involved in the F_2 mapping population.

Example 11.6. Revisit Example 3.1. Two inbred lines, semi-dwarf IR64 and tall Azucena, were crossed to generate a heterozygous F_1. The haploid chromosomes for pollens (gametes) of the F_1 were doubled to produce 123 doubled haploid (DH) plants. These DH plants, equivalent to a backcross progeny, were genotyped for 135 RFLP and 40 isozyme and RAPD markers, from which a linkage map covering the entire genome of 12 chromosomes was constructed (Yan et al. 1998; Fig. 3.3). Each of the DH lines was measured for plant height at 10 weeks after the plants were grown in the field.

The number of QTL for plant height at age 10 weeks contained in the HD population is estimated. Assuming that there are $k = 0, 1, 2, \ldots, 8$ QTLs for the trait, the log-likelihood ratios (LR) are calculated under two alternative hypotheses of QTL numbers k and $k-1$. Based on the AIC and BIC information criteria, we suggest that there possibly exists one major QTL that controls 10-week plant heights in rice (Table 11.10). Further genetic mapping with a linkage map constructed by molecular markers (Fig. 3.3) can be used to locate the location of this QTL.

Table 11.10. AIC and BIC information criteria under different numbers of QTLs in a rice DH population.

QTL Number	Likelihood Value	LR	AIC	BIC
0	-483.0999		968.19989	970.86333
		2.8541		
1	-480.2458		964.49168	969.81856
		0.2296		
2	-480.0163		968.0325	978.68626
		4.3810		
3	-475.6352		967.27044	988.57796
		4.8085		
4	-470.8267		973.65336	1016.2684
		3.6248		
5	-467.2019		998.4038	1083.6338
		34.2540		
6	-432.9479		993.8958	1164.3559
		0.7343		
7	-432.2136		1120.4272	1461.3475
		9.0629		
8	-423.1507		1358.3014	2040.1418

11.6.2 Locations of Individual QTLs

With the knowledge about the number of segregating QTLs in a mapping population, the linkage map constructed from molecular markers is used to localize the positions of these QTLs. In general, a scanning approach based on a grid of evenly–spaced genomic positions from one marker to the next is used to search for the existence of all possible QTL throughout the genome. Interval mapping assuming one QTL at a time makes use of the mixture-based likelihood function

$$L(\boldsymbol{\Omega}|y) = \prod_{i=1}^{n} \left[\omega_{1|i} f_1(y_i) + \omega_{0|i} f_0(y_i)\right] \tag{11.30}$$

for the backcross and

$$L(\boldsymbol{\Omega}|y) = \prod_{i=1}^{n} \left[\omega_{2|i} f_2(y_i) + \omega_{1|i} f_1(y_i) + \omega_{0|i} f_0(y_i)\right] \tag{11.31}$$

for the F_2, where $\omega_{j|i}$'s are the conditional probabilities of QTL genotypes given a marker genotype for individual i. Thus, by estimating a maximal log-likelihood ratio across the genome, we can determine the position of a QTL. The number of QTLs detected from the genetic linkage map is smaller than that detected by a pure phenotypic analysis (with no marker) if the linkage map does not well cover the entire genome.

Theoretically, multi-QTL models can be more efficient in detecting QTL effects than a one-QTL model because possible QTL–QTL interactions are considered for the former. But this may not always be true in practice. Simultaneous analysis and modeling of multiple QTLs will be likely to lead to a computational burden.

11.7 Exercises

11.1 Compare the advantages and disadvantages of the regression analysis and maximum likelihood approaches in QTL mapping.

11.2 Use Bayes' theorem to derive Table 10.3.

11.3 (a) Verify that equation (11.3) is the full likelihood.
(b) Verify that equation (11.5) is the likelihood under H_0 and is independent of θ.

11.4 For the likelihood equations of Section 11.4.1, show that

$$E(\omega_{j|i}|y, \boldsymbol{\Omega}) = P_{j|i},$$

$$E(\omega_{j|i}^2|y, \boldsymbol{\Omega}) = P_{j|i},$$

$$E(\omega_{j_1|i}\omega_{j_2|i}|y, \boldsymbol{\Omega}) = 0,$$

where the expectation is over the distribution of $\omega_{j|i}$ conditional on the data y and the other parameters. This is the E step of an EM algorithm in which the expected conditional QTL genotypes are determined with the equation (11.19), with the M step being equations (11.20) and (11.21), and the unknown parameters are estimated using the posterior value calculated from the E step. This iterative process is repeated between the E and M steps until convergence of estimates.

11.5 Interval mapping
When the interval mapping model was constructed, three different cases for the formation of gametes during meiosis were considered: (1) there is no double recombination, $r = r_1 + r_2$; (2) recombinations are independent, $r = r_1 + r_2 - 2r_1r_2$; and (3) there is interference between different intervals, $r = r_1 + r_2 - 2(1 - I)r_1r_2$ (see Section 11.3.2

for the definitions of r_1, r_2, r, and I). Case 3 is the most general and can be reduced to case 1 when $I = 1$ and case 2 when $I = 0$. Interference, I, has limits (Ott 1991)

$$1 - \min\left(\frac{1}{r_1}, \frac{1}{r_2}\right) \le I \le 1 - \max\left[0, \frac{r_1 + r_2 - 1/2}{2r_1 r_2}\right].$$

(a) Perform simulation studies to compare these cases by assuming different I values with regard to power and the precision of parameter estimation.
(b) Provide guidance for QTL mapping practitioners to select an appropriate analytical model for their data.

11.6 Epistatic mapping
Epistasis is defined as the dependence of the expression of one gene upon the expression of other genes. Show how the likelihood functions (11.1) and (11.15) can be extended to map two epistatic QTL for a mapping population.

12

Threshold and Precision Analysis

12.1 Introduction

In the preceding chapter, we described the basic principle for interval mapping of QTLs within the maximum likelihood context, but two fundamental questions should be addressed toward the QTL analysis of complex traits. First, what is the critical threshold of the test statistic that can be used to declare the statistical significance of a QTL? Second, after the significant QTL is determined, how are the estimates of likelihoods and QTL parameters adequately precise to make a scientific inference about QTL position and effect? Statistically, these two questions present different aspects of QTL mapping, but we will describe them here in a single chapter as two important follow-ups of QTL mapping.

All the statistical methods for QTL mapping rely upon the determination of appropriate significance thresholds (or critical values) with which the test statistics estimated from a particular data set are compared to declare whether a significant QTL exists. It is often difficult to determine critical thresholds because this relies heavily upon the assumption about the distribution of test statistics (LOD scores or likelihood ratios) under the null hypothesis that this is no QTL. The distributional properties of test statistics are affected by many factors, such as sample sizes, the distribution of the quantitative trait studies, etc. Because of the importance of this issue, there is a wealth of statistical literature on the determination of critical thresholds (Lander and Botstein 1989; Rebai et al 1994; Doerge and Rebai 1996; Piepho 2001; Churchill and Doerge 1994; Dupuis and Siegmund 1999; Zou et al. 2004). In the first part of this chapter, we will introduce several approaches for the determination of critical thresholds and discuss their advantages and disadvantages.

For a significant QTL detected in terms of the test statistics beyond the predetermined critical threshold, it is important to assess the accuracy and precision of the estimation of its genetic effects. The assessment of the precision of a statistical method for QTL mapping can be made through simulation studies. For a set of parameters, each with a given value, artificial data are simulated on the basis of the structure of the model, with multiple replicates in each of which the parameters are estimated. The means and mean square errors are then calculated using the estimates from all the simulation replicates. However, for a real data set, the precision of parameter estimation can be examined by calculating the estimation of asymptotic variance-covariances of the estimates of the parameters. Theories have been developed for estimating asymptotic variance-covariances of the parameter estimates within a mixture-model context (Louis 1982; Meng and Rubin 1991). Kao and Zeng (1997) and Chen (2005) extended these theories to evaluate the precision of the estimation of QTL parameters. In the second part of this chapter, we will describe key technical issues for estimating the asymptotic variance-covariances of QTL effect and position parameters. Interested readers are referred to the original papers cited above.

12.2 Threshold Determination

12.2.1 Background

In a scientific experiment, it is always crucial to know the probability of arriving at the wrong conclusions. For a QTL mapping study, these conclusions are: (1) there is a segregating QTL when in fact it is not present, and (2) a QTL is not detected that actually exists. The first type of error results in a false positive (type I), whereas the second in a false negative (type II). The probability of false positives (i.e., the significance level) can be controlled by choosing the appropriate significance threshold. The rate of false negatives is determined by the setting of an experiment and the magnitudes of the QTL effects (Jansen 1994; van Ooijen 1999).

Despite its importance for the declaration of significant QTLs, the characterization of critical thresholds has been considered one of the most difficult issues in QTL mapping. There are two issues that make threshold determination thorny (Lander and Schork 1994; Churchill and Doerge 1994). First, the distribution of test statistics (LOD or LR) under the null hypothesis cannot be well determined because the regularity conditions for an asymptotic χ^2 distribution for the LR test statistic are violated. This arises from the fact that, under the null hypothesis of no QTL, the QTL position is not identifiable and becomes a nuisance parameter. Also, the reliability of asymptotic approximations can be affected by other factors, such as finite sample size and distribution of the trait studied. Second, when the test is performed in the entire genome, as is usually done, a multiple-test problem will arise because the tests across the length of a linkage group are not mutually independent owing to the nature of linkage. It is difficult to characterize the dependence structure of such series of tests. To control the genome-wide type I error rate, critical threshold values of the test statistic therefore need to be adjusted.

Currently, three different approaches are available to calculate the threshold value throughout a genome; i.e., (1) analytical methods, (2) simulation studies, and (3) permutation tests. Piepho (2001) proposed a quick approximate approach for calculating the threshold for assessing genome-wide significance. More recently, Zou et al. (2004)) derived a score statistic aimed at increasing the computational efficiency of threshold determination. In this section, we will introduce each of these approaches. Some methodological comparisons will be made.

12.2.2 Analytical Approximations

The analytical approach depends on the distribution of the underlying test statistics. Typically, the profile of LR test statistics is constructed over the grid of possible QTL locations in a linkage group or an entire genome and the maximum of the LR (MLR) is used as a global test statistic. At a given position of the QTL, the LR test statistic is asymptotically χ^2-distributed under the null hypothesis with degrees of freedom equal to the number of associated QTL effects. However, under the null hypothesis H_0: no QTL, as mentioned above, the QTL position is unidentified and therefore the LR test statistic does not follow the standard χ^2–distribution asymptotically. Based on the results of Davies (1977, 1987), several authors have derived approximate formulas to determine critical thresholds for a particular design, where closed form thresholds are not available (Rebai et al. 1994; Doerge and Rebai 1996; Piepho 2001).

When a QTL is tested at a particular position, the significance level is the probability that we will reject the null hypothesis assuming that the null hypothesis is true. Letting LR(x) be the likelihood ratio test statistic at position x, the nominal significance level for this test would be

(12.1) $$\alpha = \text{Prob}(\text{LR}(x) > T_\alpha^P | \text{no QTL at position } x),$$

where T_α^P is the critical threshold for the α-level significance test of a QTL at a point. The probability specified by equation (12.1) is called the comparison-wise error rate. The comparison-wise error rate can be controlled by determining the distribution of the test statistic at a specific point in the genome. This point-specific test is possible in the practical case in which we are only interested in testing if a specific QTL detected in one population also exists in a different population. When the test is performed at a single point in the genome, the test statistic typically follows an asymptotic χ^2 distribution with one degree of freedom.

In QTL mapping, we need to do many tests for each marker interval across the genome. If there are no QTLs in a tested interval, we need to control the probability of falsely identifying any QTL in that interval. The probability of finding at least one QTL in an interval when in fact none exists is called the experiment-wide error rate. For a backcross design, the MLR for an entire marker interval is suggested to have a distribution between χ_1^2 and χ_2^2,

(12.2) $$\chi_{1,\alpha}^2 < T_\alpha^I < \chi_{2,\alpha}^2,$$

where T_α^I is the critical threshold for the α-level significance test of a QTL in the interval. The χ^2–distribution with two degrees of freedom results from the fact that two parameters in the backcross, the QTL position and QTL (additive) effect, are mixed under the null hypothesis. Intuitively, the distribution of the MLR is closer to χ_1^2 for a smaller interval than for a larger one.

In an experimental genomic study, we are more interested in the existence of a QTL in the entire genome. Thus, it is important to determine the distribution and appropriate threshold for the MLR where the maximum is over the entire genome. Under the null hypothesis (i.e., there is no QTL in the entire genome of total length L), the chance (α) of the MLR exceeding T_α somewhere in the genome is

$$\alpha = \text{Prob}(\text{MLR}_{0\leq x\leq L} > T_\alpha | \text{no QTL in the genome}).$$

A number of approximated analytical methods have been proposed to compute the threshold T_α for any significance level α (Lander and Botstein 1989; Dupuis and Siegmund 1999; Rebai et al. 1994). These methods are basically suitable for the extremes of very dense and very sparse genetic maps. We will discuss how to characterize T_α separately for these two extreme cases, mostly based on Lander and Botstein's (1989) argument. Further approximations for dense (< 1 cM) and sparse maps are given by Dupuis and Siegmund (1999).

Genome-wide Threshold for a Sparse Map

In QTL mapping, an experiment-wide significance level is usually desirable unless one wants to examine if the previously determined QTL exists in a new experiment. For this reason, the genome-wide significance–the probability of obtaining a test statistic above the threshold somewhere on the whole genome by chance–should be used. A genome-wide threshold will depend not only on the number and length of the chromosomes but also on the numbers of markers (i.e., density) on the chromosomes. When just a few markers are tested per chromosome (i.e., the so-called sparse map case), a lower threshold is needed at the same genome-wide significance level than when many markers are tested per chromosome (i.e., the so-called dense map case) (Lander and Botstein 1989; van Ooijen 1999).

For a sparse map in which markers are widely separated over the genome, we can safely assume that the probabilities of no QTLs within different marker intervals are approximately independent. Let α and p be the genome- and interval-wide significance levels for declaring a significant QTL, respectively. If m marker intervals are considered, we have

$$\begin{aligned} 1 - \alpha &= \text{Prob}(\text{no QTL in the genome}) \\ &= \prod_{i=1}^{m} \text{Prob}(\text{no QTL in interval } i) \\ &= \prod_{i=1}^{m} (1-p) \\ &= (1-p)^m \\ &\approx 1 - mp, \end{aligned}$$

or

$$p \approx \frac{\alpha}{m}. \tag{12.3}$$

The interval-wide significance level required to declare the genome-wide existence of a QTL over the genome of m intervals with equation (12.3) is conservative. This can be proven using the simple Bonferroni inequality, which states

$$\text{(12.4)} \qquad \text{Prob}(\text{MLR}_{0\le x\le L} > T_\alpha) \le \prod_{i=1}^{m} \text{Prob}(\text{MLR}_{0\le x\le l} > T_\alpha^I)$$

assuming each marker interval has length l. This is equivalent to $\alpha \le \prod_{i=1}^{m} p = mp$, or

$$p \ge \frac{\alpha}{m}.$$

Based on equation (12.3), the approximate threshold for the significance test of a QTL in the entire genome for a backcross is given by

$$\text{(12.5)} \qquad \chi^2_{1,\frac{\alpha}{m}} < T_\alpha < \chi^2_{2,\frac{\alpha}{m}}.$$

Example 12.1. Huang et al. (1997) founded a doubled haploid (DH) population of 123 lines with two inbred lines, semi-dwarf IR64 and tall Azucena. This DH population was formed by doubling haploid chromosomes of the gametes derived from the heterozygous F_1 and thus it is equivalent to a backcross population because its marker segregation follows a 1:1 ratio. A linkage map (Fig. 3.3) was constructed with a total of 175 polymorphic markers (including 146 RFLPs, 8 isozymes, 14 RAPDs, and 12 cloned genes), representing a good coverage of 12 rice chromosomes. The constructed map is 2005 cM long with 163 marker intervals, having an average distance of 11.5 cM, with 6 gaps larger than 35 cM.

Three grain traits (grain length, grain width, and the ratio of length to width) were measured for each DH line. Interval mapping was used to detect the QTLs that determine these traits. Huang et al. (1997) reported 12 such QTLs located on five different chromosomes. Table **??** tabulates the QTLs for grain traits, their locations, LR values, interval-wide significance levels (p) based on $\chi^2_{1,p}$ and $\chi^2_{2,p}$ (see equation (12.2)), and adjusted genome-wide significance levels (α).

Genome-wide Threshold for a Dense Map

Suppose there is an infinitely dense map in which markers are located everywhere over the genome. For such a dense map, occurrences of spuriously high test statistics at nearby intervals are no longer independent. For an infinitely dense map (i.e., the number of intervals $(m) \to 0$), the required nominal significance level for each interval test tends to be a nonzero limit independent of m. If the sample size used is large, Lander and Botstein (1989) argued that the change of the LOD score obeys the square of an Orenstein-Uhlenbeck (OU) diffusion process in the infinitely dense map.

According to the central limit theorem, the log-likelihood ratio (LR) test statistic at each position (x) is shown to be asymptotically proportional to the square of a random normal variable $z^2(x)$,

$$\text{LR}(x) \sim z^2(x),$$

where $z(x) \sim N(0,1)$. For two positions, x_1 and x_2, $\text{LR}(x_1)$ and $\text{LR}(x_2)$ are correlated, with

$$\text{Corr}[z(x_1), z(x_2)] \sim 1 - 2r = e^{-2d},$$

Table 12.1. Chromosomal location and LR values of QTLs for grain traits measured in the DH population. Adapted from Huang et al. (1997).

		Chromo-		p-value		α value	
Trait	Peak Interval	some	MLR	$\chi^2_{1,p}$	$\chi^2_{2,p}$	$\chi^2_{1,\alpha}$	$\chi^2_{2,\alpha}$
Length	RZ730–RZ801	1	28.6	$8.90e^{-8}$	$6.16e^{-7}$	$1.45e^{-5}$	$1.00e^{-4}$
	RZ519–RZ448	3	19.2	$1.18e^{-5}$	$6.77e^{-5}$	$1.92e^{-3}$	$1.10e^{-2}$
	RZ337A–CDO337	3	28.6	$8.90e^{-8}$	$6.16e^{-7}$	$1.45e^{-5}$	$1.00e^{-4}$
	G2155–RG134	10	26.2	$3.08e^{-7}$	$2.04e^{-6}$	$5.02e^{-5}$	$3.33e^{-4}$
Width	RG810–RG331	1	18.2	$1.99e^{-5}$	$1.12e^{-4}$	$3.24e^{-3}$	$1.82e^{-2}$
	RZ318–RZ58	2	14.0	$1.83e^{-4}$	$9.12e^{-4}$	$2.98e^{-2}$	$1.49e^{-1}$
	CDO87–Pgi-1	3	14.4	$1.48e^{-4}$	$7.46e^{-4}$	$2.41e^{-2}$	$1.22e^{-1}$
	RG134–RZ500	10	14.1	$1.73e^{-4}$	$8.67e^{-4}$	$2.82e^{-2}$	$1.41e^{-1}$
	RZ536–G186	11	15.7	$7.42e^{-5}$	$3.90e^{-4}$	$1.21e^{-2}$	6.35^{e-2}
Length/width	RG157-RZ318	2	19.6	$9.55e^{-6}$	$5.54e^{-5}$	$1.56e^{-3}$	$9.04e^{-3}$
	RZ519–Pgi-1	3	18.5	$1.70e^{-5}$	$9.61e^{-5}$	$2.77e^{-3}$	$1.57e^{-2}$
	RG179–CDO337	3	23.2	$1.4600e^{-6}$	$9.1661e^{-6}$	$2.38e^{-4}$	$1.49e^{-3}$

where r and d are the recombination fraction and genetic distance (measured in Morgans) between the two genomic positions, respectively, and the relationship between r and d is specified by the Haldane map function. It can be seen that $z(x)$ is a stationary normal process with covariance function e^{-2d}, which can be described by the OU diffusion process.

Lander and Botstein (1989) further derived appropriate genome-wide critical values for the backcross based on the asymptotic theory by the OU process. Considering a genome with C chromosomes and a total genetic length L (measured in Morgans), the probability that the LOD score exceeds a high level T in the case where there is no QTL is approximated by

$$(C + 2Lt)\chi^2(t), \tag{12.6}$$

where $T = (2\ln 10)^{-1}t$ and $\chi^2(t)$ denotes the inverse cumulative distribution function of the χ^2 distribution with one degree of freedom.

To control the probability of a false positive at the genome-wide significance level (α), we approximate the appropriate LOD threshold by

$$T_\alpha = (2\ln 10)^{-1}t_\alpha, \tag{12.7}$$

where t_α solves the equation $\alpha = (C + 2Lt_\alpha)\chi^2(t_\alpha)$.

Lander and Botstein (1989) suggested that a typical LOD score threshold should be between 2 and 3 to ensure a 5 percent overall false positive error for detecting a QTL. Lander and Botstein's OU diffusion-process–based theory was used for the F_2 (Dupuis and Siegmund 1999), in which the LOD score follows a χ^2 process with two degrees of freedom, because of the fitting of both the additive and dominance components and more general mating designs (Zou et al. 2004).

12.2.3 Simulation Studies

For many practical data sets, we cannot know the distribution of the test statistics. The best approach in this case is to directly estimate the false-positive rate by simulation. Based on a particular genetic setting, one can randomly simulate both the marker data according to the laws of Mendelian inheritance and phenotypic data following a normal distribution under the null hypothesis of no QTL. The simulated data are analyzed by a given statistical method, and the value of the test statistic is calculated for each simulation replicate. The distribution of the LR values over a number of simulation replicates (say 1000) can be approximated by a χ^2 distribution. The 95th and 99th percentiles of the distribution of the maximum are used as empirical critical values to declare the existence of a QTL for a quantitative trait at the significance levels $\alpha = 0.05$ and 0.01.

12.2.4 Permutation Tests

Churchill and Doerge (1994 proposed a data-based numerical method, based on the concept of a permutation test, to estimate empirical critical values for mapping QTL for a given data set. By randomly reshuffling the relationships between the phenotypic and marker data across individuals, permutation tests generate a new data set in which the original marker–QTL association was destroyed. The great advantages of this approach are its conceptual simplicity, its distribution-free nature, and its general applicability in different population structures, although its computational workload is heavy. The method proceeds as follows:

1. Randomly pair individual marker genotypes with an individual trait phenotype to generate a permuted sample of the data (this simulates the null hypothesis of no association between the QTL and phenotype).
2. Perform an interval mapping analysis on the permuted sample.
3. Repeat steps 1 and 2 a number of times to obtain an empirical distribution of the test statistic at the null hypothesis for determining an appropriate critical value for the test statistic in the original data analysis.

More specifically, let $\mathbf{X} = \{X_1, X_2, ..., X_n\}$ denote a random sample of size n from a population. In QTL mapping, each observation point X_i is composed of a trait value y_i and marker genotype $\mathbf{M}_i = \{M_{i1}, M_{i2}, ..., M_{im}\}$ for m markers; i.e.,

$$X_i = \{y_i, \mathbf{M}_i\}.$$

A permuted sample of size n, denoted $\mathbf{X}' = \{X'_1, X'_2, ..., X'_n\}$, is a sample from the original sample without replacement, such that there are random matches between y_i and $\mathbf{M}_i$. That is,

$$X'_k = \{y_{i_k}, \mathbf{M}_{i_k}\},$$

where the y_{i_k} is chosen without replacement from $(y_1, \ldots, y_n)$ and, independently, $\mathbf{M}_{i_k}$ is chosen without replacement from $(\mathbf{M}_1, \ldots, \mathbf{M}_n)$. Thus, in permuted samples, $\mathbf{X}'$, y', and $\mathbf{M}'$ do not have any intrinsic relationship, thus simulating the null hypothesis of no QTL.

For N permuted samples, let LR'_p be the maximum likelihood test statistic for a particular genomic position or maximum value of the test statistic for a marker interval or for the entire genome in the pth permuted sample. The $\alpha \times 100$ percent threshold of the test statistic under the null hypothesis can be estimated empirically as

$$\hat{T}_\alpha = \alpha \times 100 \text{ percent of } \{\mathrm{LR}'_1, \mathrm{LR}'_2, ..., \mathrm{LR}'_N\}.$$

How large should N be for practical data analysis? Churchill and Doerge (1994) suggest that for $\alpha = 0.05$, N should be at least 1000, and for $\alpha = 0.01$, N should be at least 5000.

12.2.5 A Quick Approach

To overcome the drawback of empirical approaches caused by their computational load, Piepho (2001) proposed a quick method to compute approximate threshold values that control the genome-wide type I error rate of tests for QTL detection based on Davies's (1977) results. Given a mixture of normal distributions with constant variance and location parameters depending on QTL effects, LR test statistics at different QTL positions (θ) are calculated. Assuming that LR values are a continuous function of θ, expressed as $L(\theta)$, and conditional on the QTL position, $L(\theta)$ follows a χ^2–distribution with k degrees of freedom, where k is the number of genetic effects for a putative QTL. The upper bound of the chromosome-wide type I error rate (α) is estimated (Piepho 2001) by

$$\alpha = Pr(\chi^2_k > T) + \frac{VT^{\frac{1}{2}(k-1)}e^{-\frac{1}{2}T}2^{-(\frac{1}{2})^k}}{\Gamma(\frac{1}{2}k)}, \tag{12.8}$$

where $Pr(\chi^2_k > T)$ is the cumulative distribution function of χ^2 with k degrees of freedom, T is the critical threshold value for the LR test statistic, $\Gamma(\cdot)$ is the Gamma function, and

$$\begin{aligned} V &= \int_0^\ell \left| \frac{\partial\sqrt{L(\theta)}}{\partial\theta} \right| d\theta \\ &= \left|\sqrt{L(0)} - \sqrt{L(\theta_1)}\right| + \left|\sqrt{L(\theta_1)} - \sqrt{L(\theta_2)}\right| + \cdots + \left|\sqrt{L(\theta_s)} - \sqrt{L(\theta_\ell)}\right|, \end{aligned}$$

where $\theta_1, \cdots, \theta_s$ are the successive turning points (points of inflection) of $\sqrt{L(\theta)}$ (i.e., the values of θ where the first derivative $\partial\sqrt{L(\theta)}/\partial\theta$ changes sign). This change of sign occurs at the local minima and maxima of $\sqrt{L(\theta)}$.

For a given α, T may be found from equation (12.8) by numerical methods. The problem in practice is to find the turning points $\theta_1, \cdots, \theta_s$. In most cases, this will have to be done numerically. Usually a grid search is done over all θ, so the turning points can only be determined to the accuracy given by the step size of the grid. We therefore suggest using a relatively fine grid (e.g., between 1 and 2 cM). The analysis is simplified by pretending that every point on the grid is a turning point.

Using the Bonferroni inequality, Piepho (2001) showed that the genome-wide type I error rate is calculated by

$$\gamma = CPr(\chi_k^2 > T) + \frac{(\sum_{c=1}^{C} V_c)T^{\frac{1}{2}(k-1)}e^{-\frac{1}{2}T}2^{-(\frac{1}{2})^k}}{\Gamma(\frac{1}{2}k)}, \tag{12.9}$$

where C is the number of chromosomes and V_c is the value of V for the cth chromosome. Instead of choosing the same α_c for each chromosome, it is suggested that a common critical value T be used for all chromosomes, while α_c may be different on each chromosome.

Example 12.2. (**Empirical Determination of Critical Thresholds**). Examples 12.4 and 11.5 report the results for QTL mapping in the backcross-like DH population of rices and the F_2 population of mice, respectively. Here, we calculate the critical thresholds for these two examples using analytical approaches, Piepho's (2001) quick approach, simulation studies, and permutation tests as below.

	DH of Rices		F_2 of Mice	
Approach	$\alpha = 0.05$	$\alpha = 0.01$	$\alpha = 0.05$	$\alpha = 0.01$
Quick approach	12.96	16.18	16.45	19.86
Simulation studies	9.14	12.64	10.60	14.58
Permutation tests	8.20	12.79	9.71	13.49

It can be seen that the three approaches provide broadly consistent results for the critical thresholds. A quick approach tends to be more conservative than the two others, especially when the markers are relatively dense (Piepho 2001).

12.2.6 A Score Statistic

More recently, Zou et al. (2004) have proposed a score statistic to test the significance of genome-wide QTL mapping for experimental crosses. The proposed method based on a resampling procedure is computationally much less intensive than the permutation procedure (on the order of 10^2 or higher) and is applicable to complex breeding designs and sophisticated genetic models that cannot be handled by the permutation and theoretical methods.

The idea of this approach is derived from the argument that the likelihood ratio test statistic for testing the existence is equivalent to the score test statistic in large samples (see Cox and Hinkley 1974). The score statistic can be approximated by a sum of independent random vectors and, as a result, its distribution for large samples can be readily derived. In interval mapping, we denote the vectors of QTL effect parameters by β and these of nuisance parameters (including the overall mean and residual variance) by η. These two types of parameters are arrayed as $\mathbf{\Phi} = (\beta, \eta)$. The log-likelihood for $\mathbf{\Phi}$ at a given position x for individual i can be expressed as

$$L_i(\mathbf{\Phi}; x) = \log \left[\sum_{j=1}^{J} \omega_{j|i}(x) f_j(y_i; \mathbf{\Phi}) \right], \tag{12.10}$$

assuming there are J QTL genotypes for a mapping population. Let LR(x) be the LR value at a position x over the genome or a specific region. Because x can be anywhere in the genome, the distribution of LR(x) can be regarded as a stochastic process.

The score functions of individual i for β and η are calculated, respectively, by

$$S_{\beta i}(\mathbf{\Phi}; x) = \frac{\partial}{\partial \beta} L_i(\mathbf{\Phi}; x),$$
$$S_{\eta i}(\mathbf{\Phi}; x) = \frac{\partial}{\partial \eta} L_i(\mathbf{\Phi}; x).$$

Under the null hypothesis $H_0 : \beta = 0$, we estimate the nuisance parameters denoted as $\widetilde{\eta}$. Let $S_i(x)$ be the score function of an individual i for β evaluated at $H_0 : \beta = 0$ and $\eta = \widetilde{\eta}$. Based on Taylor series expansions and the law of large numbers, we derive the expression of $S_i(x)$ (Zou et al. 2004) and estimate it by

$$\widehat{S}_i(x) = S_{\beta i}(0, \widetilde{\eta}; x) - \left[\frac{\partial^2}{\partial \beta \partial \eta} L_i(0, \widetilde{\eta}; x) \right] \left[\frac{\partial^2}{\partial \eta^2} L_i(0, \widetilde{\eta}; x) \right]^{-1} S_{\eta i}(0, \widetilde{\eta}; x).$$

The score test statistic for $H_0 : \beta = 0$ against $H_1 : \beta \neq 0$ at position x takes the form

$$W(x) = \widehat{S}^{\mathrm{T}}(x) \widehat{V}^{-1}(x) \widehat{S}(x), \tag{12.11}$$

where

$$\widehat{S}(x) = \sum_{i=1}^{n} \widehat{S}_i(x),$$
$$\widehat{V}(x) = n \widehat{\Psi}(x, x),$$

with $\widehat{\Psi}(x, x)$ being a special case of

$$\widehat{\Psi}(x_1, x_2) = n^{-1} \sum_{i=1}^{n} \widehat{S}_i(x_1) \widehat{S}_i^{\mathrm{T}}(x_2)$$

as the covariance between $n^{-1/2} S(x_1)$ and $n^{-1/2} S(x_2)$ at any two given positions x_1 and x_2. As shown in Cox and Hinkley (1974), $W(x)$ is asymptotically equivalent to LR(x). The genome-wide significance of a QTL can be tested by evaluating the distribution of $\max_x W(x)$ through a resampling method. This is described by Zou et al. (2004) as follows:

1 Sample an independent standard normal random variable G_i $(i = 1, \ldots, n)$ from $N(0, 1)$.
2 Calculate

$$S^*(d) = \sum_{i=1}^{n} \widehat{S}_i(x) G_i,$$

$$W^*(d) = S^{*\mathrm{T}}(d) \widehat{V}^{-1}(d) S^*(d),$$

$$M^* = \max_x W^*(x).$$

3 Repeat steps 1 and 2 R times (R is a large number).
4 For a given genome-wide type I error rate α, calculate the $100(1-\alpha)$th percentile of the R values of the M^*. If the observed value of the LR exceeds this threshold, then reject the null hypothesis.

Example 12.3. Zou et al. (2004) used a published *Drosophila* data set (Zeng et al. 2000) to demonstrate the usefulness of the score test statistic. A linkage map composed of 42 markers was constructed for chromosomes X, 2, and 3 in a backcross of 299 flies between *D. simulans* and *D. mauritiana*. Interval mapping was used to search for QTL that affect a morphometric trait. Figure 12.1 illustrates the profile of LOD scores calculated at every 1 cM across the three chromosomes. The genome-wide critical threshold was determined by permutation tests and the score statistic approach, both giving similar values, 10.08 and 9.96, at the 5 percent significance level based on 10,000 permutations and resamples (Fig. 12.1). But these two approaches differ dramatically in computing time, with the score statistic using 13 seconds whereas permutation tests use 6000 seconds.

12.3 Precision of Parameter Estimation

Precise estimates of QTL parameters are crucial for genetic mapping. The estimation precision of parameters can be assessed by estimating the sampling variances of the estimates derived from the asymptotic variance-covariance matrix. An alternative is to simulate multiple data sets via mimicking the data structure of the example and estimate the means and sampling variance of the simulation replicates.

12.3.1 Asymptotic Variance-Covariance Matrix

After the point estimates of parameters are obtained by the EM algorithm, we need to derive the asymptotic variance-covariance matrix and evaluate the sampling errors of the estimates. In this section, we provide a procedure for deriving the asymptotic variance-covariance matrix for QTL positions and effects. The techniques for so doing

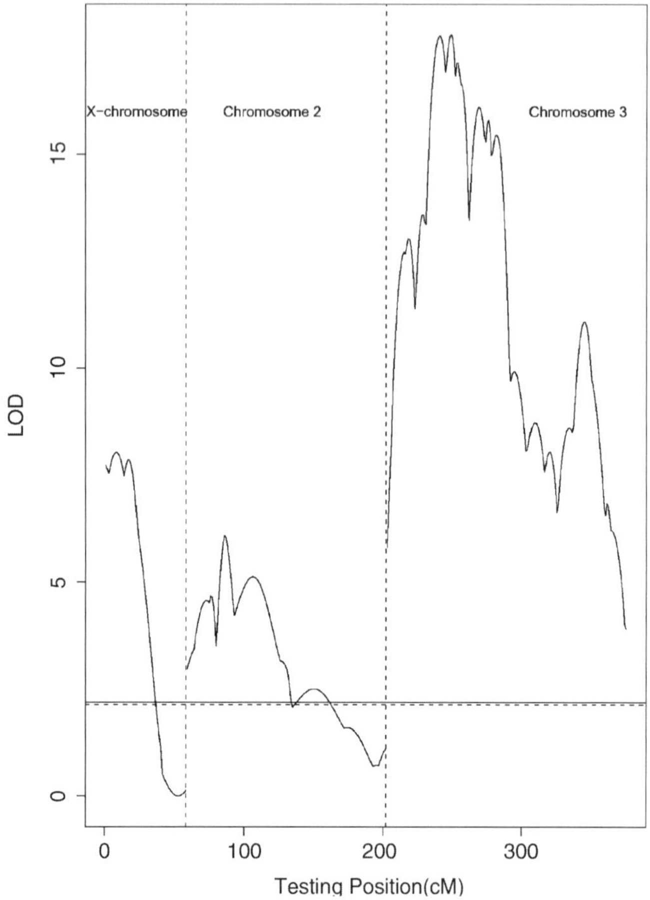

Fig. 12.1. The profile of the LOD score across chromosomes X, 2, and 3. The horizontal lines represent the 95 percent resampling (solid) and permutation (dashed) thresholds. Adapted from Zou et al. (2004).

involve calculation of the incomplete-data information matrix which is the negative second-order derivative of the incomplete-data log-likelihood. An in-depth discussion about this procedure was given in Kao and Zeng (1997). As seen from above, the EM algorithm does not automatically generate the variance-covariance matrix for the estimates, which thus suggests that some extra steps are necessary to do so. Louis (1982) and Meng and Rubin (1991) derived a general procedure for estimating the variance-covariance matrix within the mixture-model context. The basic idea of this procedure is that incomplete-data (observed) information can be obtained by extracting the missing-data information from the complete-data information.

Equations (9.2) and (9.3) are the mixture models in which the phenotype (y_i) and marker information ($\mathbf{M}$) are the *observed* or *incomplete* data, denoted by Y_{obs}, whereas the QTL genotype information contained in the conditional probabilities ($\omega_{j|i}$) presents the *unobserved* or *missing* data, denoted by Y_{mis}. The combination of incomplete and missing data is the *complete* data denoted by Y_{com}.

The joint probability distribution function of the complete-data can be factored as

$$
\begin{aligned}
f(Y_{com}|\boldsymbol{\Omega}) &= f(Y_{obs}, Y_{mis}|\boldsymbol{\Omega}) \\
&= f(Y_{obs}|\boldsymbol{\Omega}) f(Y_{mis}|Y_{obs}, \boldsymbol{\Omega}),
\end{aligned} \tag{12.12}
$$

where $f(Y_{obs}|\boldsymbol{\Omega})$ is the density of the observed data, which is the mixture likelihood, and $f(Y_{mis}|Y_{obs}, \boldsymbol{\Omega})$ is the density of missing data given observed data. The log-likelihood corresponding to $f(Y_{com}|\boldsymbol{\Omega})$ is

$$
\log L(\boldsymbol{\Omega}|Y_{com}) = \log L(\boldsymbol{\Omega}|Y_{obs}) + \log f(Y_{mis}|Y_{obs}, \boldsymbol{\Omega});
$$

that is,

$$
\log L(\boldsymbol{\Omega}|Y_{obs}) = \log L(\boldsymbol{\Omega}|Y_{com}) - \log f(Y_{mis}|Y_{obs}, \boldsymbol{\Omega}). \tag{12.13}
$$

By taking second derivatives and expectations over $f(Y_{mis}|Y_{obs}, \boldsymbol{\Omega})$ and evaluating at $\boldsymbol{\Omega} = \widehat{\boldsymbol{\Omega}}$ for equation (10.18), Louis (1982) found the observed information

$$
I_{obs}(\widehat{\boldsymbol{\Omega}}|Y_{obs}) = I_{com} - I_{mis}, \tag{12.14}
$$

where the missing information is

$$
\begin{aligned}
I_{mis} &= E\left[-\frac{\partial^2 \log f(Y_{mis}|Y_{obs}, \boldsymbol{\Omega})}{\partial \boldsymbol{\Omega}^2}|Y_{obs}, \boldsymbol{\Omega}\right]_{\boldsymbol{\Omega}=\widehat{\boldsymbol{\Omega}}} \\
&= E[S(Y_{com}, \boldsymbol{\Omega}) S^{\mathrm{T}}(Y_{com}, \boldsymbol{\Omega})|Y_{obs}, \boldsymbol{\Omega}]_{\boldsymbol{\Omega}=\widehat{\boldsymbol{\Omega}}},
\end{aligned}
$$

and the complete information is

$$
I_{com} = E\left[-\frac{\partial^2 \log f(Y_{com}|\boldsymbol{\Omega})}{\partial \boldsymbol{\Omega}^2}|Y_{obs}, \boldsymbol{\Omega}\right]_{\boldsymbol{\Omega}=\widehat{\boldsymbol{\Omega}}},
$$

with s being the gradient vector of the log-likelihood.

Based on equation (10.19), the observed information matrix for the independent but not necessarily identically distributed case can be expressed as

$$
\begin{aligned}
&I_{obs}(\widehat{\boldsymbol{\Omega}}|Y_{obs}) \\
&= \sum_{i=1}^{n} E\left[-\frac{\partial^2 \log f_i(Y_{com}|\boldsymbol{\Omega})}{\partial \boldsymbol{\Omega}^2}|Y_{obs}, \boldsymbol{\Omega}\right]_{\boldsymbol{\Omega}=\widehat{\boldsymbol{\Omega}}} \\
&\quad - \sum_{i=1}^{n} E[S_i(Y_{(com,i)}, \boldsymbol{\Omega}) S_i^{\mathrm{T}}(Y_{(com,i)}, \boldsymbol{\Omega})|Y_{(obs,i)}, \boldsymbol{\Omega}]_{\boldsymbol{\Omega}=\widehat{\boldsymbol{\Omega}}} \\
&\quad - \sum_{i_1 \neq i_2} E[S_{i_1}(Y_{(com,i_1)}, \boldsymbol{\Omega})|Y_{(obs,i_1)}, \boldsymbol{\Omega}]_{\boldsymbol{\Omega}=\widehat{\boldsymbol{\Omega}}} E[S_{i_2}^{\mathrm{T}}(Y_{(com,i_2)}, \boldsymbol{\Omega})|Y_{(obs,i_2)}, \boldsymbol{\Omega}]_{\boldsymbol{\Omega}=\widehat{\boldsymbol{\Omega}}}.
\end{aligned}
$$

Kao and Zeng (1997) provided the expressions of each element in both the complete and missing information matrices.

12.3.2 Simulation Studies

From a real data set, interval mapping has been used to estimate the position and effects of a QTL and the residual variance behind it. The precision of parameter estimation can be investigated through simulation studies. Here, a basic simulation scheme used to serve this purpose is described.

Consider a backcross in which a total of m markers were genotyped to construct a linkage map and a quantitative trait was phenotyped for each of n individuals. Interval mapping has successfully detected a QTL ($\mathbf{Q}$) located at a position ($\widehat{\theta}$) between a pair of flanking markers $\mathbf{M}_l$ and $\mathbf{M}_{l+1}$ with the recombination fraction of r_l ($l = 1, \ldots, m-1$) on a chromosome. The additive effect of the QTL was estimated as $\widehat{a}$ and two nuisance parameters estimated as $\widehat{\mu}$ for the overall mean and $\widehat{\sigma}^2$ for the residual variance. Based on these estimates, we need to simulate the same linkage map composed of m markers and the phenotypic values determined by $\mathbf{Q}$ and residual errors for this backcross of size n. The marker genotype observations of n individuals at m known markers are sampled from a polynomial distribution,

$$[(1-r_1)(1-r_2)\ldots(1-r_{m-1})]^{N_1}[(1-r_1)(1-r_2)\ldots r_{m-1}]^{N_2}\ldots[r_1r_2\ldots r_{m-1}]^{N_{2^m}},$$

where $n = N_1 + N_2 + \ldots + N_{2^m}$.

The statistical model used to simulate the phenotypic value for individual i takes the form

$$y_i = \widehat{\mu} + x_i\widehat{a} + e_i, \tag{12.15}$$

where x_i is the indicator variable, defined as 1 or 0 depending on the QTL genotype, and e_i is the residual error for individual i, following $N(0, \widehat{\sigma}^2)$. For any individual i whose marker genotype is known, the probability of its x_i taking 1 or 0 is determined by the conditional probability of a QTL genotype given the marker genotype at $\mathbf{M}_l$ and $\mathbf{M}_{l+1}$ (Table 10.3 or Table 10.4).

The simulated data that completely mimicked the real data set are analyzed by interval mapping in which the same set of parameters (μ, a, σ^2, θ) can be estimated. This procedure is repeated a large number of times (e.g., 1000), which allows the estimates of the means and the sampling errors of the estimates. Based on the means and sampling errors, one can determine the accuracy and precision of each parameter estimation.

Another use of simulation studies is to examine the power of interval mapping to analyze this given real data set. For each simulation replicate, the critical threshold for declaring the existence of a QTL can be determined using the approaches above. This threshold is compared with the LR value calculated from the simulated data to detect whether the QTL is significant at a given significance level. The proportion of the number of simulation replicates in which the QTL is detected to be significant over the total number of simulation replicates can be empirically used as the power for QTL detection in the given real data set.

Example 12.4. Revisit Example 3.1. Two inbred lines, semi-dwarf IR64 and tall Azucena, were crossed to generate a doubled haploid (DH) population in which a linkage map was constructed and plant heights were measured at age 10 weeks. In Example 12.4, a significant QTL was detected between markers RG810 and RG331 on chromosome 1 (Fig. 11.2).

Table 12.2 gives the MLEs of the QTL parameters and the estimates of the sampling errors (SE) of the MLEs by the asymptotic variance-covariance matrix for the detected QTL at the peak of the LR profile. The estimated values are used to simulate the data by mimicking the data structure of the real example. Simulated data are estimated for the parameters given in Table 12.2 and the same procedure is repeated 1000 times. The means and sampling errors of all simulation replicates for each parameter, along with the power to detect this QTL, are given in Table 12.2. It is found that genetic parameters for this DH population can be estimated with good precision, and the power to detect significant QTLs with this population is very high. The asymptotic and simulation approaches provide consistent estimation precision for the parameters of QTL position and effect and two nuisance parameters (overall mean, μ, and residual variance, σ^2).

Table 12.2. Precision analysis of QTL mapping in a rice DH population by the asymptotic and simulation approaches.

	Asymptotic		Simulation		
Parameter	MLE	SE	Mean	SE	Power
μ	109.32	1.7	109.34	1.8	
a	30.79	3.4	31.03	3.6	0.99
σ^2	295.46	29.5	278.55	45.2	
Position	215.98	1.6	216.04	1.0	

12.4 Confidence Intervals for the QTL Location

In the preceding section, we attempted to estimate the asymptotic variance-covariance matrix for all the QTL parameters, including the QTL location (θ). To avoid non-identifiability between the QTL effect and position under the null hypothesis, we often treat the QTL location as a constant. If the QTL location is fixed, the above-mentioned asymptotic derivation is only needed for the QTL effects and residual variance, arrayed in $\mathbf{\Omega}$. In this case, we need to estimate the sampling error and confidence interval for the QTL location based on other approaches.

Lander and Botstein (1989) proposed a dropoff method to estimate the "support intervals" or confidence intervals (CI) for the QTL location based on the likelihood ratio test. Using this method, the CI is calculated by finding the location at either side of the estimated QTL location that corresponds to a decrease in the LOD score of 1 or 2 units. The total width corresponding to a 1 or 2 LOD dropoff is then taken as the CI and, asymptotically, these should be approximately equivalent to 96.6 percent and 99.8 percent CI, respectively (Mangin et al. 1994). However, they noted that the LOD dropoff method of Lander and Botstein (1989) consistently underestimated the CI, especially for small QTL effects. They derived a complicated test statistic that accurately estimated the CI for small QTL effects, but the distribution of this test statistic must be computed empirically (Mangin et al. 1994).

A confidence interval can also be estimated by the simulation of a large number of samples. If 1000 samples are generated, the 95 percent confidence limits are obtained by determining the 25 lowest and 25 highest estimates for each parameter. Darvasi et al. (1993) estimated QTL parameter estimation error variances based on the Fisher information matrix and by repeated simulation for the backcross design with marker brackets. The 95 percent CI was then estimated as a ± 2 estimation SE for each parameter. Darvasi et al. (1993) also directly estimated the 95 percent CI for each parameter by repeated simulation. All methods were very accurate for estimation of QTL effect variances. Estimates based on the second derivative tended to slightly overestimate the SE for QTL means relative to the empirical estimates, especially for a large spacing between markers.

For the QTL map location, the estimates based on the empirical 95 percent CI and estimates based on four times the empirical SE were generally similar. However, estimates based on the second derivative matrix tended to underestimate the CI for small marker intervals and overestimate the CI for large marker intervals. Differences were in some cases more than two-fold. Clearly, for this parameter, the asymptotic properties of the second derivative matrix do not hold.

Empirical methods for estimating the CI also include parametric and nonparametric bootstrap and jackknife methods (Efron and Tibshirani 1993). In the parametric bootstrap method, parameter estimates are first obtained by any of the methods considered. In the second step, a large number of bootstrap samples are then derived from the assumed theoretical distribution, assuming that the original parameter estimates are the parameter values. Parameter estimates are then obtained for each sample. The CI for each parameter is then derived from the empirical distributions of the parameter estimates from the samples generated.

In nonparametric bootstrapping (Visscher et al. 1996), a large number of bootstrap samples are generated by sampling with repeats from the original data. Thus, in a particular sample some of the actual records will appear more than once, while other observations will be missing. If the actual data consist of at least several hundred points, it will be possible to draw a virtually unlimited number of different samples in this method. The parameter estimates are then derived for each sample, and as in parametric bootstrapping, the distribution of these estimates is used to derive empirical CI limits. This method is not strictly nonparametric because assumptions about the distribution are still employed to derive parametric estimates for each sample. This method is more robust to violations of assumptions used to derive parameter estimates.

12.5 Exercises

12.1 There has been an increasing number of data sets for QTL mapping in various organisms. Try to work on a real example and practice how a critical threshold can be determined using different approaches.

12.2 By fixing a putative QTL at a given chromosomal position, Kao and Zeng (1997) derived general formulas for estimating the MLEs of QTL effects and nuisance parameters and the asymptotic variance-covariance matrix of the MLEs. Chen (2005) derived the simultaneous estimation of the asymptotic variance-covariance matrix for the QTL position and effect and nuisance parameters with the EM context. The difference between these two studies lies in the estimation strategy for the QTL position. Kao and Zeng used a grid approach, whereas Chen made use of a closed form of the position estimation. Chen's approach made it possible to directly estimate the sampling variance of the QTL position.

In his derivations for the backcross, however, Chen (2005) used simplified but approximated conditional probabilities (Table 10.4). As shown before, this approximated form works only when marker intervals are reasonably small, which is hard in many practical mapping studies. It would be a nice addition to the mapping literature if Chen's approach could be incorporated by exact conditional probabilities (Table 10.3). Please try to answer the following questions:

(a) Show how the MLE of the QTL position and its sampling error can be estimated by considering the conditional probabilities of Table 10.3.

(b) Derive the asymptotic variance-covariance matrix for the QTL effect, position, and nuisance parameters for the F_2.

(c) Consider how the exact conditional probabilities of the F_2 can be incorporated for the estimation of the asymptotic variance-covariance matrix.

13

Composite QTL Mapping

13.1 Introduction

Lander and Botstein's interval method has an advantage for mapping QTLs genome-wide by scanning for the position of a QTL throughout the genome. But this method can lead to biased estimates of QTL positions and effects when multiple QTLs occur on the same linkage group because it makes use of one single-marker interval at a time and has no mechanism to alleviate the impact of other QTLs outside the interval. For this reason, if a real QTL is located near a marker interval with no QTL, interval mapping may still detect a "ghost" QTL due to the linkage between the real QTL and the interval being tested (Martinez and Curnow 1992). Although a simultaneous search for multiple QTLs on different intervals can overcome this problem, this will bring about new difficulties in parameter estimation and model identifiability.

Referring to Fig. 10.1, if the interest is in mapping a QTL effect in the interval $(\mathbf{M}, \mathbf{N})$, there could be confounding effects from markers (or QTLs) that are outside the interval. A natural way to eliminate the influences of genetic background is to attempt to remove this confounding information using covariates or cofactors. This is the approach of *composite interval mapping*, which constructs test statistics by combining interval mapping on two flanking markers and multiple regression analysis on the other markers. This composite interval mapping strategy, proposed independently by Zeng (1993, 1994) and Jansen and Stam (1994), removes some of the interference from QTLs outside the interval being tested. Zeng (1993) particularly demonstrates the advantages and disadvantages of composite interval mapping.

In this chapter, we introduce the basic theory for composite interval mapping and its algorithm for the estimation of QTL position and effect parameters. We will compare the advantages of composite interval mapping over interval mapping through simulation studies and real examples. The areas in which composite interval mapping can be improved toward a complete characterization of the genetic architecture of a quantitative trait are mentioned.

13.2 Composite Interval Mapping for a Backcross

We start the description of the analysis of composite interval mapping with a backcross design toward the homozygous parent carrying unfavorable alleles. Assume that this backcross has a size n in which a number of markers $(\mathbf{M}_1, \ldots, \mathbf{M}_m)$ have been genotyped to construct a linkage map. At a putative QTL bracketed by a particular pair of markers, $\mathbf{M}_t$ and $\mathbf{M}_{t+1}$, there are two classes of genotypes, heterozygote Qq (1) and homozygote qq (0), in the backcross.

A linear model for combining the testing interval formed by the two flanking markers and other markers is expressed as

$$y_i = \mu + (x^*_{1|i} - x^*_{0|i})a + \sum_{k=1}^{m-2} b_k x_{ki} + e_i, \tag{13.1}$$

where y_i is the phenotypic trait of the ith progeny in the backcross, μ is the overall mean, a is the additive genetic effect of the QTL, $x^*_{1|i}$ and $x^*_{0|i}$ are the indicator variables that define the QTL genotype (Qq or qq) of progeny i at a fixed location in $\mathbf{M}_t$–$\mathbf{M}_{t+1}$ based on the interval information, x_{ki} is the genotype of progeny i on the kth marker outside of the interval and defined as 1 for the marker heterozygote and 0 for the marker homozygote, $b_1, \ldots, b_{m-2}$ are the partial regression coefficients (additive marker effects) to be estimated, and e_i is the residual error, which includes effects exerted by other QTLs and nongenetic factors, $e_i \sim N(0, \sigma^2)$.

This approach is essentially interval mapping but controlling for the effects outside the interval. Thus, the methodologies are very similar to those in Section 11.2, except they are complicated by the necessity of estimating the additional effects.

13.2.1 The Likelihood

Following the development in Section 11.2, similar to equation (11.3), we have the likelihood

$$L(a, \mathbf{b}, \sigma^2, \omega_{1|i}, \omega_{0|i}|\mathbf{y}) = \prod_{i=1}^{n}[\omega_{1|i} f_1(y_i|a, \mathbf{b}, \sigma^2) + \omega_{0|i} f_0(y_i|\mathbf{b}, \sigma^2)] \tag{13.2}$$

or

$$\begin{aligned} L(a, \mathbf{b}, \sigma^2, \theta|\mathbf{y}) = & \prod_{i=1}^{n_1} f_1(y_i|a, \mathbf{b}, \sigma^2) \\ & \times \prod_{i=1}^{n_2}[(1-\theta) f_1(y_i|a, \mathbf{b}, \sigma^2) + \theta f_0(y_i|\mathbf{b}, \sigma^2)] \\ & \times \prod_{i=1}^{n_3}[\theta f_1(y_i|a, \mathbf{b}, \sigma^2) + (1-\theta) f_0(y_i|\mathbf{b}, \sigma^2)] \\ & \times \prod_{i=1}^{n_4} f_0(y_i|\mathbf{b}, \sigma^2) \end{aligned} \tag{13.3}$$

The mixture-model–based likelihood in equation (13.2) is specified by a general form of mixture proportions, $\omega_{1|i}$ and $\omega_{0|i}$, the conditional probabilities of QTL genotypes Qq and qq, respectively, given the marker interval genotype of progeny i. The conditional probabilities in the likelihood of equation (13.3) are a simple form assuming no double recombination (Table 10.4).

The normal density probability of the trait value within each QTL genotype class is defined as

$$f_1(y_i|a, \mathbf{b}, \sigma^2) = N\left(a + \mathbf{X}_i\mathbf{b}^\mathrm{T}, \sigma^2\right),$$
$$f_0(y_i|\mathbf{b}, \sigma^2) = N\left(\mathbf{X}_i\mathbf{b}^\mathrm{T}, \sigma^2\right),$$

because

$$y_i = \mu + a + \sum_{k=1}^{m-2} b_k x_{ki} + e_i = a + \mathbf{X}_i\mathbf{b}^\mathrm{T} + e_i, \text{ for } Qq\ (x_i = 1),$$
$$y_i = \mu + \sum_{k=1}^{m-2} b_k x_{ki} + e_i = \mathbf{X}_i\mathbf{b}^\mathrm{T} + e_i, \text{ for } qq\ (x_i = 0),$$

where $\mathbf{b} = (\mu, b_1, \ldots, b_{m-2})$ and $\mathbf{X}_i = (1, x_{1i}, x_{2i}, \ldots, x_{(m-2)i})$.

13.2.2 Maximizing the Likelihood

The parameters contained in the likelihood (13.2) are estimated by implementing the EM algorithm. This can be done using a procedure similar to that described in Section 11.2.2. In the E step, we define the posterior probability that progeny i carries a particular QTL genotype given its marker genotypes and phenotypic value as

$$P_{1|i} = \frac{\omega_{1|i} f_1(y_i|a, \mathbf{b}, \sigma^2)}{\omega_{1|i} f_1(y_i|a, \mathbf{b}, \sigma^2) + \omega_{0|i} f_0(y_i|\mathbf{b}, \sigma^2)},$$
$$P_{0|i} = \frac{\omega_{0|i} f_0(y_i|\mathbf{b}, \sigma^2)}{\omega_{1|i} f_1(y_i|a, \mathbf{b}, \sigma^2) + \omega_{0|i} f_0(y_i|\mathbf{b}, \sigma^2)}. \quad (13.4)$$

In the M step, the parameters are estimated in terms of the posterior probabilities by differentiating the log-likelihood (13.2) with respect to each parameter, setting the derivatives equal to zero, and solving the log-likelihood equations. This procedure is described as follows.

For a, we have

$$\frac{\partial}{\partial a} \ln L(a, \mathbf{b}, \sigma^2, \theta|\mathbf{y}) \propto \sum_{i=1}^{n} P_{1|i} \left(\frac{y_i - a - \mathbf{X}_i\mathbf{b}^\mathrm{T}}{\sigma^2}\right) = 0,$$

which leads to

$$\hat{a} = \frac{\sum_{i=1}^{n}(y_i - \mathbf{X}_i\mathbf{b}^{\mathrm{T}})P_{1|i}}{\sum_{i=1}^{n} P_{1|i}}$$

(13.5)
$$= \frac{1}{c}(\mathbf{y} - \mathbf{X}\mathbf{b}^{\mathrm{T}})^{\mathrm{T}}\mathbf{P}_1,$$

where $\mathbf{y} = \{y_i\}_{n\times 1}$, $\mathbf{X} = \{X_i\}_{n\times 1}$, $\mathbf{P}_1 = \{P_{1|i}\}_{n\times 1}$, and $c = \sum_{i=1}^{n} P_{1|i}$.

For $\mathbf{b}$ and σ^2, we have

$$\frac{\partial}{\partial \mathbf{b}} \ln L(a, \mathbf{b}, \sigma^2, \theta|\mathbf{y}) \propto \sum_{i=1}^{n}[P_{1|i}\mathbf{X}_i(y_i - a - \mathbf{X}_i\mathbf{b}^{\mathrm{T}}) + P_{0|i}\mathbf{X}_i(y_i - \mathbf{X}_i\mathbf{b}^{\mathrm{T}})]\frac{1}{\sigma^2} = 0$$

and

$$\frac{\partial}{\partial \sigma^2} \ln L(a, \mathbf{b}, \sigma^2, \theta|\mathbf{y}) \propto \sum_{i=1}^{n}[P_{1|i}\mathbf{X}_i(y_i - a - \mathbf{X}_i\mathbf{b}^{\mathrm{T}})^2 + P_{0|i}\mathbf{X}_i(y_i - \mathbf{X}_i\mathbf{b}^{\mathrm{T}})^2]\frac{1}{2\sigma^4} - \frac{n}{2\sigma^2} = 0,$$

leading to

(13.6)
$$\hat{\mathrm{b}} = (\mathbf{X}^{\mathrm{T}}\mathbf{X})^{-1}\mathbf{X}^{\mathrm{T}}(\mathbf{y} - \mathbf{P}\hat{b}),$$

(13.7)
$$\hat{\sigma}^2 = (\mathbf{y} - \mathbf{X}\hat{\mathbf{b}}^{\mathrm{T}})^{\mathrm{T}}(\mathbf{y} - \mathbf{X}\mathbf{b}^{\mathrm{T}}) - \hat{a}^2 c.$$

For the estimation of the QTL position θ, two approaches can be used, one based on its closed form, like equation (11.7), derived from likelihood (13.3), and the second based on a grip scan for a series of fixed positions within a marker interval. Given the initial values for the unknown parameters $(a, \mathbf{b}, \sigma^2)$, the E and M steps are calculated with equations (13.4) and (13.5)–(13.7), respectively, and repeated until the estimates are stable.

13.2.3 Hypothesis Testing

The significance of the genetic effect of a QTL detected by composite interval mapping can be tested by formulating the hypotheses

$$H_0 : a = 0 \text{ vs. } H_1 : a \neq 0.$$

The likelihood value under the alternative hypothesis is calculated by plugging the MLEs of parameters into likelihood (13.2). The likelihood under the null hypothesis is constructed as

$$L(\mathbf{b}, \sigma^2|\mathbf{y}) = \prod_{i=1}^{n} f(y_i|\mathbf{b}, \sigma^2),$$

from which the MLEs are derived as

$$\widetilde{\mathbf{b}} = (\mathbf{X}^{\mathrm{T}}\mathbf{X})^{-1}\mathbf{X}^{\mathrm{T}}\mathbf{Y}$$
$$\sigma^2 = \frac{1}{n}(\mathbf{y} - \mathbf{X}\widetilde{\mathbf{b}}^{\mathrm{T}})^{\mathrm{T}}(\mathbf{y} - \mathbf{X}\widetilde{\mathbf{b}}^{\mathrm{T}}).$$

The likelihood ratio test statistic is then calculated using

$$\text{(13.8)} \qquad \mathrm{LR} = -2\ln\left[\frac{L(a=0,\widetilde{\mathbf{b}},\widetilde{\sigma}^2|\mathbf{y}}{L(\widehat{a},\widehat{\mathbf{b}},\widehat{\sigma}^2|\mathbf{y},\mathbf{M})}\right].$$

Note that the calculation of the likelihood under the alternative hypothesis involves marker information ($\mathbf{M}$), whereas under the null hypothesis marker information is not needed. The critical threshold for declaring the existence of a QTL is determined using the approaches proposed in Chapter 12. In practice, empirical approaches based on permutation tests, although they are computationally expensive, are widely used because they do not rely upon the distribution of the test statistics.

13.3 Composite Interval Mapping for an F_2

A similar procedure for composite interval mapping can be developed for the F_2 in which three genotypes at a QTL are denoted as QQ (2), Qq (1), and qq (0), respectively. The linear model of the phenotypic value of progeny i over the tested interval $\mathbf{M}_t$–$\mathbf{M}_{t+1}$, similar to equation (13.1), which explicitly models additive and dominance effects, can be written as

$$\text{(13.9)} \qquad y_i = \mu + (x^*_{2|i} - x^*_{0|i})a + x^*_{1|i}d + \sum_{k=1}^{m-2} x_{ki}b_k + \sum_{k=1}^{m-2} z_{ki}h_k + e_i,$$

where $x^*_{2|i}$, $x^*_{1|i}$, and $x^*_{0|i}$ are the indicator variables for the QTL genotypes for progeny i determined by its marker interval genotype, a and d are the additive and dominant effects of the QTL, x_{ki} and z_{ki} are the indicator variables for marker genotypes at $\mathbf{M}_k$,

$$x_{ki} = \begin{cases} \frac{1}{2} & \text{for } M_kM_k \\ 0 & \text{for } M_km_k \\ -\frac{1}{2} & \text{for } m_km_k \end{cases}$$

and

$$z_{ki} = \begin{cases} 1 & \text{for } M_km_k \\ 0 & \text{for the other,} \end{cases}$$

and b_k and h_k are the additive and dominance effects associated with marker $\mathbf{M}_k$.

The likelihood function of the observed phenotypic ($\mathbf{y}$) and marker data ($\mathbf{M}$) in the F_2 is

$$\text{(13.10)} \qquad L(\mathbf{\Omega}|\mathbf{y},\mathbf{M}) = \prod_{i=1}^{n}[\omega_{2|i}f_2(y_i) + \omega_{1|i}f_1(y_i) + \omega_{0|i}f_0(y_i)],$$

where the $\omega_{j|i}$'s ($j = 2, 1, 0$) are the F_2 weights for each QTL genotype given the marker genotype of progeny i, which are determined by the QTL position in the marker interval (see Table 11.7), and the normal distributions are given by

$$f_2(y_i) = \frac{1}{\sqrt{2\pi}\sigma} \exp\left[-\frac{(y_i - \mathbf{X}_i\mathbf{b}^{\mathrm{T}} + a)^2}{2\sigma^2}\right],$$

$$f_1(y_i) = \frac{1}{\sqrt{2\pi}\sigma} \exp\left[-\frac{(y_i - \mathbf{X}_i\mathbf{b}^{\mathrm{T}} + d)^2}{2\sigma^2}\right],$$

$$f_0(y_i) = \frac{1}{\sqrt{2\pi}\sigma} \exp\left[-\frac{(y_i - \mathbf{X}_i\mathbf{b}^{\mathrm{T}} - a)^2}{2\sigma^2}\right],$$

with

$$\mathbf{X}_i\mathbf{b}^{\mathrm{T}} = \mu + \sum_{k=1}^{m-2} x_{ki}b_k + \sum_{k=1}^{m-2} z_{ki}h_k,$$

$$\mathbf{X}_i = (1, x_{1i}, \ldots, x_{(m-2)i}, z_{1i}, \ldots, z_{(m-2)i}),$$

$$\mathbf{b} = (\mu, b_1, \ldots, b_{m-2}, h_1, \ldots, h_{m-2}).$$

The likelihood (13.10) contains an unknown vector $\mathbf{\Omega} = (a, d, \mathbf{b}, \sigma^2, \omega_{j|i})$. As in the backcross, the EM algorithm can be implemented to obtain the MLEs of the parameters ($\mathbf{\Omega}$) that maximize the likelihood (13.10).

In the E step, we define the posterior probability of QTL j for progeny i by

$$\mathrm{P}_{j|i} = \frac{\omega_{j|i}f_j(y_i)}{\sum_{j'=0}^{2}[\omega_{j'|i}f_{j'}(y_i)]}, \tag{13.11}$$

with $\mathrm{P}_{2|i} + \mathrm{P}_{1|i} + \mathrm{P}_{0|i} = 1$. Then, in the M step, we obtain the MLEs of the unknown parameters as follows:

$$\begin{aligned}
\widehat{a} &= \frac{(\mathbf{y} - \mathbf{X}\widehat{\mathbf{b}}^{\mathrm{T}})^{\mathrm{T}}\mathrm{P}_2}{2\mathbf{1}^{\mathrm{T}}\mathrm{P}_2}, \\
\widehat{d} &= \frac{(\mathbf{y} - \mathbf{X}\widehat{\mathbf{b}}^{\mathrm{T}})^{\mathrm{T}}\mathrm{P}_1}{\mathbf{1}^{\mathrm{T}}\mathrm{P}_1} - \widehat{a}, \\
\widehat{\mathbf{b}} &= (\mathbf{X}^{\mathrm{T}}\mathbf{X})^{-1}\mathbf{X}^{\mathrm{T}}[\mathbf{y} - (2\mathrm{P}_2 + \mathrm{P}_1)\widehat{a} - \mathrm{P}_1\widehat{d}], \\
\widehat{\sigma}^2 &= \frac{1}{n}[(\mathbf{Y} - \mathbf{X}\widehat{\mathbf{b}}^{\mathrm{T}})^{\mathrm{T}}(\mathbf{y} - \mathbf{X}\widehat{\mathbf{b}}^{\mathrm{T}}) - 4(\mathbf{1}^{\mathrm{T}}\mathrm{P}_2)\widehat{a}^2 - (1^{\mathrm{T}}\mathrm{P}_1)(a+d)^2],
\end{aligned} \tag{13.12}$$

where $\mathrm{P}_2 = (\mathrm{P}_{2|1}, \ldots, \mathrm{P}_{2|n})^{\mathrm{T}}$ and $\mathrm{P}_1 = (\mathrm{P}_{1|1}, \ldots, \mathrm{P}_{1|n})^{\mathrm{T}}$.

Equations (13.11) and (13.12) construct an iterative procedure to solve for the MLEs of the unknown parameters.

For the F_2, hypothesis tests include three steps. First, the existence of a QTL can be teste on the basis of

(13.13) $H_0 : a = d = 0$ vs. H_1 : at least one of them is not equal to zero.

Second, the significance of the additive effect of the QTL (a) can be tested by

$$H_0 : a = 0 \text{ vs. } H_1 : a \neq 0. \tag{13.14}$$

Third, the dominance effect of the QTL (d) is tested using

$$H_0 : d = 0 \text{ vs. } H_1 : d \neq 0. \tag{13.15}$$

In each case, the log-likelihood ratios under the null and alternative hypotheses are calculated. The critical threshold for the first hypothesis (13.13) can be determined empirically from permutation tests. Those for the second (13.14) and third hypotheses (13.15) can be determined empirically by simulation studies. In fact, hypotheses (13.14) and (13.15) do not contain nonidentifiable parameters in their null hypothesis, and therefore their log-likelihood ratios can each be thought to follow a χ^2 distribution with one degree of freedom.

13.4 A Statistical Justification of Composite Interval Mapping

Zeng (1993) has shown that, under the assumption of no crossover inference and no epistasis, the true partial regression coefficient (true parameter) of a trait on a marker depends only on those QTLs that are located on the interval bracketed by these two flanking markers, independent of any other QTL.

Ideally, the test statistic constructed from a marker interval hypothesized to carry a QTL is independent of the effects of any possible QTL at the chromosomal region outside the interval. This relationship is a property of the parameters in a regression-based composite interval mapping model. Much of the discussion on the statistical foundations of composite interval mapping is from Zeng (1993).

13.4.1 Conditional Marker (Co)variances

Suppose there are two markers, $\mathbf{M}_j$ with two alleles M_j and m_j and $\mathbf{M}_k$ with two alleles M_k and m_k, genotyped for a backcross population. These two markers are linked with the recombination fraction r_{jk}. We denote by 1 and 0 the "values" for the heterozygote and homozygote, respectively, at each marker in the backcross. Thus, the frequencies and "value" of the two backcross genotypes are expressed as

	$M_jm_jM_km_k$	$M_jm_jm_km_k$	$m_jm_jM_km_k$	$m_jm_jm_km_k$
Frequency	$\frac{1}{2}(1-r_{jk})$	$\frac{1}{2}r_{jk}$	$\frac{1}{2}r_{jk}$	$\frac{1}{2}(1-r_{jk})$
"Value" at $\mathbf{M}_1$	1	1	0	0
"Value" at $\mathbf{M}_2$	1	0	1	0

The variances at each marker can be calculated as

$$\sigma_j^2 = \sigma_k^2 = \frac{1}{4} \tag{13.16}$$

and the covariance between the two markers calculated as

$$\sigma_{jk} = \frac{1}{4}(1 - 2r_{jk}), \tag{13.17}$$

with a correlation of $1 - 2r_{ij}$. Based on equations (13.16) and (13.17), the conditional variance of marker $\mathbf{M}_k$ given marker $\mathbf{M}_j$ is

$$\begin{aligned} \sigma_{k|j}^2 &= \sigma_k^2 - \frac{\sigma_{kj}^2}{\sigma_j^2} \\ &= \frac{1}{4} - \frac{[\frac{1}{4}(1-2r_{jk})]^2}{4} \\ &= r_{jk}(1 - r_{jk}). \end{aligned} \tag{13.18}$$

Considering three markers, $\mathbf{M}_j$, $\mathbf{M}_k$ and $\mathbf{M}_l$, we similarly derive the covariance between markers $\mathbf{M}_j$ and $\mathbf{M}_k$ conditional upon marker $\mathbf{M}_l$ as

$$\begin{aligned} \sigma_{jk|l} &= \sigma_{jk} - \frac{\sigma_{jl}\sigma_{kl}}{\sigma_l^2} \\ &= \frac{1}{4}[(1-2r_{jk}) - (1-2r_{jl})(1-2r_{kl})] \\ &= \begin{cases} 0 & \text{for order } \mathbf{M}_j\mathbf{M}_l\mathbf{M}_k \text{ or } \mathbf{M}_k\mathbf{M}_l\mathbf{M}_j \\ r_{kl}(1-r_{kl})(1-2r_{jk}) & \text{for order } \mathbf{M}_j\mathbf{M}_k\mathbf{M}_l \text{ or } \mathbf{M}_l\mathbf{M}_k\mathbf{M}_j \\ r_{jl}(1-r_{jl})(1-2r_{jk}) & \text{for order } \mathbf{M}_l\mathbf{M}_j\mathbf{M}_k \text{ or } \mathbf{M}_k\mathbf{M}_j\mathbf{M}_l, \end{cases} \end{aligned} \tag{13.19}$$

because, without inference, $(1 - 2r_{ik}) = (1 - 2r_{il})(1 - 2r_{kl})$ for order $\mathbf{M}_i\mathbf{M}_l\mathbf{M}_k$ or $\mathbf{M}_k\mathbf{M}_l\mathbf{M}_i$. Equation (13.19) states that, conditional on an intermediate marker, the covariance between two flanking markers is expected to be zero.

Using equations (13.18) and (13.19), we derive the variance of marker $\mathbf{M}_j$ conditional on markers $\mathbf{M}_k$ and $\mathbf{M}_l$ as

$$\begin{aligned} \sigma_{j|kl}^2 &= \sigma_{j|k}^2 - \frac{\sigma_{jl|k}^2}{\sigma_{l|k}^2} \\ &= \sigma_{j|l}^2 - \frac{\sigma_{jk|l}^2}{\sigma_{k|l}^2} \end{aligned}$$

$$= \begin{cases} \sigma_{j|k}^2 & \text{for order } \mathbf{M}_j\mathbf{M}_k\mathbf{M}_l \text{ or } \mathbf{M}_l\mathbf{M}_k\mathbf{M}_j \\ \sigma_{k|j}^2 & \text{for order } \mathbf{M}_j\mathbf{M}_l\mathbf{M}_k \text{ or } \mathbf{M}_k\mathbf{M}_l\mathbf{M}_j \\ \dfrac{r_{jk}(1-r_{jk})r_{jl}(1-r_{jl})}{r_{kl}(1-r_{kl})} & \text{for order } \mathbf{M}_k\mathbf{M}_j\mathbf{M}_l \text{ or } \mathbf{M}_l\mathbf{M}_j\mathbf{M}_k. \end{cases} \tag{13.20}$$

According to equation (13.20), the conditional variance of a marker conditional upon all other markers is only dependent on the markers that are right next to the marker under consideration. This can be stated generally as follows. The variance of marker $\mathbf{M}_j$ conditional on all other markers is expressed as

$$\sigma^2_{j|s_{-j}} = \sigma^2_{j|(j-1)(j+1)},$$

where s_{-j} denotes a set that includes all markers except for marker $\mathbf{M}_j$.

Similarly, considering four markers with $\mathbf{M}_j < \mathbf{M}_k$ and $\mathbf{M}_l < \mathbf{M}_m$, the conditional covariance between marker $\mathbf{M}_j$ and $\mathbf{M}_k$ given $\mathbf{M}_l$ and $\mathbf{M}_m$ is derived as

$$\begin{aligned}
\sigma^2_{jk|lm} &= \sigma_{jk|l} - \frac{\sigma_{jm|l}\sigma_{km|l}}{\sigma^2_{m|l}} \\
&= \sigma_{jk|m} - \frac{\sigma_{jl|m}\sigma_{kl|m}}{\sigma^2_{l|m}} \\
&= \begin{cases} 0 & \text{for order } \mathbf{M}_j\mathbf{M}_l\mathbf{M}_k\mathbf{M}_m \\ & \text{or } \mathbf{M}_j\mathbf{M}_m\mathbf{M}_k\mathbf{M}_l \\ \sigma^2_{jk|l} & \text{for order } \mathbf{M}_j\mathbf{M}_k\mathbf{M}_l\mathbf{M}_m \\ \sigma^2_{jk|m} & \text{for order } \mathbf{M}_l\mathbf{M}_m\mathbf{M}_j\mathbf{M}_k \\ \dfrac{r_{lk}(1-r_{lk})r_{km}(1-r_{km})(1-2r_{jk})}{r_{lm}(1-r_{lm})} & \text{For order } \mathbf{M}_l\mathbf{M}_j\mathbf{M}_k\mathbf{M}_m. \end{cases}
\end{aligned} \tag{13.21}$$

Equation (13.21) contains two important conclusions. First, as long as we condition on one intermediate marker, the covariance between any two markers will be zero. Second, the covariance between two flanking markers given all other markers is only dependent on the two markers that are the closest to each of the flanking markers. In other words, conditioning on two flanking markers would make the covariance between two interior markers independent of all those markers that are outside the marker interval. This can be mathematically expressed as

$$\sigma^2_{jk|s_{-jk}} = \sigma^2_{jk|(j-1)(k+1)},$$

where s_{-jk} denotes a set that includes all markers except for markers $\mathbf{M}_j$ and $\mathbf{M}_k$.

13.4.2 Conditional QTL Variance

In this section, we extend the conditional marker variance to consider the conditional variance involving QTL. Consider marker $\mathbf{M}_j$ and QTL $\mathbf{Q}_u$ with two alleles, Q_u and q_u, whose genotypes, frequencies, and values are expressed in the backcross as

	$M_jm_jQ_uq_u$	$M_jm_jq_uq_u$	$m_jm_jQ_uq_u$	$m_jm_jq_uq_u$
Frequency	$\frac{1}{2}(1-r_{ju})$	$\frac{1}{2}r_{ju}$	$\frac{1}{2}r_{ju}$	$\frac{1}{2}(1-r_{ju})$
"Value" at $\mathbf{M}_j$	1	1	0	0
Value at $\mathbf{Q}_u$	a	0	a	0

where r_{ju} is the recombination fraction between the marker and QTL and a_u is the additive effect due to $\mathbf{Q}_u$. It is easy to show $\sigma_u^2 = \frac{1}{4}a^2$ and $\sigma_{ju} = \frac{1}{4}(1-2r_{ju})a_u^2$.

Considering a quantitative trait, y, controlled by U QTLs with no epistasis, we derive the covariance between trait y and marker $\mathbf{M}_j$ conditional on marker $\mathbf{M}_k$ as

$$
\begin{aligned}
\sigma_{yj|k} &= \sigma_{yj} - \frac{\sigma_{yk}\sigma_{jk}}{\sigma_k^2} \\
&= \frac{1}{4}\sum_{u=1}^{U}[(1-2r_{uj}) - (1-2r_{uk})(1-2r_{jk})]a_u \\
&= \begin{cases}
r_{jk}(1-r_{jk})\sum_{u\le j}(1-2r_{uj})a_u + \sum_{j<u<k} r_{uk}(1-r_{uk})(1-2r_{ju})a_u \\
\qquad\qquad \text{for order } \mathbf{M}_j\mathbf{M}_k \\
r_{jk}(1-r_{jk})\sum_{u\ge j}(1-2r_{uj})a_u + \sum_{j<u<k} r_{uk}(1-r_{uk})(1-2r_{ju})a_u \\
\qquad\qquad \text{for order } \mathbf{M}_k\mathbf{M}_j,
\end{cases}
\end{aligned}
\tag{13.22}
$$

which suggests that the covariance between one marker and trait conditional upon a second marker does not contain those QTLs that are beyond the conditional marker. More clearly, let us consider two conditional markers. It is easy to derive

$$
\begin{aligned}
\sigma_{yj|kl}^2 &= \sigma_{yj|k} - \frac{\sigma_{yl|k}\sigma_{jl|k}}{\sigma_{l|k}^2} \\
&= \sigma_{yj|l} - \frac{\sigma_{yk|l}\sigma_{jk|l}}{\sigma_{l|m}^2} \\
&= \begin{cases}
\sigma_{yj|k} & \text{for order } \mathbf{M}_j\mathbf{M}_k\mathbf{M}_l \\
\sigma_{yj|l} & \text{for order } \mathbf{M}_j\mathbf{M}_l\mathbf{M}_k \\
\dfrac{r_{jk}(1-r_{jk})}{r_{kl}(1-r_{kl})}\sum_{l<u\le j} r_{lu}(1-t_{lu})(1-2r_{uj})a_u + & \\
\dfrac{r_{jk}(1-r_{jk})}{r_{kl}(1-r_{kl})}\sum_{j<u\le k} r_{uk}(1-t_{uk})(1-2r_{ju})a_u & \text{for order } \mathbf{M}_l\mathbf{M}_j\mathbf{M}_k.
\end{cases}
\end{aligned}
\tag{13.23}
$$

The general expression for equation (13.23) is $\sigma_{yj|s_{-j}} = \sigma_{yj|(j-1)(j+1)}$. Based on the formulas above, the partial regression coefficient of the trait value on marker M_j conditional upon the rest of the markers is derived as

$$\begin{aligned} b_{yj|s_{-j}} &= \frac{\sigma_{yj|s_{-j}}}{\sigma^2_{j|s_{-j}}} \\ &= \frac{\sigma_{yj|(j-1)(j+1)}}{\sigma^2_{j|(j-1)(j+1)}} \\ &= \sum_{j-1<u\leq j} \frac{r_{(j-1)u}(1-r_{(j-1)u})(1-2r_{uj})}{r_{(j-1)j}(1-r_{(j-1)j})} a_u \\ &\quad + \sum_{j<u<j+1} \frac{r_{u(j-1)}(1-r_{u(j+1)})(1-2r_{ju})}{r_{j(j+1)}(1-r_{j(j+1)})} a_u . \end{aligned} \tag{13.24}$$

which contains two summations for all QTLs located between markers $\mathbf{M}_{j-1}$ and $\mathbf{M}_j$ and for all QTLs between markers $\mathbf{M}_j$ and $\mathbf{M}_{j+1}$. This suggests that the partial regression coefficient depends only on those QTLs located between markers $\mathbf{M}_{j-1}$ and $\mathbf{M}_{j+1}$, although this does not apply when QTLs display strong epistatic interactions. In any case with no epistasis, if interval mapping with a pair of flanking markers is combined with multiple regression analysis of other markers, all possible QTLs outside this testing interval will be absorbed into the partial regression coefficients. For this reason, the combination of interval mapping and regression analysis can limit test statistics for QTL localization within a specific marker interval under consideration and makes it independent of the effects of other possible QTLs outside the tested interval on a chromosome. This has laid a statistical foundation for composite interval mapping.

With further derivations, Zeng (1993) showed that the sampling variance of the partial regression coefficient can be reduced and therefore the statistical power of QTL mapping can be increased when conditional on unlinked markers. According to the analysis of the sampling variance of the partial regression coefficient, the combination of interval mapping and a multiple-regression analysis can reduce the influence of multiple linked QTLs on hypothesis testing and thus the precision of the QTL test and estimation when the conditioned markers are linked, although this possibly reduces the statistical power. The tradeoff between the power and precision of QTL mapping can be balanced for a practical problem. If one purports to separate multiple QTLs on the same genomic region, linked markers should be chosen for regression analysis. On the other hand, if one hopes to increase the power of QTL mapping, regression analysis should be constructed with unlinked markers.

Zeng (1993) derived the form of the sampling correlation between two partial regression coefficients on different markers. He showed that such a correlation is generally zero unless the two markers are adjacent. This property helps to derive the correlation of test statistics between two testing positions in two intervals for an interval test and determine the significance threshold of a test statistic under a null hypothesis for an overall test covering the entire genome.

Zeng (1993) thought that the theoretical aspects of composite interval mapping derived for the backcross are also applicable to the F_2. For the F_2 with no dominance, the underlying theory is qualitatively equivalent to the backcross. But for the F_2 with dominance, the regression analysis needs to introduce one more parameter to reflect dominance deviation.

13.4.3 Marker Selection

Composite interval mapping integrates interval mapping and multiple regression analysis. The selection of markers as cofactors is an important issue. Conditioning on linked markers helps to separate multiple linked QTLs on a region but leads to the reduction of statistical power. Although conditioning on unlinked markers can increase analytical power by reducing residual variance, it has little to do with the separation of linked QTLs. These properties of composite interval mapping make its marker selection difficult.

Unfortunately, there is no theoretical derivation for the choice of markers as cofactors. Zeng (1994) proposed forward or backward stepwise regression analysis to add or drop markers. As discussed above, marker selection should also be made empirically on the basis of the purpose of the study. The high-resolution mapping of linked QTLs favors the inclusion of more linked markers, whereas statistical power can be increased by conditioning on more unlinked markers.

13.5 Comparisons Between Composite Interval Mapping and Interval Mapping

Zeng (1994) performed extensive simulation studies to compare the advantages and disadvantages of composite interval mapping and interval mapping. He simulated four chromosomes of equal length. Ten QTLs with different effect sizes were simulated for a quantitative trait. Composite interval mapping conditional on linked markers can detect individual QTLs that are located on the same chromosome. Composite interval mapping conditional on all the markers displays greater power, but it cannot well separate linked QTLs on some chromosome. Traditional interval mapping has lower power and also is biased for the estimation of QTL positions.

We used two examples to demonstrate the differences between these two different approaches. In both examples, there are multiple linkage groups, each corresponding to a different chromosome. Marker selection includes three options: (1) linked markers on the same chromosome, (2) unlinked markers on a different chromosome, and (3) all markers on the same and different chromosomes. The results from these options are compared with those for interval mapping.

Example 13.1. Revisit Example 3.1. Two inbred lines, semi-dwarf IR64 and tall Azucena, were crossed to generate a heterozygous F_1. The haploid chromosomes for pollens (gametes) of the F_1 were doubled to produce 123 doubled haploid (DH) plants. These DH plants, equivalent to a backcross progeny, were genotyped for 135 RFLP and 40 isozyme and RAPD markers, from which a linkage map covering the entire genome of 12 chromosomes was constructed (Yan et al. 1998; Fig. 3.3). Each of the DH lines was measured for plant height at each of ten consecutive weeks.

As an example, the first chromosome was used to perform composite interval mapping (CIM) for plant weight at week 10. We first used interval mapping (IM) to scan for the existence of a QTL throughout rice chromosome 1. The log-likelihood ratio (LR) test statistics were calculated by fixing the position of a putative QTL at every 2 cM in each marker interval. After scanning for a QTL from the first to the last interval, we draw a smoothed LR profile with a series of discrete LR values (Fig. 13.1). Two LR peaks were observed by IM, a flatter one between markers RZ730 and RZ801 and a more narrow one between markers RG810 and RG331. The LR values at the two peaks (> 55) are largely beyond the critical threshold at the 5 percent significance level (10.39) determined from 100 permutation tests by destroying the original phenotype–marker relationships. It was found that these two QTLs detected by IM for height growth are highly significant.

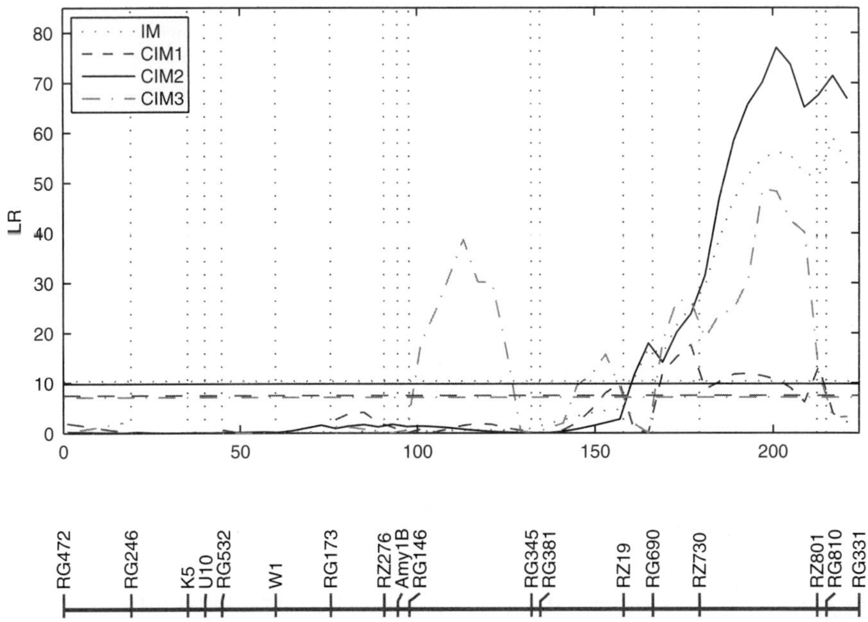

Fig. 13.1. Profile plot of likelihood–ratio statistics (LRs) as a function of map position of the putative QTL in a DH population of rice. IM, Interval Mapping; CIM, Composite Interval Mapping; CIM1, linked markers on the same chromosome; CIM2, unlinked markers on a different chromosome; and CIM3, all markers on the same and different chromosomes. The critical thresholds for IM, CIM1, CIM2 and CIM3 are 10.39, 7.47, 9.79, and 7.13, respectively, as determined from 100 permutation tests.

To test whether each of the two LR peaks detected by IM correspond to a different QTL or collectively present a QTL on chromosome 1, we carried out composite interval mapping conditional upon linked markers on the same chromosome (CIM1). According to the property of CIM found by Zeng (1993), CIM1 has greater power to separate multiple linked QTLs. As shown in Fig. 13.1, CIM1 detects three distinct LR peaks at marker intervals RG381-RZ19 and RG690-RZ730 and marker RZ801, larger than the 5 percent critical threshold (7.47). It seems that CIM1 dissolves a big flat peak by IM into three smaller but narrower ones, suggesting that CIM1 has better resolution for mapping linked QTLs than IM. However, because CIM1 has increased sampling variances, the caution should be taken for the estimation precision of QTL position and effect parameters.

Composite interval mapping conditional upon unlinked markers (CIM2) can increase the power of QTL mapping by controlling some residual variance (Zeng 1993). It is not surprising to find that CIM2 in this example identified two higher LR peaks (> 70) at positions similar to those detected by IM, leading to increased significance levels (compared to the 5 percent threshold of 9.79). However, as expected, CIM2 has no improvement for the separation of these two peaks compared with IM.

Composite interval mapping conditional upon both linked and unlinked markers (CIM3) displays results different from those for IM, CIM1, and CIM2. It detects four different peaks at marker intervals RG146-RG345, RG381-RZ19, RG690-RZ730 and RZ730-RZ801, with LR values being larger than the 5 percent threshold (7.13) (Fig. 13.1). Perhaps CIM3 preserves the advantages of CIM1 and CIM2, thus leading to the better separation of more linked QTL and the better power of QTL detection.

In summary, CIM3 conditional upon unlinked markers, although increasing the power of QTL detection, obtains results similar to those by IM. CIM1 conditional on linked markers has better resolution for multiple linked QTL, and its power can be improved when some unlinked markers are involved in the partial regression analysis. To determine the reliability of the results about QTL detection, simulation studies can be performed by mimicking the above example in terms of the genomic distribution of QTLs, their inheritance mode and effect sizes.

Example 13.2. Revisit Example 3.2. Cheverud et al. (1996) constructed a linkage map using 75 microsatellite markers in a population of 535 F_2 progeny derived from two strains, the Large (LG/J) and Small (SM/J). The F_2 progeny were measured for body mass at ten weekly intervals starting at age 7 days. The raw weights were corrected for the effects of each covariate due to dam, litter size at birth, and parity but not for the effect due to sex. We used chromosome 2 composed of nine markers to map the QTL affecting body weight at age 10 weeks with interval mapping (IM) and different types of composite interval mapping (CIM). In each case, permutation tests were used to empirically determine critical thresholds.

Interval mapping (IM) obtains a flat LR profile covering most of the chromosome (Fig. 13.2), suggesting that more than one QTL on this chromosome may exist to affect mouse body weight. Composite interval mapping conditional upon unlinked markers (CIM2) obtains a similar LR profile, although the peak detected by CIM2 is larger than that by IM. Composite interval mapping conditional upon linked markers (CIM1) detects no QTL, despite some distinct small peaks, suggesting that CIM1 tends to decrease the power in this particular example. Composite interval mapping conditional upon both linked and unlinked markers (CIM3) produces a different LR profile, with a single peak in marker interval DZMIT17–DZMIT26, showing no evidence for the presence of two QTLs. To examine the reliability of each mapping scheme above, simulation studies by mimicking the example would be helpful.

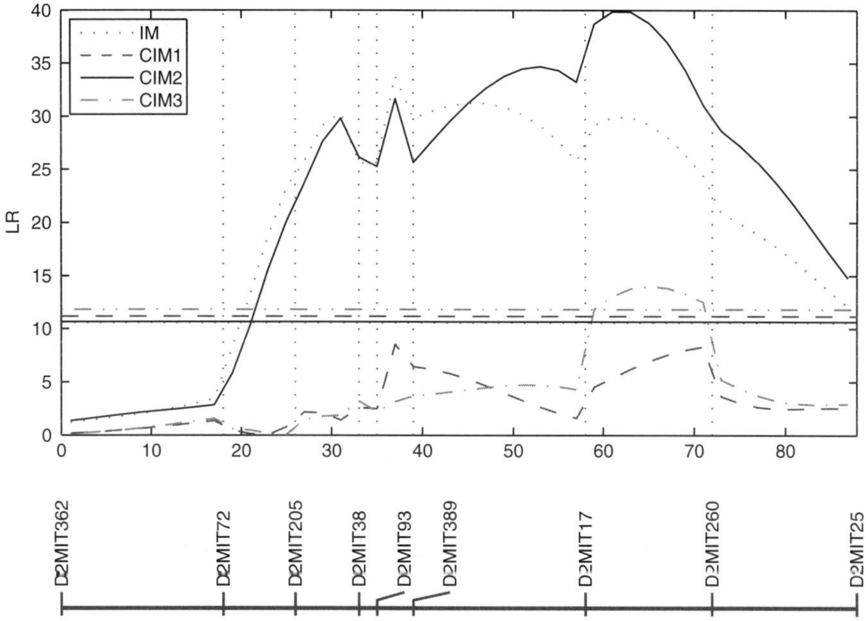

Fig. 13.2. Profile plot of likelihood–ratio statistics (LRs) as a function of map position of the putative QTL. IM, Interval Mapping; CIM, Composite Interval Mapping; CIM1, linked markers on the same chromosome; CIM2, unlinked markers on a different chromosome; and CIM3, all markers on the same and different chromosomes.

13.6 Multiple Interval Mapping

Both interval mapping and composite interval mapping model one QTL at a time. Although composite interval mapping that combines the idea of interval mapping and marker regression analysis can overcome the problem of multi-QTL linkage, it has less power to characterize the detailed genetic architecture of a quantitative trait. Because a complex trait may be controlled by a number of QTLs, it is crucial to have a mapping approach that can model multiple QTLs simultaneously and identify and locate all the QTLs responsible for quantitative variation. Such an approach has been proposed by Zeng and colleagues, and is named multiple interval mapping (Kao et al. 1999; Zeng et al. 2000). Kao and Zeng (1997) have derived general formulas for obtaining maximum likelihood estimates for the positions and effects of multiple QTLs.

Multiple interval mapping models multiply QTLs in such a way that QTLs can be directly controlled in the model through the simultaneous use of multiple marker intervals. They have proven more powerful and precise for estimating the positions and effects of QTLs than conventional interval mapping and composite interval mapping. In addition, by searching and mapping all possible QTLs in multiple marker intervals simultaneously, multiple interval mapping allows the full estimation of the genetic architecture of a quantitative trait in terms of the number of underlying QTLs, their genetic effects, pleiotropic effects, and epistatic network among different QTLs. The area of research that is open to multiple interval mapping is the procedure for the model selection of multiple QTLs, and their genomic positions and effects that collectively provides the best fit of the data observed.

13.7 Exercises

13.1 Given the likelihoods of composite interval mapping for the backcross (13.2) and F_2 (13.10), derive the MLEs of the parameters in the M step as shown by equations (13.5) (backcross) and (13.7) (F_2). (backcross) and (13.12) (F_2), respectively.

13.2 Zeng (1993, 1994) provided the statistical properties of composite interval mapping based on a simple backcross design, which are reiterated below:

(a) *For additive QTLs (ignoring epistasis), the expected partial regression coefficient of the trait on a marker depends only on those QTLs that are located on the interval bracketed by the two neighboring markers, and is unaffected by the effects of QTLs located on other intervals.*

(b) *Conditioning on unlinked markers in the multiple regression analysis will reduce the sampling variance of the test statistic by controlling some residual genetic variation and thus will increase the power of QTL mapping.*

(c) *Conditioning on linked markers in the multiple regression analysis will reduce the chance of interference of possible multiple linked QTLs on hypothesis testing and parameter estimation but with a possible increase in sampling variance.*

(d) *Two sample partial regression coefficients of the trait value on two markers in a multiple regression analysis are generally uncorrelated unless the two markers are adjacent markers.*

Show that these properties apply to an F_2 design with three different genotypes at each locus. Hint: You can denote 2, 1, and 0 for the homozygote for one parent's allele, and heterozygote and homozygote for the other parent's allele, respectively.

13.3 Perform simulation studies to show how many and what types of markers are involved in composite interval mapping as cofactors.

13.4 Read Kao and Zeng's (1997) paper to find out how general formulas for estimating the asymptotic sampling variances of the MLEs of QTL positions and effects can be derived in multiple interval mapping. Also, show how the asymptotic covariance matrix for the MLEs can be derived with Louis' (1982) mixture-model–based approach.

14

QTL Mapping in Outbred Pedigrees

14.1 Introduction

QTL mapping approaches were developed originally for experimental crosses, such as the backcross, double haploid, RILs, or F_2, derived from inbred lines. Because of the homozygosity of inbred lines, the Mendelian (co)segregation of all markers with two alternative alleles in such crosses can be observed directly. In practice, there also is a group of species called outcrossing species, such as forest trees or large animals, in which it is difficult or impossible to generate inbred lines due to long generation intervals and high heterozygosity, although experimental hybrids can be commercially used for their genetic improvement.

Many commercially available hybrid populations for outcrossing species can serve directly as mapping material. But statistical mapping models used for these populations should vary, depending on the biological characteristics of species. Some species, such as forest trees, can generate a large hybrid family (Grattapaglia and Sederoff 1994; Bradshaw and Stettler 1995), so that one single cross would be adequate to obtain the power needed. On the other hand, for some species, such as dogs, it is impossible to produce a large family and thus multiple families, each with a small size and including some from both related and unrelated parents, will be needed (Bliss et al. 2002; Todhunter et al. 2005). Linkage analysis for these two different types of outcrossing species has been described in Chapters 4 and 7, respectively. In this chapter, we will present statistical models that concern QTL mapping for outcrossing species. Based on the biological features of outcrossing species, two different types of mapping models will be described, the first dealing with a full-sib family of large size and the second type dealing with many related families, each with a small size.

For a given outbred line, some markers may be heterozygous, whereas others may be homozygous over the genome. All markers may or may not have the same allele system between any two outbred lines used for a cross. Also, for a pair of heterozygous loci, their allelic configuration along two homologous chromosomes (i.e., linkage phase) cannot be observed from the segregation pattern of genotypes in the cross. Unfortunately, a consistent number of alleles across different markers and their known linkage phases are the prerequisites for statistical mapping approaches described for the backcross or F_2 in the preceding chapters.

14.2 A Fixed-Effect Model for a Full-Sib Family

14.2.1 Introduction

Several particular statistical models have been proposed for QTL mapping in a full-sib family (Schäfer-Pregl et al. 1996; Johnson et al. 1999; Song et al. 1999). In some studies, more sophisticated statistical algorithms, such as Bayesian approaches relying on a Markov chain Monte Carlo method, have been proposed to take the complexity of full-sib family mapping into account (Hoeschele et al. 1997; Sillanpa and Arjas 1999;). Lin et al. (2003) derived a general model for full-sib mapping by integrating uncertainties about allelic numbers and configurations into the mixture-model context. In this section, this general model will be described and illustrated by an example.

14.2.2 A Mixture Model for a Parental Diplotype

For a full-sib family derived from two outbred parents, up to four marker alleles, besides a null allele, can occur at a single locus. Also, the number of alleles may vary over loci. Each of the marker alleles is dominance to the null allele. In Section 3.4, all possible cross types for a segregating marker locus were tabulated in Table 3.2.

Consider two outbred parental lines denoted as P_1 and P_2, which contain two homologous chromosomes, **12** and **34**, respectively, in a set. The cross between these two lines, $\mathbf{12} \times \mathbf{34}$, results in four possible parental chromosome pairings: **13**, **14**, **23**, and **24**. We used bold Arabic numerals to denote parental chromosomes. Although there may be many different marker types in a full-sib family derived from the two outbred parental lines, all observed markers, no matter which type they come from, are generally expressed as 1 and 2 for parent P_1 and 3 and 4 for parent P_2.

Suppose there is a QTL located between the two markers. The four alleles of the QTL are denoted by 1 and 2 for parent P_1 and 3 and 4 for parent P_2, which will be segregating to generate zygotes 13, 14, 23, and 24 following a 1:1:1:1 ratio in the family. The recombination fractions between the two markers, between marker $\mathbf{M}_1$ and the QTL and between the QTL and marker $\mathbf{M}_2$, are denoted by r, r_1, and r_2, respectively, with

$$r = \begin{cases} r_1 + r_2 & \text{for no double recombination} \\ r_1 + r_2 - 2r_1 r_2 & \text{for independent recombination} \\ r_1 + r_2 - 2c r_1 r_2 & \text{for interfered recombination,} \end{cases}$$

where c is the coefficient of coincidence between the recombinations at two different intervals. Parent-specific difference of linkage is ignored. The alleles of these two markers and the QTL are arranged between the two homologous chromosomes in each of a total of four possible linkage phases for each parent.

But the allelic linkage phases of the two markers can be known for both parents through linkage analyses of markers using a strategy proposed in Chapter 4. Thus, under a fixed marker linkage phase, we will have $2 \times 2 = 4$ parental combinations ($\mathbf{\Phi}$'s) of linkage phase (or diplotype) of the QTL relative to the two markers, which are schematically expressed, along with the order of the four QTL genotypes in the progeny, as

$$
\left\{
\begin{array}{lcl}
\mathbf{\Phi}_{11} & = & \begin{array}{c|c|}1&2\\1&2\\1&2\end{array} \times \begin{array}{c|c|}3&4\\3&4\\3&4\end{array} \rightarrow (13, 14, 23, 24), \\[2ex]
\mathbf{\Phi}_{12} & = & \begin{array}{c|c|}1&2\\1&2\\1&2\end{array} \times \begin{array}{c|c|}3&4\\4&3\\3&4\end{array} \rightarrow (14, 13, 24, 23), \\[2ex]
\mathbf{\Phi}_{21} & = & \begin{array}{c|c|}1&2\\2&1\\1&2\end{array} \times \begin{array}{c|c|}3&4\\3&4\\3&4\end{array} \rightarrow (23, 24, 13, 14), \\[2ex]
\mathbf{\Phi}_{22} & = & \begin{array}{c|c|}1&2\\2&1\\1&2\end{array} \times \begin{array}{c|c|}3&4\\4&3\\3&4\end{array} \rightarrow (24, 23, 14, 13),
\end{array}
\right. \tag{14.1}
$$

where the first and second subscripts of $\mathbf{\Phi}$ denote two possible phases of parents P_1 and P_2, respectively, and the vertical lines for each diplotype combination denote two parental chromosomes, **12** and **34**, each for a parent. Each parent, no matter which possible diplotype combination it has, will generate eight three-locus haploid gametes (haplotypes), with the haplotype frequencies depending on the recombination fractions of the three loci. Table 14.1 gives the frequencies of the eight haplotypes, denoted by p_{111}, p_{121}, p_{112}, p_{122}, p_{211}, p_{221}, p_{212}, and p_{222}, for parent P_1 along with the haplotype frequencies for the two markers, denoted by p_{11}, p_{12}, p_{21}, and p_{22}. The corresponding haplotype frequencies can also be given for parent P_2.

The marker–QTL haplotype frequencies can be expressed in terms of the recombination fractions between each marker and QTL (r_1 and r_2) when no double recombination (case 1) or independent recombination (case 2) is assumed. However, if interference exists between different intervals during meiosis (case 3), the haplotype frequencies should be expressed differently. Let g_{00}, g_{01}, g_{10} and g_{11} be the occurrence probabilities of no recombination both in the $\mathbf{M}_1$–QTL and QTL–$\mathbf{M}_2$ intervals, one recombination only in the $\mathbf{M}_1$–QTL interval, one recombination only in the QTL–$\mathbf{M}_2$ interval, and one recombination in each interval, respectively. The haplotype frequencies in the three cases are expressed as

Haplotype Frequency	Case 1	Case 2	Case 3
p_{111}	1	$\frac{1}{2}(1-r_1)(1-r_2)$	$\frac{1}{2}g_{00}$
p_{121}	0	$\frac{1}{2}r_1r_2$	$\frac{1}{2}g_{11}$
p_{112}	$\frac{1}{2}r_2$	$\frac{1}{2}(1-r_1)r_2$	$\frac{1}{2}g_{01}$
p_{122}	$\frac{1}{2}r_1$	$\frac{1}{2}r_1(1-r_2)$	$\frac{1}{2}g_{10}$
p_{211}	$\frac{1}{2}r_1$	$\frac{1}{2}r_1(1-r_2)$	$\frac{1}{2}g_{10}$
p_{221}	$\frac{1}{2}r_2$	$\frac{1}{2}(1-r_1)r_2$	$\frac{1}{2}g_{01}$
p_{212}	0	$\frac{1}{2}r_1r_2$	$\frac{1}{2}g_{11}$
p_{222}	1	$\frac{1}{2}(1-r_1)(1-r_2)$	$\frac{1}{2}g_{00}$

Table 14.1. Frequencies of haplotypes formed by two markers and a QTL that is located between the two markers in parent P_1.

Flanking Markers		QTL	
Haplotype	Frequency	1	2
11	$p_{11}=\frac{1}{2}(1-r)$	p_{111}	p_{121}
12	$p_{12}=\frac{1}{2}r$	p_{112}	p_{122}
21	$p_{21}=\frac{1}{2}r$	p_{211}	p_{221}
22	$p_{22}=\frac{1}{2}(1-r)$	p_{212}	p_{222}

For the frequencies of two-marker haplotypes, the first and second subscripts correspond to alleles from markers $\mathbf{M}_1$ and $\mathbf{M}_2$, respectively. For the frequencies of three-locus haplotypes, the second subscripts corresponds to the QTL alleles, whereas the first and second subscript corresponds to the alleles of the two markers, respectively. It can be shown that $p_{11} = p_{111}+p_{121}$, $p_{12}=p_{112}+p_{122}$, $p_{21}=p_{211}+p_{221}$, and $p_{22}=p_{212}+p_{222}$.

The eight haplotypes from parent P_1 unite randomly with the eight haplotypes from parent P_2 to generate a total of 64 zygotic genotypes. The frequencies of these genotypes in the full-sib family are calculated in terms of the haplotype frequencies (expressed by g's), which are tabulated in Table 14.1. It is noticed that the two parents for the cross have four possible combinations of parental diplotypes, as shown by display (14.1), but one and only one combination is correct. Under these different diplotype combinations, the two parents form the same set of 64 zygotic genotypes, but the formation frequencies of genotypes are phase-specific. The genotype frequencies presented in Table 14.1 correspond to different parental diplotype combinations $\mathbf{\Phi}_{11}$, $\mathbf{\Phi}_{12}$, $\mathbf{\Phi}_{21}$, and $\mathbf{\Phi}_{22}$.

Table 14.2. Genotype frequencies for the two markers and the QTL bracketed by the markers in a full-sib family generated by two outbred parents under different diplotype combinations.

			Phase	QTL genotype			
			$\mathbf{\Phi}_{11}$	13	14	23	24
			$\mathbf{\Phi}_{12}$	14	13	24	23
Flanking Markers			$\mathbf{\Phi}_{21}$	23	24	13	14
No.	Genotype	Frequency	$\mathbf{\Phi}_{22}$	24	21	14	13
				$\mathbf{G}_{\ell_1\ell_2}$			
1	13/13	p_{11}^2		p_{111}^2	$p_{111}p_{121}$	$p_{121}p_{111}$	p_{121}^2
2	13/14	$p_{11}p_{12}$		$p_{111}p_{112}$	$p_{111}p_{122}$	$p_{121}p_{112}$	$p_{121}p_{122}$
3	13/23	$p_{12}p_{11}$		$p_{111}p_{112}$	$p_{112}p_{121}$	$p_{122}p_{111}$	$p_{121}p_{122}$
4	13/24	p_{12}^2		p_{112}^2	$p_{112}p_{122}$	$p_{122}p_{112}$	p_{122}^2
5	14/13	$p_{11}p_{21}$		$p_{111}p_{211}$	$p_{111}p_{221}$	$p_{121}p_{211}$	$p_{121}p_{221}$
6	14/14	$p_{11}p_{22}$		$p_{111}p_{212}$	$p_{111}p_{222}$	$p_{121}p_{212}$	$p_{121}p_{222}$
7	14/23	$p_{12}p_{21}$		$p_{112}p_{211}$	$p_{112}p_{221}$	$p_{122}p_{211}$	$p_{122}p_{221}$
8	14/24	$p_{12}p_{22}$		$p_{112}p_{212}$	$p_{112}p_{122}$	$p_{122}p_{212}$	$p_{122}p_{222}$
9	23/13	$p_{21}p_{11}$		$p_{211}p_{111}$	$p_{211}p_{121}$	$p_{221}p_{111}$	$p_{221}p_{121}$
10	23/14	$p_{21}p_{12}$		$p_{211}p_{112}$	$p_{211}p_{122}$	$p_{221}p_{112}$	$p_{221}p_{122}$
11	23/23	$p_{22}p_{11}$		$p_{212}p_{111}$	$p_{212}p_{121}$	$p_{222}p_{111}$	$p_{222}p_{121}$
12	23/24	$p_{22}p_{12}$		$p_{212}p_{112}$	$p_{212}p_{122}$	$p_{222}p_{112}$	$p_{222}p_{122}$
13	24/13	p_{21}^2		p_{211}^2	$p_{211}p_{221}$	$p_{221}p_{211}$	p_{221}^2
14	24/14	$p_{21}p_{22}$		$p_{211}p_{212}$	$p_{211}p_{222}$	$p_{221}p_{212}$	$p_{221}p_{222}$
15	24/23	$p_{21}p_{22}$		$p_{211}p_{212}$	$p_{212}p_{221}$	$p_{222}p_{211}$	$p_{221}p_{222}$
16	24/24	p_{22}^2		p_{212}^2	$p_{212}p_{222}$	$p_{222}p_{212}$	p_{222}^2

Note: The genotype frequencies under parental diplotype combination $\mathbf{\Phi}_{\ell_1\ell_2}$ $(\ell_1, \ell_2 = 1, 2)$ are arrayed into a (16×4) matrix $\mathbf{G}_{\ell_1\ell_2}$. By adjusting the order of the QTL genotypes in the full-sib family under parental diplotype combinations $\mathbf{\Phi}_{11}$, $\mathbf{\Phi}_{12}$, $\mathbf{\Phi}_{21}$, and $\mathbf{\Phi}_{22}$, matrices $\mathbf{G}_{11}$, $\mathbf{G}_{12}$, $\mathbf{G}_{21}$, and $\mathbf{G}_{22}$ can be generated.

Let p and q be the probabilities with which the first diplotype in display (14.1) occurs for parents P_1 and P_2, respectively. Thus, the corresponding probabilities of the four parental diplotype combinations will be $\phi_{11} = pq$, $\phi_{12} = p(1-q)$, $\phi_{21} = (1-p)q$, and $\phi_{22} = (1-p)(1-q)$ for $\mathbf{\Phi}_{11}$, $\mathbf{\Phi}_{12}$, $\mathbf{\Phi}_{21}$, and $\mathbf{\Phi}_{22}$, respectively. Let $\mathbf{G}_{11}$, $\mathbf{G}_{12}$, $\mathbf{G}_{21}$, and $\mathbf{G}_{22}$ be the matrices for genotype frequencies under the corresponding diplotype combinations (Table 14.2). Thus, the observed genotype frequencies $\mathbf{G}$ in the full-sib family should be a mixture of the genotype frequencies weighted by the diplotype combination probabilities, expressed as

$$\mathbf{G} = \phi_{11}\mathbf{G}_{11} + \phi_{12}\mathbf{G}_{12} + \phi_{21}\mathbf{G}_{21} + \phi_{22}\mathbf{G}_{22}, \tag{14.2}$$

which is a (16×4) matrix with 16 rows for two-marker genotypes and four columns for QTL genotypes in the full-sib family. According to Bayes' theorem, the conditional probabilities of a QTL genotype, uv (with alleles $u = 1, 2$ inherited from parent P_1 and alleles $v = 3, 4$ inherited from parent P_2), given marker genotypes are estimated by dividing $\mathbf{G}$ by the marker genotype frequencies given in Table 14.2. The conditional probability of a particular QTL genotype given the marker genotype of individual i is correspondingly expressed as

$$\omega_{uv|i} = \phi_{11}\omega_{uv|i}^{11} + \phi_{12}\omega_{uv|i}^{12} + \phi_{21}\omega_{uv|i}^{21} + \phi_{22}\omega_{uv|i}^{22}, \tag{14.3}$$

where the right side contains parental diplotype-specific conditional probabilities.

14.2.3 Quantitative Genetic Model

In a full-sib family, a QTL generates four genotypes. Let μ_{uv} be the value of a QTL genotype inheriting allele u from parent P_1 and allele v from parent P_2. Based on quantitative genetic theory, this genotypic value can be partitioned into the additive and dominance effects as

$$\mu_{uv} = \mu + \alpha_u + \beta_v + \gamma_{uv},$$

where μ is the overall mean, α_u and β_v are the allelic (additive) effects of alleles u and v, respectively, and γ_{uv} is the interaction (dominance) effect at the QTL. Considering all possible alleles and allele combinations between the two parents, there are a total of four additive effects (α_1 and α_2 from parent P_1 and β_3, and β_4 from parent P_2) and four dominance effects (γ_{13}, γ_{14}, γ_{23}, and γ_{34}). But these additive and dominance effects are not independent and therefore are not estimable. After parameterization, there are two independent additive effects, $\alpha = \alpha_1 = -\alpha_2$ and $\beta = \beta_3 = -\beta_4$, and one dominance effect, $\gamma = \gamma_{13} = -\gamma_{14} = -\gamma_{23} = \gamma_{24}$, to be estimated.

Let $\mathbf{u} = (\mu_{uv})_{4\times1}$ and $\mathbf{a} = (\mu, \alpha, \beta, \gamma)^{\mathrm{T}}$, which can be connected by a design matrix $\mathbf{D}$. We have

$$\mathbf{u} = \mathbf{D}\mathbf{a},$$

where

$$\mathbf{D} = \begin{bmatrix} 1 & 1 & 1 & 1 \\ 1 & 1 & -1 & -1 \\ 1 & -1 & 1 & -1 \\ 1 & -1 & -1 & 1 \end{bmatrix}.$$

The MLE of $\mathbf{a}$ can be obtained from the MLE of $\mathbf{u}$ by

$$\hat{\mathbf{a}} = \mathbf{D}^{-1}\hat{\mathbf{u}}.$$

14.2.4 Likelihood Analysis

Suppose there is a full-sib family of size n derived from two outbred lines. Consider a QTL for a quantitative trait that is bracketed by two markers. The linear model of the trait value for individual i (y_i) affected by this bracketed QTL is written as

$$y_i = \sum_{u=1}^{2}\sum_{v=3}^{4} \xi_{iuv}\mu_{uv} + e_i, \tag{14.4}$$

where ξ_{iuv} is the indicator variable for QTL genotypes, defined as 1 if a particular genotype uv is considered for individual i and 0 otherwise, and e_i is the residual error normally distributed with mean 0 and variance σ^2. The probability that individual i carries QTL genotype uv can be inferred from its marker genotype, with this probability expressed as $\omega_{uv|i}$.

The log-likelihood of the trait values (y) and marker information ($\mathbf{M}$) is given by

$$\log L(\mathbf{\Omega}|y, \mathbf{M}) = \sum_{i=1}^{n} \log\left[\sum_{u=1}^{2}\sum_{v=3}^{4} \omega_{uv|i} f_{uv}(y_i)\right], \tag{14.5}$$

where $\mathbf{\Omega}$ is the vector for unknown parameters QTL that include the QTL position (r_1 and r_2), parental phase probabilities, QTL genotypic values (μ_{uv}), and the residual variance (σ^2). The first two parameters, denoted by $\mathbf{\Omega}_p$, are contained in the mixture proportions of the model above, whereas the second two, denoted by $\mathbf{\Omega}_q$, are quantitative genetic parameters.

As usual, the MLEs of the unknown vector are obtained by differentiating the log-likelihood function with respect to each unknown Ω_τ, setting the derivatives equal to zero, and solving the log-likelihood equations. This procedure is shown as follows:

$$\begin{aligned}
&\frac{\partial}{\partial \Omega_\tau} \log L(\mathbf{\Omega}|y, \mathbf{M}) \\
&= \sum_{i=1}^{n}\sum_{u=1}^{2}\sum_{v=3}^{4} \frac{f_{uv}(y_i)\frac{\partial}{\partial \mathbf{\Omega}_p}\omega_{uv|i} + \omega_{uv|i}\frac{\partial}{\partial \mathbf{\Omega}_q} f_{uv}(y_i)}{\sum_{u'=1}^{2}\sum_{v'=3}^{4} \omega_{u'v'|i} f_{u'v'}(y_i)} \\
&= \sum_{i=1}^{n}\sum_{u=1}^{2}\sum_{v=3}^{4} \left[\frac{\omega_{uv|i} f_{uv}(y_i)\frac{1}{\omega_{uv|i}}\frac{\partial}{\partial \mathbf{\Omega}_p}\omega_{uv|i}}{\sum_{u'=1}^{2}\sum_{v'=3}^{4} \omega_{u'v'|i} f_{u'v'}(y_i)} + \frac{\omega_{uv|i} f_{uv}(y_i)\frac{\partial}{\partial \mathbf{\Omega}_q}\log f_{uv}(y_i)}{\sum_{u'=1}^{2}\sum_{v'=3}^{4} \omega_{u'v'|i} f_{u'v'}(y_i)}\right] \\
&= \sum_{i=1}^{n}\sum_{u=1}^{2}\sum_{v=3}^{4} \Pi_{uv|i}\left[\frac{1}{\omega_{uv|i}}\frac{\partial}{\partial \mathbf{\Omega}_p}\omega_{u'v'|i} + \frac{\partial}{\partial \mathbf{\Omega}_q}\log f_{uv}(y_i)\right],
\end{aligned}$$

where we define

$$\Pi_{uv|i} = \frac{\omega_{uv|i} f_{uv}(y_i)}{\sum_{u'=1}^{2} \sum_{v'=3}^{4} \omega_{u'v'|i} f_{u'v'}(y_i)}, \tag{14.6}$$

which could be thought of as the posterior probability that individual i has a QTL genotype uv. We then implement the EM algorithm with the expanded parameter set $\{\mathbf{\Omega}, \mathbf{\Pi}\}$, where $\mathbf{\Pi} = \{\Pi_{uv|i}\}$. Conditional on $\mathbf{\Pi}$, we solve for the zeros of $\frac{\partial}{\partial \Omega_\ell} \log L(\mathbf{\Omega}|y, \mathbf{M})$ to get the estimates of $\mathbf{\Omega}$ (the M step). The estimates are then used to update $\mathbf{\Pi}$ (the E step), and the process is repeated until convergence. The values at convergence are the maximum likelihood estimates (MLEs).

The estimates of the genotypic values and variance in the M step are derived as

$$\begin{aligned} \hat{\mu}_{uv} &= \frac{\sum_{i=1}^{n} \Pi_{uv|i} y_i}{\sum_{i=1}^{n} \Pi_{uv|i}}, \\ \hat{\sigma}^2 &= \frac{1}{n} \sum_{i=1}^{n} \sum_{u=1}^{2} \sum_{v=3}^{4} \Pi_{uv|i} (y_i - \hat{\mu}_{uv})^2. \end{aligned} \tag{14.7}$$

The estimates of the parental diplotype probabilities will be difficult because these probabilities are contained at a lower hierarchy of the mixture proportions, as shown by equation (14.3).

Substituting equation (14.3) into the log-likelihood (14.5), we have

$$\log L(\mathbf{\Omega}|y, \mathbf{M}) = \sum_{i=1}^{n} \log[\omega_{13} f_{13}(y_i) + \omega_{14} f_{14}(y_i) + \omega_{23} f_{23}(y_i) + \omega_{24} f_{24}(y_i)]$$

$$\begin{aligned} = \sum_{i=1}^{n} \log \Big\{ & \left[pq\omega_{13|i}^{11} + p(1-q)\omega_{13|i}^{12} + (1-p)q\omega_{13|i}^{21} + (1-p)(1-q)\omega_{13|i}^{22} \right] f_{13}(y_i) \\ & + \left[pq\omega_{14|i}^{11} + p(1-q)\omega_{14|i}^{12} + (1-p)q\omega_{14|i}^{21} + (1-p)(1-q)\omega_{14|i}^{22} \right] f_{14}(y_i) \\ & + \left[pq\omega_{23|i}^{11} + p(1-q)\omega_{23|i}^{12} + (1-p)q\omega_{23|i}^{21} + (1-p)(1-q)\omega_{23|i}^{22} \right] f_{23}(y_i) \\ & + \left[pq\omega_{24|i}^{11} + p(1-q)\omega_{24|i}^{12} + (1-p)q\omega_{24|i}^{21} + (1-p)(1-q)\omega_{24|i}^{22} \right] f_{24}(y_i) \Big\}, \end{aligned}$$

from which the closed forms for the estimates of the parental diplotype probabilities are derived as

$$\begin{aligned} \hat{p} &= \frac{1}{n} \sum_{i=1}^{n} \left(\psi_1 \Pi_{13|i} + \psi_2 \Pi_{14|i} + \psi_3 \Pi_{23|i} + \psi_4 \Pi_{24|i} \right), \\ \hat{q} &= \frac{1}{n} \sum_{i=1}^{n} \left(\psi_1' \Pi_{13|i} + \psi_2' \Pi_{14|i} + \psi_3' \Pi_{23|i} + \psi_4' \Pi_{24|i} \right), \end{aligned} \tag{14.8}$$

where

$$\psi_1 = \frac{pq\omega_{13|i}^{11} + p(1-q)\omega_{13|i}^{12}}{\omega_{13|i}},$$

$$\psi_2 = \frac{pq\omega^{11}_{14|i} + p(1-q)\omega^{12}_{14|i}}{\omega_{14|i}},$$

$$\psi_3 = \frac{pq\omega^{11}_{23|i} + p(1-q)\omega^{12}_{23|i}}{\omega_{23|i}},$$

$$\psi_4 = \frac{pq\omega^{11}_{24|i} + p(1-q)\omega^{12}_{24|i}}{\omega_{24|i}},$$

$$\psi'_1 = \frac{pq\omega^{11}_{13|i} + (1-p)q\omega^{12}_{13|i}}{\omega_{13|i}},$$

$$\psi'_2 = \frac{pq\omega^{11}_{14|i} + (1-p)q\omega^{12}_{14|i}}{\omega_{14|i}},$$

$$\psi'_3 = \frac{pq\omega^{11}_{23|i} + (1-p)q\omega^{12}_{23|i}}{\omega_{23|i}},$$

$$\psi'_4 = \frac{pq\omega^{11}_{24|i} + (1-p)q\omega^{12}_{24|i}}{\omega_{24|i}}.$$

There are two approaches for estimating the recombination fraction between the markers and QTL (r_1 or r_2) that describes the QTL position when no double recombination (case 1) or independent recombination (case 2) is assumed. The first is based on genome-wide scanning where the QTL position is estimated by treating r_1 (and therefore r_2) as fixed. Using a grid search, we can obtain the MLE of the QTL position from the peak of the profile of the log-likelihood ratio test statistics across a chromosome. The second is based on closed-form estimates of haplotype frequencies p's (and therefore the recombination fractions). If interference is assumed between different intervals (case 3), only the closed-form estimate is used. Combining Table 14.2 and equation (14.3) will generate a mixture of conditional probabilities in terms of p. Closed forms can be derived for the MLEs of p's, with constraints $p_{121} = p_{212} = 0$ for case 1 and $p_{121} + p_{212} = (p_{121} + p_{212} + p_{122} + p_{211})(p_{121} + p_{212} + p_{112} + p_{221})$ for case 2 and no constraint for case 3. In each case, the MLEs of p's are used to obtain the MLEs of the recombination fractions by

MLE	Case 1	Case 2	Case 3
$\hat{r}_1$	$\hat{p}_{122} + \hat{p}_{211}$	$\hat{p}_{121} + \hat{p}_{212} + \hat{p}_{122} + \hat{p}_{211}$	$\hat{p}_{122} + \hat{p}_{211} + \hat{p}_{121} + \hat{p}_{212}$
$\hat{r}_2$	$\hat{p}_{112} + \hat{p}_{221}$	$\hat{p}_{121} + \hat{p}_{212} + \hat{p}_{112} + \hat{p}_{221}$	$\hat{p}_{112} + \hat{p}_{221} + \hat{p}_{121} + \hat{p}_{212}$

Note that the MLEs of the recombination fractions may be different among the three cases, with the degrees depending on how much these cases are consistent for a given data set.

14.2.5 Fitting Marker Phenotypes

We have built a general framework for QTL mapping in a full-sib family based on marker zygote genotypes. But in practice only the phenotypes of the marker zygotes can be observed. The number of zygote phenotypes of a marker is 4, 3, or 2, depending on marker cross types (see Fig. 3.3). We can design different incidence matrices **I** to connect zygotic genotypes to zygotic phenotypes for all different marker types listed in Table 3.2. Thus, the joint frequency matrix of two markers and a QTL for particular marker types can be derived by using the corresponding incidence matrices, which are expressed as

$$\dot{\mathbf{G}} = (\mathbf{I}_{\mathbf{M}_1} \otimes \mathbf{I}_{\mathbf{M}_2})\mathbf{G},$$

where $\otimes$ is the Kronecker product and $\mathbf{I}_{\mathbf{M}_1}$ and $\mathbf{I}_{\mathbf{M}_2}$ are the incidence matrices for markers $\mathbf{M}_1$ and $\mathbf{M}_2$. The frequency vector of two-marker genotypes can also be collapsed in a similar way.

The pattern and structure of an incidence matrix (**I**) relating the zygotic genotypes to phenotypes for a marker depend on the cross type of this marker. Let a, b, c, and d be alleles at a marker between two outbred parents $\mathrm{P}_1 \times \mathrm{P}_2$, all dominance to a null allele o. We have

$$\mathbf{I} = \begin{bmatrix} 1\,0\,0\,0 \\ 0\,1\,0\,0 \\ 0\,0\,1\,0 \\ 0\,0\,0\,1 \end{bmatrix}$$

and for marker cross type $ab \times cd$,

$$\mathbf{I} = \begin{cases} \begin{bmatrix} 1\,1\,0\,0 \\ 0\,0\,1\,0 \\ 0\,0\,0\,1 \end{bmatrix} & \text{for } ab \times ab \\[2em] \begin{bmatrix} 1\,0\,0\,0 \\ 0\,1\,0\,0 \\ 0\,0\,1\,1 \end{bmatrix} & \text{for } ab \times ao, \end{cases}$$

$$\mathbf{I} = \begin{cases} \begin{bmatrix} 1\,0\,1\,0 \\ 0\,1\,0\,0 \\ 0\,0\,0\,1 \end{bmatrix} & \text{for } ao \times ab \\[2em] \begin{bmatrix} 1\,0\,0\,0 \\ 0\,1\,0\,1 \\ 0\,0\,0\,1 \end{bmatrix} & \text{for } ab \times ao, \end{cases}$$

$$\mathbf{I} = \begin{cases} \begin{bmatrix} 1\,0\,0\,0 \\ 0\,1\,1\,0 \\ 0\,0\,0\,1 \end{bmatrix} & \text{for } ab \times ao \\ \\ \begin{bmatrix} 1\,0\,0\,1 \\ 0\,1\,0\,0 \\ 0\,0\,0\,1 \end{bmatrix} & \text{for } ab \times ao, \end{cases}$$

$$\mathbf{I} = \begin{cases} \begin{bmatrix} 1\,1\,1\,0 \\ 0\,0\,0\,1 \end{bmatrix} & \text{for } ab \times ao \\ \\ \begin{bmatrix} 1\,1\,0\,1 \\ 0\,0\,1\,0 \end{bmatrix} & \text{for } ab \times ao \\ \\ \begin{bmatrix} 1\,0\,1\,1 \\ 0\,1\,0\,0 \end{bmatrix} & \text{for } ab \times ao \\ \\ \begin{bmatrix} 1\,0\,0\,0 \\ 0\,1\,1\,1 \end{bmatrix} & \text{for } ab \times ao, \end{cases}$$

$$\mathbf{I} = \begin{bmatrix} 1\,1\,0\,0 \\ 0\,0\,1\,1 \end{bmatrix},$$

$$\mathbf{I} = \begin{bmatrix} 1\,0\,1\,0 \\ 0\,1\,0\,1 \end{bmatrix}.$$

With the marker-QTL joint frequencies and marker frequencies for a given marker type, the conditional probability ($\omega_{uv|i}$) of a QTL genotype given the marker genotype of individual i can be calculated. This is used as a basis for QTL mapping in outcrossing species.

14.2.6 Hypothesis Tests

The existence of a QTL of significant effect within a marker interval can be tested by calculating a log-likelihood ratio (LR) test statistic under the null (H_0: There is no QTL) and alternative hypotheses (H_1: There is a QTL) expressed as

$$\text{LR} = -2[\log L_0(\mu_{uv} = \tilde{\mu}, \tilde{\sigma}^2, \tilde{p}, \tilde{q}) - \log L_1(\hat{\mathbf{\Omega}})].$$

Because the position of a QTL under the null hypothesis is not identifiable, the LR under the null hypothesis may not be asymptotically χ^2-distributed with four degrees of freedom. Churchill and Doerge (1994) proposed a permutation test approach to determine a critical threshold for declaring the existence of a QTL at a given type I error rate.

Hypotheses can be made regarding the significance of genetic additive ($H_0 : \alpha = \beta = 0$) and dominance effects ($H_0 : \gamma = 0$). The log-likelihood ratio (LR) test statistics are calculated for each test, in which the critical thresholds can be determined either by simulation studies or from the χ^2 distribution with two or one degrees of freedom, respectively, if the sample size used is sufficiently large.

In a full-sib family derived from two outbred parents, it is possible that a putative QTL does not segregate in a 1:1:1:1 ratio. In practice, the segregation pattern of a significant QTL should be tested because this is important for designing an efficient breeding strategy. We can test if the QTL detected is diallelic, segregating 1:2:1 or 1:1. The hypothesis that a significant QTL conforms to segregation type $ab \times ab$, for example, can be tested by formulating

$$\begin{aligned} H_0 &: \alpha = \beta, \\ H_1 &: \alpha \neq \beta. \end{aligned}$$

Similarly, the hypothesis for testing for the consistency of the QTL segregation to type $ab \times bb$ can be formulated as

$$\begin{aligned} H_0 &: \ \alpha = 0 \text{ or } \beta = 0, \\ H_1 &: \ \text{Neither of them equals zero.} \end{aligned}$$

As usual, the critical values for claiming the significance of these effects are determined on the basis of simulation studies.

We can also make a hypothesis test for the occurrence of double or interfered recombinations by formulating the null hypotheses

$$H_0 : p_{121} = p_{212} = 0 \tag{14.9}$$

and

$$H_0 : c = \frac{p_{121} + p_{212}}{(p_{121} + p_{212} + p_{122} + p_{211})(p_{121} + p_{212} + p_{112} + p_{221})} = 1, \tag{14.10}$$

respectively. The LR values under each of the two null hypotheses (14.9) and (14.10) and its alternative are calculated, which can each be thought to asymptotically follow a χ^2 distribution with one degree of freedom.

Example 14.1. The first examples of using a controlled cross to map QTL for outbred trees include studies in poplar (Bradshaw and Stettler 1995), eucalyptus (Grattapaglia et al. 1995) and loblolly pine (Groover et al. 1994). Here, we use an example from the hybridization between two poplar species, *Populus deltoides* and *P. euramericana*. A genetic linkage map was constructed using a so-called pseudo-testcross strategy (Grattapaglia and Sederoff 1994) based on 90 genotypes randomly selected from the F_1 interspecific hybrid family with random amplified polymorphic DNAs (RAPDs), amplified fraction length polymorphisms (AFLPs), and inter-sample sequence repeats (ISSR) (Yin et al. 2002). This map comprises the 19 largest linkage groups for each parental map, which roughly represent 19 pairs of chromosomes. The 90 hybrid genotypes used for map construction were measured for wood density with wood samples collected from 11-year-old stems in a field trial. The measurement for each genotype was repeated 4–6 times to reduce measurement errors. The means of these genotypes were calculated and used for QTL mapping here.

A significant QTL for wood density is detected on linkage group D17 reported in Yin et al. (2002). In this example, the empirical estimate of the critical value is obtained from 1000 permutation tests. It is found that the critical value for declaring the existence of a QTL on the whole linkage group under consideration is 6.9 at the significance level $p = 0.05$. The profile of the LRs of the full vs. reduced model across the length of linkage group D17 has a steep peak between a narrow marker interval AG/CGA-480–AG/CGA-330 (Fig. 14.1). The LR value at this peak is 11.7, well beyond the empirical critical threshold at the significance level $p = 0.05$.

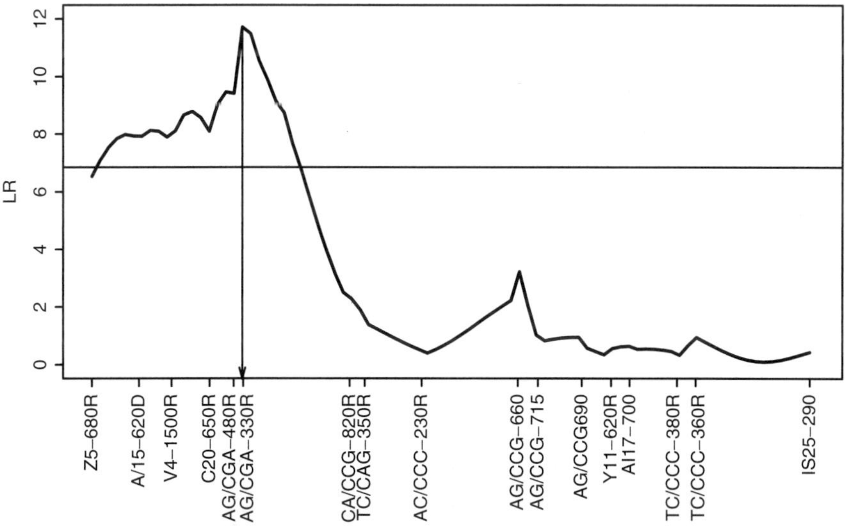

Fig. 14.1. The profile of the log-likelihood ratio (LR) test statistic for QTL detection across linkage group D17 in Yin et al. (2002) using the mixed-phase analysis. The empirical threshold based on permutation tests (Churchill and Doerge 1994) is indicated at the horizontal line. The marker names across the linkage group are given below the profile. Adapted from Lin et al. (2003).

The additive effect of this significant QTL detected is 0.033, or equivalent to 7 percent relative to the overall mean. This QTL was found to explain about 30 percent of the phenotypic variance for wood density in hybrid poplars. The MLE of phase probability p is 0.82, thus suggesting that there is quite a high probability of having a linkage phase $\mathbf{\Phi}_{11}$. This indicates that the positive allele of this QTL that increases wood density is, at a probability of 0.82, in coupling phase with dominance alleles of the two markers AG/CGA-480 and AG/CGA-330 flanking the QTL.

The same material was analyzed using a traditional interval mapping approach that assumes a possible QTL-marker linkage phase at one time. This phase-separate approach can also identify a significant QTL for wood density but cannot determine a correct linkage phase because the maximums of the LR values are identical between two possible linkage phases. Our method provides important information about non-allelic arrangements on the homologous chromosomes.

14.2.7 The Influence of Linkage Phases

A simulation was performed to test the influence of incorrectly characterizing a linkage phase on QTL detection and parameter estimation. A full-sib family is simulated for six equally spaced (20 cM) fully informative markers, forming five intervals. A QTL is hypothesized at 26 cM from the first marker (located within the second interval). The phenotypic values for this full-sib family are simulated by giving a particular set of unknown QTL effect parameters under the parental phase combination $\mathbf{\Phi}_{11}$ for the two parents. The sample size simulated is 400 and the heritability of the trait is 0.4.

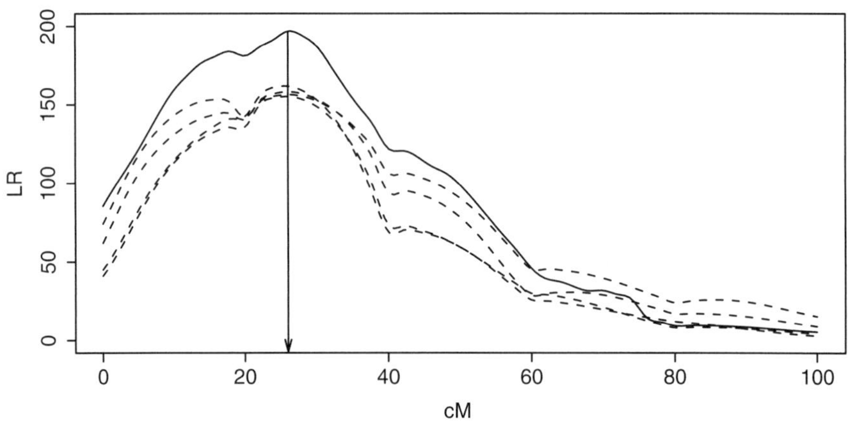

Fig. 14.2. The profiles of the log-likelihood ratio (LR) test statistic from one random simulation replicate for QTL detection across a linkage group under one mixed- (solid curve) and four separate (dot curves) phase analyses. The heritability for the trait hypothesized is $H^2 = 0.4$ with a sample size of 400. It should be noted that the same simulated data set given $\alpha = 0.5, \beta = 0.5$, and $\gamma = 0.5$) is used for all of these five different (one mixed- and four separate-phase) analyses. Adapted from Lin et al. (2003).

The simulated data under linkage phase combination $\mathbf{\Phi}_{11}$ were analyzed using models based on this phase and three other different phases, $\mathbf{\Phi}_{12}$, $\mathbf{\Phi}_{21}$, and $\mathbf{\Phi}_{22}$. Because different linkage phases only change the order of the parental chromosomal pairings, the maximums of the LR values from the correct linkage phase $\mathbf{\Phi}_{11}$ and the three incorrect linkage phases $\mathbf{\Phi}_{12}$, $\mathbf{\Phi}_{21}$, and $\mathbf{\Phi}_{22}$ will be identical (see Fig. 14.2), suggesting that phase-separate analyses have no power to select a most likely linkage phase. Also, as shown by flat, crooked curves, the maximum LR value from a single linkage phase model cannot be used to precisely determine the QTL position. Figure 14.2 also illustrates the LR values across the linkage group calculated when all linkage phase combinations are considered simultaneously based on the same simulated data set. A higher peak of the curve for a mixed-phase analysis indicates that the mixed model is better at detecting a significant QTL than the usual phase-separate analyses. When an incorrect linkage phase is used, the signs of the MLEs of the additive and dominance effects of a QTL will reverse.

Example 14.2. Wullschleger et al. (2005) performed QTL mapping for biomass partitioning in a backcross, (**T** × **D**) × **D**, derived from two poplar species, *Populus trichocarpa* Torr. & Gray (**T**) and *P. deltoids* Bartr. (**D**). The mapping population includes a total of 171 backcross trees that were genotyped for microsatellite (SSR) and AFLP markers. Because of the heterozygous nature of poplars, multiple types of markers were observed. Of them, the testcross markers that are segregating in parent F_1 but not in parent D include 92 SSR and 556 AFLP markers. A genetic linkage map based on the F_1 parent was then constructed from these markers with the pseudo-testcross strategy (Yin et al. 2004). This map is composed of 19 linkage groups, equivalent to the *Populus* chromosome number.

If we are going to map an intercross QTL ($Qq \times Qq$) using a pair of flanking testcross markers ($TD \times DD$), the F_1 parent may have two different diplotypes, $[TQT][DqD]$ and $[TqT][DDD]$. These two diplotypes with a probability of p and $1-p$, respectively, are incorporated into the model for QTL mapping, leading to the detection of a few significant QTLs that affect biomass partitioning traits in poplar (Table 14.2).

An intercross QTL for stem biomass was mapped to marker P_204_C2 on linkage group 6, whereas an intercross QTL near marker $T4\_10$ on linkage group 13 was observed for stem biomass, leaf biomass percentage, above-ground biomass, and total biomass.

14.3 Random-Effect Mapping Model for a Complicated Pedigree

14.3.1 Introduction

In the mixture model (9.4), we assume that different normal components are characterized by a known or unknown number of QTL genotypes. Genetic effects of putative QTLs on the phenotype, which are embedded within normal distributions, can be directly estimated by incorporating the fixed-effect model approach. This approach is

Table 14.3. The detection of intercross QTLs and the parental origin of favorable QTL alleles for biomass partitioning at year 2 in the field for a hybrid poplar family.

Trait	Group	Position	Phase	$\hat{p}$	$\hat{a}$	$\hat{d}$	LR
Stem biomass	6	P_204_C2	DQT	0.71	0.15	-0.02	29.51
Branch biomass	13	$T4\_10$	DQT	0.82	0.92	0.23	35.03
Leaf%	13	$T4\_10$	DQT	0.81	0.79	0.18	31.03
Above-ground biomass	13	$T4\_10$	DQT	0.82	0.83	0.19	32.84
Total biomass	13	$T4\_10$	DQT	0.81	0.80	0.19	31.47

Note: a and d are the additive and dominance genetic effects of an outcrossed QTL, respectively.

useful if the underlying genetic effects can be readily specified, as for controlled crosses of large size derived from inbred or outbred lines (Lander and Botstein 1989; Haley and Knott 1992; Zeng 1994; Lin et al. 2003). However, for many outcrossing populations, the size of a single family can be limited, so that the combination analysis of multiple families derived from unrelated or related parents is merely a choice. Because these parents are heterozygous, in which the number of alleles at a locus and the linkage phase of different loci are unknown, it would be difficult to specify the genetic effects. A robust method based on a random-effect model approach (Xu and Atchley 1995), which is not dependent on the parental diplotypes and the number of QTL alleles, can be used.

Under the fixed model, all effects contained in the expected mean of the normal distribution are fixed and can be estimated directly from the mixture model. Since there are no variances for the fixed effects, the phenotypic variance only contains the residual variance within a QTL genotype. However, in the random-effects model, all effects contained in the expected mean cannot be estimated because their expectations are zero. But their variances can be estimated by partitioning the total phenotypic (co)variances into the corresponding components. Thus, where the fixed model approach estimates the effect of allelic substitution (or allelic effect), the random model mapping approach estimates the segregating variance of the QTL.

The theoretical basis of the random model approach is the phenotypic resemblance (or covariance) between genetically related individuals. The phenotypic variance of an individual from a noninbred full-sib family is partitioned into the genetic variance due to the putative QTL, the variance due to the family-specific effect, and the variance due to the residual effect. Yet, the phenotypic covariance of two different individuals within the same family can be partitioned into the genetic covariance due to the QTL and the covariance due to the family-specific effect since the residual effects can be assumed to be independent between the sibs.

14.3.2 Statistical Model

Consider multiple related families, each with a different number of sibs. A quantitative trait, y, is measured for each sib within each family. The phenotypic value of sib j ($j = 1, \ldots, m_i$) within family i ($i = 1, \ldots, n$) is expressed as a linear function of K QTLs and other fixed covariates,

$$y_{ij} = \mu + \sum_{k=1}^{K} \alpha_k + \sum_{l=1}^{L} \beta_l X_l + e_{ij}, \tag{14.11}$$

where μ is the grand mean, α_k is the effect of the kth QTL, X_l's are some covariates such as sex or age, β_l is the effect of the lth covariate that is assumed to be uncorrelated with genetic and environmental errors, and e_{ij} represents a random environmental error term. The total sample size is $N = \sum_{i=1}^{n} m_i$.

Suppose there is a QTL of interest with an additive effect on the trait. Such a one-QTL model can be written as

$$y_{ij} = \mu + a_{ij} + d_{ij} + g_{ij} + \sum_{l=1}^{L} \beta_l X_l + e_{ij}, \tag{14.12}$$

where $a_{ij} \sim \mathcal{N}(0, \sigma_a^2)$ is the additive genetic effect, $d_{ij} \sim \mathcal{N}(0, \sigma_d^2)$ is the dominance genetic effect, $g_{ij} \sim \mathcal{N}(0, \sigma_g^2)$ is the polygenic additive effect that reflects the effects of unlinked genes or other familial influences, including environmental factors shared by families (excluding the hypothesized QTL), and $e_{ij} \sim \mathcal{N}(0, \sigma_e^2)$ is the environmental error.

Assuming that a_{ij}, d_{ij}, g_{ij} and e_{ij} are uncorrelated random variables, each with expectation 0, the total variance for a single observation (y_{ij}) becomes

$$\text{var}(y_{ij}) = \sigma_a^2 + \sigma_d^2 + \sigma_g^2 + \sigma_e^2.$$

The covariance between two sibs j and j' from family i is

$$\text{cov}(y_{ij}, y_{ij'}) = \pi_{ia}\sigma_a^2 + \pi_{id}\sigma_d^2 + \phi_{ig}\sigma_g^2,$$

where π_{ia} is the proportion of alleles identical–by–descent (IBD) shared by family members j and j', π_{id} is a binary variable indicating whether j and j' share both alleles IBD, and ϕ_{ig} is the expected proportion of shared alleles IBD and is, by expectation, equal to 0.5 for full-sib pairs. Therefore, the total variance-covariance matrix for y is given by

$$\mathbf{\Sigma} = \mathbf{\Pi}_a\sigma_a^2 + \mathbf{\Pi}_d\sigma_d^2 + \mathbf{\Phi}_g\sigma_g^2 + \mathbf{I}\sigma_e^2, \tag{14.13}$$

where $\mathbf{\Pi}_a$ is the matrix of the proportion of shared marker alleles IBD, $\mathbf{\Pi}_d$ is the matrix of binary variable π_{id}, and $\mathbf{\Phi}_g$ is a matrix of the expected proportion of shared alleles IBD, and $\mathbf{I}$ is the identity matrix.

We write the covariance matrix for two sibs j and j' in the same family i as

$$\Sigma_i = \text{Var}\begin{bmatrix} y_{ij} \\ y_{ij'} \end{bmatrix} = \sigma^2 \begin{bmatrix} 1 & \rho_i \\ \rho_i & 1 \end{bmatrix} = \sigma^2 \mathbf{R}_i, \tag{14.14}$$

where σ^2 is the total phenotypic variance, and

$$\begin{aligned} \rho_i &= \pi_{ia}\left(\frac{\sigma_a^2}{\sigma^2}\right) + \pi_{id}\left(\frac{\sigma_d^2}{\sigma^2}\right) + \phi_{ig}\left(\frac{\sigma_g^2}{\sigma^2}\right) \\ &= \pi_{ia}h_a^2 + \pi_{id}h_d^2 + \phi_{ig}h_g^2, \end{aligned} \tag{14.15}$$

where σ_a^2, σ_d^2, and σ_g^2, as defined, are the additive and dominance genetic variances of the QTL and the polygenic genetic variance, respectively, and h_a^2, h_d^2, and h_g^2 are the additive and dominance heritabilities of the putative QTL and the proportion of the phenotypic variance accounted for by the polygenic effect, respectively. The IBD variables have discrete distributions

$$\pi_{ia} = \begin{cases} 0 & \text{if no allele is IBD} \\ \frac{1}{2} & \text{if one allele is IBD} \\ 1 & \text{if both alleles are IBD} \end{cases}$$

and

$$\pi_{id} = \begin{cases} 1 & \text{if both alleles are IBD} \\ 0 & \text{otherwise.} \end{cases}$$

Equation (14.15) can be extended to relate individuals from different but related families i and i', in which case ϕ_{ig} should be denoted as $\phi_{ii'g}$. The determination of $\phi_{ii'g}$ should be based on quantitative genetic theory. For example, $\phi_{ii'g} = 1/4$ for two half-sibs, 1/8 for two cousins, and so on (Falconer and Mackay 1996).

14.3.3 IBD at a QTL

Since genotypes of a QTL cannot be directly observed, we need to use observed marker information to infer QTL genotypes linked with the markers. It has been shown from the joint probability for two linked loci that the expected IBD of one locus can be expressed as a linear function of the IBD of another locus (Haseman and Elston 1972). More recently, a linear model for expressing the IBD of a putative QTL (π_{ia}) in terms of the IBD of two flanking markers $\mathbf{M}_1$ and $\mathbf{M}_2$ has been developed (Fulker and Cardon 1994). Let r_1, r_2, and r be the recombination fractions between marker $\mathbf{M}_1$ and the QTL, the QTL and marker $\mathbf{M}_2$, and the two markers, respectively. Fulker and Cardon (1994) found the relationship

$$\pi_{ia} = a + b_1\pi_1 + b_2\pi_2, \tag{14.16}$$

where π_1 and π_2 are the IBDs of the two markers, respectively,

$$b_1 = \frac{(1-2r_1)^2 - (1-2r_2)^2(1-2r)^2}{1-(1-2r)^4},$$

$$b_2 = \frac{(1-2r_2)^2 - (1-2r_1)^2(1-2r)^2}{1-(1-2r)^4},$$

$$a = (1 - b_1 - b_2)/2.$$

The IBD of the QTL considered in equation (14.15) is then substituted by equation (14.16).

Example 14.3. Consider two outbred parents with genotypes 12/12 and 34/34 at two markers, respectively. These two parents are crossed to generate two offspring, j and j', with two-marker genotypes 13/24 and 14/24 (see Fig. 14.3). The IBD coefficients of alleles shared between sibs j and j' are $\pi_1 = 0.5$ for marker **A** and $\pi_2 = 1$ for marker **B**. Suppose there is a QTL between the two markers with the recombination fractions $r_1 = 0.05$ and $r_2 = 0.10$. The recombination fraction between the two markers is then 0.14, assuming no interference for crossover at meioses. Based on equation (14.16), we calculate the IBD of the QTL as follows:

$$\begin{aligned}
\pi_{ia} = & \left[1 - \frac{(1-2\times 0.05)^2 - (1-2\times 0.10)^2(1-2\times 0.14)^2}{1-(1-2\times 0.14)^4} \right. \\
& \left. - \frac{(1-2\times 0.10)^2 - (1-2\times 0.05)^2(1-2\times 0.14)^2}{1-(1-2\times 0.14)^4}\right] \\
& + \frac{(1-2\times 0.05)^2 - (1-2\times 0.10)^2(1-2\times 0.14)^2}{1-(1-2\times 0.14)^4} \times 0.5 \\
& + \frac{(1-2\times 0.10)^2 - (1-2\times 0.05)^2(1-2\times 0.14)^2}{1-(1-2\times 0.14)^4} \times 1 \\
= & \ 0.6730
\end{aligned}$$

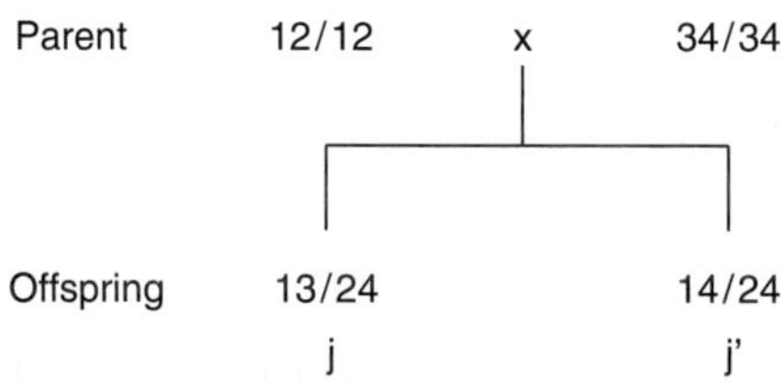

Fig. 14.3. Diagram for marker segregation between two outbred parents.

14.3.4 The Likelihood

Unrelated Families

The likelihood of observations for a pedigree can be constructed in terms of its structure. We will consider two types of pedigrees, one in which families are unrelated to each other and one in which families are related to different extents. Consider one (i) of n unrelated families in which there are m_i sibs. As defined in equation (14.14), $\mathbf{R}_i$ is now the $m_i \times m_i$ matrix with 1 on the diagonal and ρ_i on the off-diagonals. Then, under the normality assumption, we have a density function

$$f_i(\mathbf{y}_i) = \frac{1}{(2\pi\sigma^2)^{n/2}|\mathbf{R}_i|}\exp\left[-\frac{1}{2\sigma^2}\left(\mathbf{Z}_i^{\mathrm{T}}\mathbf{R}_i^{-1}\mathbf{Z}_i\right)\right], \tag{14.17}$$

where $\mathbf{Z}_i = \mathbf{y}_i - \mathbf{1}\mu - \sum_{l=1}^{L}\beta_l X_l$, $\mathbf{y}_i = (y_1, , y_{m_i})^{\mathrm{T}}$ is an $(m_i \times 1)$ vector of phenotypes, and $\mathbf{1}$ is an $(m_1 \times 1)$ vector with all entries equal to 1. For n independent families, we have the overall log-likelihood

$$L(\mathbf{\Omega}|\mathbf{y}) = \log[f_1(\mathbf{y}_i)] + ... + \log[f_n(\mathbf{y}_n)]. \tag{14.18}$$

The unknown parameters $\mathbf{\Omega} = (\mu, \beta_l, \sigma^2, h_a^2, h_d^2, h_g^2, r_1, r_2)$ contained in the likelihood (14.18) can be estimated using a maximum likelihood method. As usual, the estimate of the QTL position can be based on a grid approach by treating r_1 or r_2 as a known constant and varying it throughout the marker interval. By taking the derivative of the log-likelihood function with respect to μ, β_l, and σ^2, we obtain their MLEs as

$$\begin{aligned}
\hat{\mu} &= \left[\sum_{i=1}^{n}\mathbf{1}^{\mathrm{T}}\mathbf{R}_i^{-1}\mathbf{1}\right]^{-1}\left[\sum_{i=1}^{n}\mathbf{1}^{\mathrm{T}}\mathbf{R}_i^{-1}\mathbf{y}_i\right], \\
\hat{\beta}_l &= (\mathbf{1}^{\mathrm{T}}\mathbf{R}^{-1}\mathbf{1})^{-1}(\mathbf{1}^{\mathrm{T}}\mathbf{R}^{-1}\mathbf{y}), \\
\hat{\sigma}^2 &= \frac{1}{N}\sum_{i=1}^{n}\hat{\mathbf{Z}}^{\mathrm{T}}\mathbf{R}_i^{-1}\hat{\mathbf{Z}}_i,
\end{aligned}$$

where $\hat{\mathbf{Z}}_i = \mathbf{y}_i - \mathbf{1}\hat{\mu} - \sum_{l=1}^{L}\hat{\beta}_l X_l$ and $N = \sum_{i=1}^{n} m_i$. By plugging $\hat{\mu}$, $\hat{\beta}_l$, and $\hat{\sigma}^2$ into the log-likelihood function, the MLEs of h_a^2, h_d^2, and h_g^2 are estimated using the simplex algorithm (Nelder and Mead 1965).

Related Families

If all families in the pedigree are related to each other, the likelihood function of phenotypic data $(\mathbf{y})$ is given, under the assumption of multivariate normality, by

$$L(\mathbf{\Omega}|\mathbf{y}) = (2\pi)^{\frac{N}{2}}|\mathbf{\Sigma}|^{\frac{1}{2}}\exp\left[-\frac{1}{2}\left(\mathbf{Z}^{\mathrm{T}}\mathbf{\Sigma}^{-1}\mathbf{Z}\right)\right], \tag{14.19}$$

where N is the total number of offspring from all families, $\mathbf{Z} = \{Z_i\}_{i=1}^{n}$, $\mathbf{y} = \{y_{im_i}\}_{i=1}^{n}$ is an $(N \times 1)$ vector of all phenotypes, $\mathbf{1}$ is an $(N \times 1)$ vector with all entries equal to 1, and

$$\begin{aligned}
\mathbf{\Sigma} &= \sigma^2(\mathbf{\Pi}_a h_a^2 + \mathbf{\Pi}_d h_d^2 + \mathbf{\Pi}_g h_g^2 + \mathbf{I}) \\
&= \sigma^2\mathbf{R},
\end{aligned} \tag{14.20}$$

where $\mathbf{\Pi}_a$, $\mathbf{\Pi}_d$, and $\mathbf{\Pi}_g$ are the $(N \times N)$ matrices for the coefficients of IBD alleles at the QTL shared with any pair of individuals, the binary variables indicating whether the IBD alleles are shared between any pair of individuals, and the coefficients of IBD alleles generally shared with any pair of individuals in the pedigree, respectively. By taking the derivative of the log-likelihood with respect to μ, β_l, and σ_e^2, their MLEs can be obtained as

$$\begin{aligned}\hat{\mu} &= (\mathbf{1}^{\mathrm{T}}\mathbf{H}^{-1}\mathbf{1})^{-1}(\mathbf{1}^{\mathrm{T}}\mathbf{H}^{-1}\mathbf{y}),\\ \hat{\beta}_l &= (\mathbf{1}^{\mathrm{T}}\mathbf{H}^{-1}\mathbf{1})^{-1}(\mathbf{1}^{\mathrm{T}}\mathbf{H}^{-1}\mathbf{y}),\\ \hat{\sigma}^2 &= \frac{1}{N}(\mathbf{y}-\mathbf{1}\mu)^{\mathrm{T}}\mathbf{H}^{-1}(\mathbf{y}-\mathbf{1}\mu).\end{aligned}$$

By plugging $\hat{\mu}$, $\hat{\beta}_l$, and $\hat{\sigma}^2$ into the log-likelihood function, the MLEs of h_a^2, h_d^2, and h_g^2 are estimated using the simplex algorithm.

14.3.5 Hypothesis Testing

After the parameters are estimated, the hypothesis regarding the existence of QTLs can be tested. This can be done by formulating the hypotheses

$$\begin{aligned}&H_0: \ h_a^2 = h_d^2 = 0,\\ &H_1: \ \text{At least one of the heritabilities above is not equal to zero.}\end{aligned} \tag{14.21}$$

The likelihoods under the null $(L_0(\tilde{\mathbf{\Omega}}|\mathbf{y})$ and alternative hypotheses $(L_1(\hat{\mathbf{\Omega}}|\mathbf{y}))$ are calculated, with which the log-likelihood ratio is calculated

$$\mathrm{LR} = -2[\ln L_0(\tilde{\mathbf{\Omega}}|\mathbf{y}) - \ln L_1(\hat{\mathbf{\Omega}}|\mathbf{y})], \tag{14.22}$$

where $\tilde{\mathbf{\Omega}}$ and $\hat{\mathbf{\Omega}}$ are the MLEs of parameters under the H_0 and H_1, with the former not affected by marker genotypes. The critical value for the declaration of the existence of a QTL can be empirically determined by permutation tests. Similar tests for the additive or dominance variances explained by the QTL can also be made. The polygenic variance is important in explaining the results, whose significance can be tested by constructing the corresponding log-likelihood ratio.

Example 14.4. (**The Canine Genome Project**). A canine pedigree was developed to map the QTL responsible for canine hip dysplasia (CHD) using molecular markers. Seven founding Greyhounds and six founding Labrador retrievers were intercrossed, followed by backcrossing F_1's to the Greyhounds and Labrador retrievers and intercrossing the F_1's. A series of subsequent intercrosses among the progeny at different generation levels led to a complex network pedigree structure (Fig. 14.4) that maximized phenotypic ranges in CHD-related quantitative traits and the chance of detecting a segregating QTL (Bliss et al. 2002; Todhunter et al. 2005). A total of 148 dogs from this outbred population were genotyped for 240 microsatellite markers located on 38 pairs of autosomes and 1 pair of sex chromosomes (Breen et al. 2001). A linkage map of the canine genome constructed with these markers displays good coverage of each chromosome. The distances between adjacent markers were estimated in cM for the linkage map (Breen et al. 2001). Age at detection of femoral capital ossification (OSS), one of the important criteria for evaluating CHD, was measured for each of the dogs studied at its left and right side.

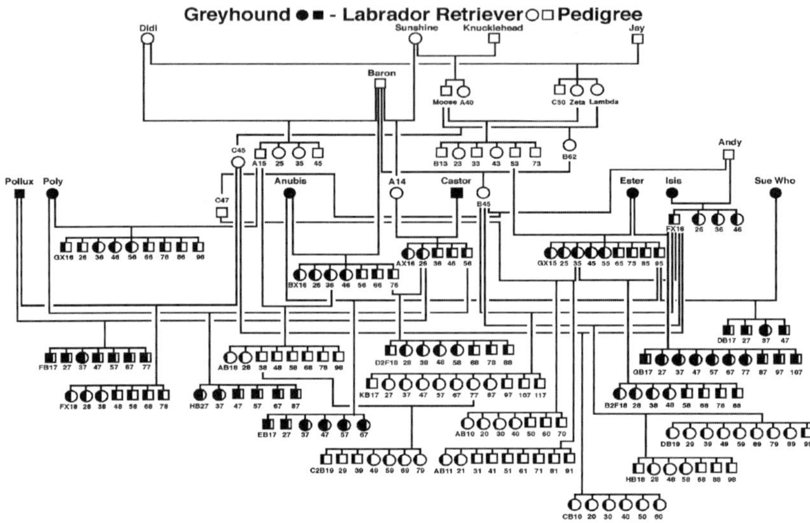

Fig. 14.4. Diagram of an outbred pedigree in a dog. Squares and circles represent males and females, respectively. Filled and open portions of each symbol represent the proportion of Greyhound and Labrador Retriever alleles, respectively, possessed by that dog.

We will use the random-effect Mendelian model (14.12) to map OSS based on the linkage map constructed by the 240 markers. For simplicity, this model ignores the dominance effect of the QTL and covariate effects, which is expressed as

$$y_{ij} = \mu + a_{ij} + g_{ij} + e_{ij},$$

allowing the estimates of the overall mean (μ), the genetic variance contributed by the QTL (σ_a^2), polygenic variance (σ_g^2), and residual variance (σ^2).

We identified 13 QTLs for OSS at the 0.1 percent genome-wide significance level, as shown by the peaks of the genome-wide log-likelihood ratio profile that indicate the MLEs of the QTL positions (Fig. 14.5). There are different estimation values of σ_a^2, suggesting that QTLs contribute differently to the genetic variance. Of the QTLs detected, six detected on CFA1, CFA3, CFA5, CFA8, CFA9, and CFA28 are "generalist" in that they affect OSS for both the left and right sides of a hip (Fig. 14.5). In most cases, consistent σ_a^2 estimates at the left and right imply that these QTLs play equally important roles in the susceptibility of CHD for the two sides. Other QTLs are "specialist", with two, on CFA17 and CFAX, being responsible for the left side, while five, on CFA7, CFA10, CFA18, CFA22, and CFA37, being responsible for the right side. As shown, these specialist QTLs have no contribution to CHD at the opposite hip.

14.4 Exercises

14.1 The model for full-sib mapping assumes that marker phases were known prior to QTL analysis. In other words, given the data set of a full-sib family, two steps are used for QTL mapping: (*i*) the determination of marker phases using approaches described in Chapter 4, and (*ii*) mapping QTL with known marker phases. Theoretically, the full-sib QTL model can incorporate the uncertainty of marker phases into the mixture model (14.4).

(a) Show the procedure for jointly modeling marker phases and marker-QTL phases.

(b) Compare the unifying and separate models. Is the former more advantageous than the latter?

(c) Find a real example in full-sib mapping. Use the statistical models described in this chapter to analyze the example and make statistical inferences from your analysis.

14.2 Mapping epistatic QTLs in outbred crosses

Mapping approaches for inbred crosses can be readily extended to map epistatic QTLs by including more QTL genotypes within the mixture model. Similar modeling can be done for epistatic mapping in outbred crosses. For fully informative loci, there are four different alleles at a locus between two outbred parents. Consider two QTL, each of which has four different genotypes, 13, 14, 23, and 24, in the outbred progeny population of size n. Let $\mu_{u_1v_1/u_2v_2}$ be the genotypic value for QTL genotype u_1v_1/u_2v_2 for $u_1, u_2 = 1, 2$ and $v_1, v_2 = 3, 4$ and let $\mathbf{u} = (\mu_{u_1v_1/u_2v_2})$ be the corresponding mean vector. Genetic effect parameters for two interacting QTLs are arrayed in $\mathbf{a} = (\mu, \alpha_1, \beta_1, \gamma_1, \alpha_2, \beta_2, \gamma_2, I_{\alpha\alpha}, I_{\alpha\beta}, I_{\beta\alpha}, I_{\beta\beta}, J_{\alpha\gamma}, J_{\beta\gamma}, K_{\gamma\alpha}, K_{\gamma\beta}, L_{\gamma\gamma})^{\mathrm{T}}$, where

(a) μ is the overall mean;

(b) α_1 is the additive effect due to the substitution from allele 1 to 2 at the first QTL;

(c) β_1 is the additive effect due to the substitution from allele 3 to 4 at the first QTL;

(d) γ_1 is the dominance effect due to the interaction between alleles from different parents;

(e) α_2 is the additive effect due to the substitution from allele 1 to 2 at the second QTL;

(f) β_2 is the additive effect due to the substitution from allele 3 to 4 at the second QTL;

(g) γ_2 is the dominance effect due to the interaction between alleles from different parents;

(h) $I_{\alpha\alpha}$ is the additive $\times$ additive epistatic effect due to the interaction between the substitutions from allele 1 to 2 at the first and second QTLs;

(i) $I_{\alpha\beta}$ is the additive $\times$ additive epistatic effect due to the interaction between the substitutions from allele 1 to 2 at the first QTL and from allele 3 to 4 at the second QTL;

(j) $I_{\beta\alpha}$ is the additive $\times$ additive epistatic effect due to the interaction between the substitutions from allele 3 to 4 at the first QTL and from allele 1 to 2 at the second QTL;

(k) $I_{\alpha\beta}$ is the additive $\times$ additive epistatic effect due to the interaction between the substitutions from allele 3 to 4 at the first and second QTLs;

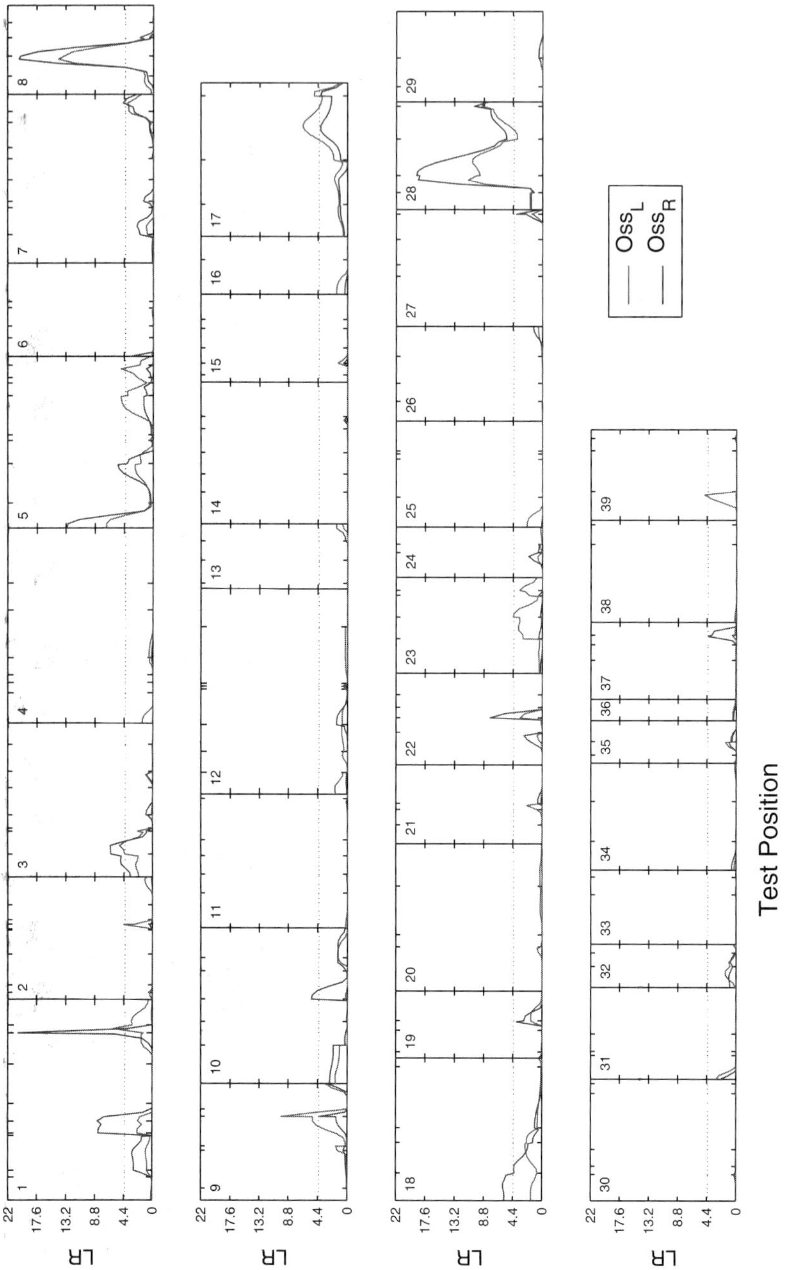

Fig. 14.5. The profiles of the log-likelihood ratios (LR) between the full (there is a QTL) and reduced (there is no QTL) models estimated from the interval model for OSS measured at the left (OSS_L, red) and right (OSS_R, blue) of a canine hip across the entire genome from chromosomes 1 to 39 using the linkage map constructed from microsatellite markers. The horizontal line indicates the critical threshold at $p = 0.001$ determined from permutation tests. Adapted from Liu et al. (2006).

(l) $J_{\alpha\beta}$ is the additive $\times$ dominance epistatic effect due to the interaction between the substitutions from allele 1 to 2 at the first QTL and the dominance effect at the second QTL;

(m) $J_{\alpha\beta}$ is the additive $\times$ dominance epistatic effect due to the interaction between the substitutions from allele 3 to 4 at the first QTL and the dominance effect at the second QTL;

(n) $K_{\alpha\beta}$ is the dominance $\times$ additive epistatic effect due to the interaction between the dominance effect at the first QTL and the substitutions from allele 1 to 2 at the second QTL;

(o) $K_{\alpha\beta}$ is the dominance $\times$ additive epistatic effect due to the interaction between the dominance effect at the first QTL and the substitutions from allele 3 to 4 at the second QTL;

(p) $K_{\alpha\beta}$ is the dominance $\times$ dominance epistatic effect due to the interaction between the dominance effects at the first and second QTLs.

We relate the genotypic value vector and genetic-effect vector by

$$\mathbf{u} = \mathbf{D}\mathbf{a},$$

where the design matrix is

$$\mathbf{D} = \begin{bmatrix}
1 & 1 & 1 & 1 & 1 & 1 & 1 & 1 & 1 & 1 & 1 & 1 & 1 & 1 & 1 & 1 \\
1 & 1 & 1 & 1 & 1 & -1 & -1 & 1 & -1 & 1 & -1 & -1 & -1 & 1 & -1 & -1 \\
1 & 1 & 1 & 1 & 1 & 1 & -1 & -1 & 1 & -1 & 1 & -1 & -1 & -1 & 1 & -1 \\
1 & 1 & 1 & 1 & -1 & -1 & 1 & -1 & -1 & -1 & -1 & 1 & 1 & -1 & -1 & 1 \\
1 & 1 & -1 & -1 & 1 & 1 & 1 & 1 & 1 & -1 & -1 & 1 & -1 & -1 & -1 & -1 \\
1 & 1 & -1 & -1 & 1 & -1 & -1 & 1 & -1 & -1 & 1 & -1 & 1 & -1 & 1 & 1 \\
1 & 1 & -1 & -1 & -1 & 1 & -1 & -1 & 1 & 1 & -1 & -1 & 1 & 1 & -1 & 1 \\
1 & 1 & -1 & -1 & -1 & -1 & 1 & -1 & -1 & 1 & 1 & 1 & -1 & 1 & 1 & -1 \\
1 & -1 & 1 & -1 & 1 & 1 & 1 & -1 & -1 & 1 & 1 & -1 & 1 & -1 & -1 & -1 \\
1 & -1 & 1 & -1 & 1 & -1 & -1 & -1 & 1 & 1 & -1 & 1 & -1 & -1 & 1 & 1 \\
1 & -1 & 1 & -1 & -1 & 1 & -1 & 1 & -1 & -1 & 1 & 1 & -1 & 1 & -1 & 1 \\
1 & -1 & 1 & 1 & -1 & -1 & 1 & 1 & 1 & -1 & -1 & -1 & 1 & 1 & 1 & -1 \\
1 & -1 & -1 & 1 & 1 & 1 & 1 & -1 & -1 & -1 & -1 & -1 & -1 & 1 & 1 & 1 \\
1 & -1 & -1 & 1 & 1 & -1 & -1 & -1 & 1 & -1 & 1 & 1 & 1 & 1 & -1 & -1 \\
1 & -1 & -1 & 1 & -1 & 1 & -1 & 1 & -1 & 1 & -1 & 1 & 1 & -1 & 1 & -1 \\
1 & -1 & -1 & 1 & -1 & -1 & 1 & 1 & 1 & 1 & 1 & -1 & -1 & -1 & -1 & 1
\end{bmatrix}.$$

By estimating the genotypic value vector, the genetic-effect vector can be readily estimated as $\hat{\mathbf{a}} = \mathbf{D}^{\mathrm{T}}\hat{\mathbf{u}}$. The likelihood for two interacting "outbred" QTLs can be formulated as

$$L(y) = \prod_{i=1}^{n} \sum_{u_1=1}^{2} \sum_{v_1=3}^{4} \sum_{u_2=1}^{2} \sum_{v_2=3}^{4} \omega_{u_1v_1/u_2v_2|i} f_{u_1v_1/u_2v_2}(y_i),$$

where $\omega_{u_1v_1/u_2v_2|i}$ is the conditional probability of a QTL genotype given the marker genotype of individual i and $f_{u_1v_1/u_2v_2}(y_i)$ is a normal distribution density with mean $\mu_{u_1v_1/u_2v_2}$ and variance σ^2. The conditional probabilities can be derived in terms of the recombination fractions between the markers and QTL.

The standard EM algorithm can be developed to estimate the genotypic values and residual variance. By defining

$$\Pi_{u_1v_1/u_2v_2|i} = \frac{\omega_{u_1v_1/u_2v_2|i} f_{u_1v_1/u_2v_2}(y_i)}{\sum_{u_1'=1}^{2} \sum_{v_1'=3}^{4} \sum_{u_2'=1}^{2} \sum_{v_2'=3}^{4} \omega_{u_1'v_1'/u_2'v_2'|i} f_{u_1'v_1'/u_2'v_2'}(y_i)}$$

in the E step, we derived the MLEs of the unknown parameters in the M step as

$$\hat{\mu}_{u_1v_1/u_2v_2} = \frac{\sum_{u_1=1}^{2}\sum_{v_1=3}^{4}\sum_{u_2=1}^{2}\sum_{v_2=3}^{4}\Pi_{u_1v_1/u_2v_2|i}y_i}{\sum_{u_1=1}^{2}\sum_{v_1=3}^{4}\sum_{u_2=1}^{2}\sum_{v_2=3}^{4}\Pi_{u_1v_1/u_2v_2|i}},$$

$$\hat{\sigma}^2 = \frac{1}{n}\sum_{i=1}^{n}\sum_{u_1=1}^{2}\sum_{v_1=3}^{4}\sum_{u_2=1}^{2}\sum_{v_2=3}^{4}(y_i - \hat{\mu}_{u_1v_1/u_2v_2})^2\Pi_{u_1v_1/u_2v_2|i}.$$

The E and M steps are iterated until the estimates are stable. After the parameters are estimated, a number of hypothesis tests can be made.

(a) Show how you test for the existence of a QTL and determine the critical threshold for declaring the existence of a QTL.

(b) Hypothesis tests for different genetic effects, including the additive, dominance, and epistatic effects, can be formulated with the respective null hypotheses:

$H_0 : \alpha_1 = 0,$
$H_0 : \beta_1 = 0,$
$H_0 : \gamma_1 = 0,$
$H_0 : \alpha_2 = 0,$
$H_0 : \beta_2 = 0,$
$H_0 : \gamma_2 = 0,$
$H_0 : I_{\alpha\alpha} = 0,$
$H_0 : I_{\alpha\beta} = 0,$
$H_0 : I_{\beta\alpha} = 0,$
$H_0 : I_{\beta\beta} = 0,$
$H_0 : J_{\alpha\gamma} = 0,$
$H_0 : J_{\beta\gamma} = 0,$
$H_0 : K_{\gamma\alpha} = 0,$
$H_0 : K_{\gamma\beta} = 0,$
$H_0 : L_{\gamma\gamma} = 0.$

Under $H_0 : \alpha_1 = 0$, the genotypic values should be constrained by

$$\begin{aligned} &\mu_{13/13} + \mu_{13/14} + \mu_{13/23} + \mu_{13/24} \\ &+ \mu_{14/13} + \mu_{14/14} + \mu_{14/23} + \mu_{14/24} \\ &= \mu_{23/13} + \mu_{23/14} + \mu_{23/23} + \mu_{23/24} \\ &+ \mu_{24/13} + \mu_{24/14} + \mu_{24/23} + \mu_{24/24}. \end{aligned}$$

This constraint is implemented with the EM algorithm as described above, which will lead to the MLEs of the genotypic values with α_1 restricted to 0. Provide the constraints and algorithms for parameter estimation under each of the other null hypotheses.

(c) Use the algorithms you develop as a practical example for QTL mapping in outcrossing populations.

14.3 Show how the relationship between the IBDs of the markers and QTL, as described by equation (14.16), is derived.

14.4 To map QTLs in a complicated pedigree, we need the matrices IBD between each pair of dogs. Using Fig. 14.4 as an example:

(a) Write down a (148 × 148) matrix for the proportion of alleles IBD at a QTL.

(b) Write down a (148 × 148) matrix for the binary variables that indicate whether any pair of dogs are IBD at the QTL.
(c) Write down a (148 × 148) matrix for the coefficients of IBD between each pair of dogs for all genes.

14.5 Mapping epistatic QTLs in a complicated pedigree
If epistatic QTLs should be mapped for a pedigree as in Fig. 14.4, we need to incorporated epistatic variances, i.e., additive × additive ($\sigma^2_{i_{aa}}$), additive × dominance ($\sigma^2_{i_{ad}}$), dominance × additive ($\sigma^2_{i_{da}}$), and dominance × dominance ($\sigma^2_{i_{dd}}$) into the random-effects model. In this case, the total variance for a single observation (y_{ij}) is expressed as

$$\text{var}(y_{ij}) = \sigma^2_{a_1} + \sigma^2_{d_1} + \sigma^2_{a_2} + \sigma^2_{d_2} + \sigma^2_{i_{aa}} + \sigma^2_{i_{ad}} + \sigma^2_{i_{da}} + \sigma^2_{i_{dd}} + \sigma^2_g + \sigma^2_e.$$

The covariance between two sibs j and j' from family i is

$$\begin{aligned}\text{cov}(y_{ij}, y_{ij'}) = {} & \pi_{ia_1}\sigma^2_{a_1} + \pi_{id_1}\sigma^2_{d_1} + \pi_{ia_2}\sigma^2_{a_2} + \pi_{id_2}\sigma^2_{d_2} \\ & + \pi_{ia_1}\pi_{ia_2}\sigma^2_{i_{aa}} + \pi_{ia_1}\pi_{id_2}\sigma^2_{i_{ad}} + \pi_{id_1}\pi_{id_2}\sigma^2_{i_{da}} + \pi_{id_1}\pi_{id_2}\sigma^2_{i_{dd}} + \phi_{ig}\sigma^2_g,\end{aligned}$$

where the second-order subscripts of π stand for the identification of a QTL.
(a) Formulate the likelihood for two interactive QTLs.
(b) Derive the algorithms for the estimates of the additive ($\sigma^2_{a_1}$ and $\sigma^2_{a_2}$) and dominance variances ($\sigma^2_{d_1}$ and $\sigma^2_{d_2}$) at two different QTLs and their epistatic variances.

A

General Statistical Results and Algorithms

A.1 Likelihood Asymptotics

The following material is adapted from the book *Statistical Inference, Second Edition*, by Casella and Berger (2001). The material is a bit advanced, and is included for those who want to see a more complete picture of the asymptotics of likelihood.

After the data $\mathbf{X} = \mathbf{x}$ are observed, the likelihood function, $L(\theta|\mathbf{x})$, is a completely defined function of the variable θ, and the LR statistic is

$$\lambda(\mathbf{x}) = \frac{\sup_{\Theta_0} L(\theta|\mathbf{x})}{\sup_{\Theta} L(\theta|\mathbf{x})}.$$

Even if the two suprema of $L(\theta|\mathbf{x})$, over the sets Θ_0 and Θ, cannot be analytically obtained, they can usually be computed numerically. Thus, the test statistic $\lambda(\mathbf{x})$ can be obtained for the observed data point even if no convenient formula defining $\lambda(\mathbf{x})$ is available.

To define a level α test, the constant c must be chosen so that

$$\sup_{\theta \in \Theta_0} P_\theta\left(\lambda(\mathbf{X}) \leq c\right) \leq \alpha. \tag{A.1}$$

Theorem A.1 (Asymptotic Distribution of the LRT–Simple H_0). *For testing $H_0 : \theta = \theta_0$ versus $H_1 : \theta \neq \theta_0$, suppose $X_1, \ldots, X_n$ are iid $f(x|\theta)$, $\hat{\theta}$ is the MLE of θ, and $f(x|\theta)$ satisfies the usual regularity conditions. Then, under H_0, as $n \to \infty$,*

$$-2 \log \lambda(\mathbf{X}) \to \chi_1^2 \textit{ in distribution,}$$

where χ_1^2 is a χ^2 random variable with one degree of freedom.

Theorem A.1 can be extended to the case where the null hypothesis concerns a vector of parameters. The following generalization, which we also state without proof, allows us to ensure equation (A.1) is true, at least for large samples.

Theorem A.2. *Let $X_1, \ldots, X_n$ be a random sample from a pdf or pmf $f(x|\theta)$. Under the usual regularity conditions, if $\theta \in \Theta_0$, then the distribution of the statistic $-2\log\lambda(\mathbf{X})$ converges to a chi-squared distribution as the sample size $n \to \infty$. The degrees of freedom of the limiting distribution is the difference between the number of free parameters specified by $\theta \in \Theta_0$ and the number of free parameters specified by $\theta \in \Theta$.*

Rejection of $H_0 : \theta \in \Theta_0$ for small values of $\lambda(\mathbf{X})$ is equivalent to rejection for large values of $-2\log\lambda(\mathbf{X})$. Thus,

$$H_0 \text{ is rejected if and only if } -2\log\lambda(\mathbf{X}) \geq \chi^2_{\nu,\alpha},$$

where ν is the degrees of freedom specified in Theorem A.2. The type I error probability will be approximately α if $\theta \in \Theta_0$ and the sample size is large. In this way, equation (A.1) will be approximately satisfied for large sample sizes, and an *asymptotic size α test* has been defined. Note that the theorem will actually imply only that

$$\lim_{n\to\infty} P_\theta(\text{reject } H_0) = \alpha \quad \text{for each } \theta \in \Theta_0,$$

not that the $\sup_{\theta\in\Theta_0} P_\theta(\text{reject } H_0)$ converges to α. This is usually the case for asymptotic size α tests.

Theorem A.3. *Let $X_1, \ldots, X_n$ be a random sample from a pdf or pmf $f(x|\theta)$, let $\hat\theta$ denote the MLE of θ, and let $h(\theta)$ be a continuous function of θ. Under regularity conditions,*

$$\sqrt{n}[h(\hat\theta) - h(\theta)] \to n[0, v(\theta)],$$

where $\mathrm{Var}(h(\hat\theta)) = v(\theta)$ can be approximated by

$$\mathrm{Var}(h(\hat\theta)) \approx \frac{[h'(\theta)]^2|_{\theta=\hat\theta}}{-\frac{\partial^2}{\partial\theta^2}\log L(\theta|\mathbf{X})|_{\theta=\hat\theta}},$$

where the denominator is $\hat I_n(\hat\theta)$, the observed information number.

Regularity Conditions. Theorems A.1, A.2, and A.3 refer to "usual regularity conditions". These are mathematically technical conditions, rather boring, and usually satisfied in most reasonable problems. But they are a necessary evil.

These conditions mainly relate to differentiability of the density and the ability to interchange differentiation and integration. For more details and generality, see Casella and Berger (2001), Stuart, Ord and Arnold (1999, Chapter 18), Ferguson (1996, Part 4), or Lehmann and Casella (1998, Section 6.3).

A.2 General Form of the EM Algorithm

Although we will give a detailed description of the workings of the EM algorithm, we will suppress some of the more mathematical details. For those, we refer the reader to Casella and Berger (2001, Chapter 7).

We start the statistical formulation in the usual manner, having data $\mathbf{y}$ with likelihood function $L(\theta|\mathbf{y})$. Then the "missing data" z are introduced (sometimes called the *augmented data*), and we refer to $(\mathbf{y}, z)$ as the *complete data* and $L(\theta|\mathbf{y}, z)$ as the *complete-data likelihood.* (In the problems that we will encounter, it will usually be clear how to choose $L(\theta|\mathbf{y}, z)$, as it will be clear what type of missing data will make the computation easier.) Analogously, $L(\theta|\mathbf{y})$ is sometimes called the *incomplete-data likelihood* or the *observed-data likelihood.* From these two likelihoods, the entire EM algorithm follows.

The critical piece of the algorithm is the definition of the missing-data density k, which is given by

$$k(z|\theta, \mathbf{y}) = \frac{L(\theta|\mathbf{y}, z)}{L(\theta|\mathbf{y})}, \tag{A.2}$$

which we then rearrange and take logs to get

$$\log L(\theta|\mathbf{y}) = \log L(\theta|\mathbf{y}, z) - \log k(z|\theta, \mathbf{y}). \tag{A.3}$$

The EM algorithm is an iterated algorithm. At each step $t = 1, 2, \ldots$, we calculate an estimate of θ, $\hat{\theta}^{(t)}$ with the property that as $t \to \infty$, $\hat{\theta}^{(t)} \to \hat{\theta}$, the true MLE. To get this sequence we work with the expected value of equation (A.3) in the following way. At iteration t, we have $\hat{\theta}^{(t)}$, and we are ready to calculate $\hat{\theta}^{(t+1)}$, which we denote by θ for now. To do so, we first take the expected value of both sides of equation (A.3) (*The E step*) with respect to $k(z|\hat{\theta}^{(t+1)}, \mathbf{y})$ to get

$$\begin{aligned} \mathrm{E}_{\hat{\theta}^{(t)}}[\log L(\theta|\mathbf{y})] &= E_{\hat{\theta}^{(t)}}[\log L(\theta|\mathbf{y}, z)] - E_{\hat{\theta}^{(t)}}[\log k(z|\theta, \mathbf{y})] \\ &\text{or} \\ \log L(\theta|\mathbf{y}) &= E_{\hat{\theta}^{(t)}}[\log L(\theta|\mathbf{y}, z)] - E_{\hat{\theta}^{(t)}}[\log k(z|\theta, \mathbf{y})]. \end{aligned} \tag{A.4}$$

As the expectation is over the distribution of z, and $\log L(\theta|\mathbf{y})$ is free of z, we do not need to take expectations.

Now here is the beauty of the EM algorithm. We choose $\hat{\theta}^{(t+1)}$ to be the value of θ that maximizes $E_{\hat{\theta}^{(t)}}[\log L(\theta|\mathbf{y}, z)]$, the *expected complete-data likelihood.* By the EM algorithm theory, this value of θ automatically decreases the last term in equation (A.4), $E_{\hat{\theta}^{(t)}}[\log k(z|\theta, \mathbf{y})]$. Thus, it follows that

$$\log L(\hat{\theta}^{(t)}|\mathbf{y}) \leq \log L(\hat{\theta}^{(t+1)}|\mathbf{y})$$

and we do not have to calculate $E_{\hat{\theta}^{(t)}}[\log k(z|\theta, \mathbf{y})]$!

Thus, the EM sequence $\hat{\theta}^{(1)}, \hat{\theta}^{(2)}, \ldots, \hat{\theta}^{(t)}\hat{\theta}^{(t+1)}$ results in an increasing likelihood $L(\theta|\mathbf{y})$ and under a few mild conditions will converge to the MLE of θ.

B

R Programs

B.1 Chapter 2

(1) *R* program for Example 2.4

```
y<-c(79,82,85,87,100,101,102,103,124,125,126,127)
n<-length(y);nit<-20;
MAA<-array(max(y),dim=c(nit,1));
MAa<-array(mean(y),dim=c(nit,1));
Maa<-array(min(y),dim=c(nit,1));
S<-array(sd(y),dim=c(nit,1));S2<-array(sy,dim=c(nit,1));
S2[1]<-5;
for(i in 2:nit){
temp1<-.25*dnorm(y,mean=MAA[i-1],sd=S[i-1]);
temp2<-.5*dnorm(y,mean=MAa[i-1],sd=S[i-1]);
temp3<-.25*dnorm(y,mean=Maa[i-1],sd=S[i-1]);
PAA<-temp1/(temp1+temp2+temp3);
PAa<-temp2/(temp1+temp2+temp3);
Paa<-1-PAA-PAa;
MAA[i]<-sum(y*PAA)/sum(PAA);
MAa[i]<-sum(y*PAa)/sum(PAa);
Maa[i]<-sum(y*Paa)/sum(Paa);
#S[i]<-sqrt((1/nit)*(sum((y-MAA[i])^2)
    +sum((y-MAa[i])^2)+sum((y-Maa[i])^2)))
S2[i]<-(sum(PAA*(y-MAA[i])^2)+sum(PAa*(y-MAa[i])^2)
    +sum(Paa*(y-Maa[i])^2))/(n*sum(PAA+PAa+Paa))
S[i]<-sqrt(S2[i])
}
par(mfrow=c(2,2))
par(mar=c(4,4,1,1))
plot(MAA,type="l",lwd=2,main="",
  ylab="Mean AA",xlab="Iteration")
```

```
plot(MAa,type="l",lwd=2,main="",
  ylab="Mean Aa",xlab="Iteration")
plot(Maa,type="l",lwd=2,main="",
  ylab="Mean aa",xlab="Iteration")
plot(S2,type="l",lwd=2,main="",
  ylab="Variance",xlab="Iteration")
MAA[nit];MAa[nit];Maa[nit];S2[nit]
```

(2) *R* program for Example 2.12

```
y<-c(1339,154,151,1195);
m<-length(y);n<-sum(y);nsim<-10000;
#-----Log likelihood function---------
lambda<-function(p1,p2,p3,x1,x2,x3)
    (x1*log(p1)+x2*log(p2)+x3*log(p3)
      +(n-x1-x2-x3)*log(1-p1-p2-p3));
p1hat<-(y[1]+y[4])/(2*n);p2hat<-(y[2]+y[3])/(2*n);
Lnull<-lambda(p1hat,p2hat,p2hat,y[1],y[2],y[3])
Lmax<-lambda(y[1]/n,y[2]/n,y[3]/n,y[1],y[2],y[3]);
Tobs<- -2*(Lnull-Lmax);
Tsim<-array(0,dim=c(nsim,1));
for(i in 1:nsim)
{
x<-rmultinom(1,n,c(y[1]/n,y[2]/n,y[3]/n,y[4]/n))
Tsim[i]<- -2*(lambda((x[1]+x[4])/(2*n),
    (x[2]+x[3])/(2*n),(x[2]+x[3])/(2*n),
    x[1],x[2],x[3])-lambda(x[1]/n,
    x[2]/n,x[3]/n,x[1],x[2],x[3]))
}
hist(Tsim,main=expression(-2(log)(lambda)),
    xlab="Simulated Observations",xlim=c(0,40),
    freq=F,col="green",breaks=50)
mean(Tsim>Tobs)
#-----------------------------------------------------
#The function rmultinom returns a random multinomial vector
#n=number of variables desired
#size=sum of cells
#prob=vector of cell probabilities
rmultinom <- function(n,size,prob){
K <- length(prob) # #{classes}
matrix(tabulate(sample(K, n*size,
      repl = TRUE, prob)+K *0:(n-1),
nbins=n*K),
nrow=n, ncol=K, byrow=TRUE)}
```

B.2 Chapter 8

(1) *R* program for Example 8.6

```
#This gets MLEs for the Tomato data
y1<-c(79,82,100,102,124);y2<-c(85,87,101,103,125,126,127)
n1<-length(y1);n2<-length(y2);n<-n1+n2
#-----Initialize Estimates-------
r<-.25;m1<-mean(y1);m2<-mean(y2);s<-sqrt(var(c(y1,y2)))
m1plot<-m1;m2plot<-m2;splot<-s;rplot<-r
#-----Start Iteration------------------------------
nit<-100
for(i in 1:nit)
{
w1<-P1(y1,m1,m2,s,r);w2<-P2(y2,m1,m2,s,r)
m1<-(sum(w1*y1)+sum(w2*y2))/(sum(w1)+sum(w2))
m2<-(sum((1-w1)*y1)+sum((1-w2)*y2))/(sum((1-w1))+sum((1-w2)))
s<-sqrt((sum(w1*(y1-m1)^2+(1-w1)*(y1-m2)^2)
    +sum(w2*(y2-m1)^2+(1-w2)*(y2-m2)^2))/n)
r<-(sum(1-w1)+sum(w2))/n
r<-min(r,.5)
m1plot<-c(m1plot,m1);m2plot<-c(m2plot,m2)
splot<-c(splot,s);rplot<-c(rplot,r)
}
#----------------Plot---------------------------
par(mfrow=c(2,2))
par(mar=c(4,4,1,1))
plot(m1plot,type="l",lwd=2,main="",
  ylab="mean 1",xlab="Iteration")
plot(m2plot,type="l",lwd=2,main="",
  ylab="mean 2",xlab="Iteration")
plot(splot,type="l",lwd=2,main="",
  ylab="std",xlab="Iteration")
plot(rplot,type="l",lwd=2,main="",
  ylab="r",xlab="Iteration")

#----------Define Functions for P1 and P2
P1<-function(y,t1,t2,w,r)
{
temp1<-(1-r)*dnorm(y,mean=t1,sd=w)
temp2<-r*dnorm(y,mean=t2,sd=w)
return(temp1/(temp1+temp2))
}
P2<-function(y,t1,t2,w,r)
{
temp1<-r*dnorm(y,mean=t1,sd=w)
```

```
temp2<-(1-r)*dnorm(y,mean=t2,sd=w)
return(temp1/(temp1+temp2))
}
```

(2) *R* program for Example 8.6

```
#This gets permutation distribution for the Tomato data
y1<-c(79,82,100,102,124);y2<-c(85,87,101,103,125,126,127)
n1<-length(y1);n2<-length(y2);n<-n1+n2
y<-c(y1,y2)
#-----Initialize Estimates-------
r<-.25;m1<-mean(y1);m2<-mean(y2);s<-sd(y)
Lplot<- logL(y1,y2,m1,m2,s,r)
#-----Start Iteration------------------------------
nperm<-5000         #number of permutation statistics
nit<-50          #number of iterations to find MLEs
for(j in 1:nperm)
{
yp<-sample(y);y1<-yp[1:n1];y2<-yp[(n1+1):n]
for(i in 1:nit)
{
w1<-P1(y1,m1,m2,s,r);w2<-P2(y2,m1,m2,s,r)
m1<-(sum(w1*y1)+sum(w2*y2))/(sum(w1)+sum(w2))
m2<-(sum((1-w1)*y1)+sum((1-w2)*y2))/(sum((1-w1))+sum((1-w2)))
s<-sqrt((sum(w1*(y1-m1)^2+(1-w1)*(y1-m2)^2)
    +sum(w2*(y2-m1)^2+(1-w2)*(y2-m2)^2))/n)
r<-(sum(1-w1)+sum(w2))/n
r<-min(r,.5)
}
L<- logL(y1,y2,m1,m2,s,r);Lplot<-c(Lplot,L)
}
#----------------Calculate Statistics-----------
m<-mean(y);s<-sd(y)
LH0<-logL(y1,y2,m,m,s,.5)
Lplot<- -2*(LH0-Lplot)
sort(Lplot)[.95*nperm]
sort(Lplot)[.05*nperm]
#----------------Plot---------------------------
hist(Lplot,main="Permutation Distribution",
    freq=F,xlab="-2 log lambda")
#----------Define Functions for P1 and P2
P1<-function(y,t1,t2,w,r)
{
temp1<-(1-r)*dnorm(y,mean=t1,sd=w)
temp2<-r*dnorm(y,mean=t2,sd=w)
return(temp1/(temp1+temp2))
```

```
}
P2<-function(y,t1,t2,w,r)
{
temp1<-r*dnorm(y,mean=t1,sd=w)
temp2<-(1-r)*dnorm(y,mean=t2,sd=w)
return(temp1/(temp1+temp2))
}
#------------Define Functions for log likelihood---------
logL<-function(y1,y2,t1,t2,w,r)
{
S1<-sum(log((1-r)*dnorm(y1,t1,s)+r*dnorm(y1,t2,s)))
S2<-sum(log(r*dnorm(y2,t1,s)+(1-r)*dnorm(y2,t2,s)))
return(S1+S2)
}
```

(3) *R* program for Example 11.2

```
#This gets MLEs for the Poplar data
data<-read.table("PoplarMarker3.txt",sep = ",",header=F)
M<-data[,1:4];nm<-dim(M);
Y<-data[,5];n<-length(Y)
mY<-mean(Y);sY<-sd(Y)
LL0<-sum(dnorm(Y,mY,sY,log=T))
#----------Create Classes-------------------
#----------Class1=11, Class2=10,Class3=01,Class4=00
Cl<-array(1,c(nm[1],(nm[2]-1)));nc<-dim(Cl)
for(i in 1:nc[1])
{
for(j in 1:nc[2])
{
c1<-1*(M[i,j]==2)*(M[i,(j+1)]==2);
c2<-2*(M[i,j]==2)*(M[i,(j+1)]==1);
c3<-3*(M[i,j]==1)*(M[i,(j+1)]==2);
c4<-4*(M[i,j]==1)*(M[i,(j+1)]==1);
Cl[i,j]<-c1+c2+c3+c4
}
}
ML<-array(0,c(nc[2],4));LL<-array(0,c(nc[2],1))
#---------Define the Y classes for each Interval-----------
for(j in 1:nc[2])
{
 y1<-(Cl[,j]==1)*Y;y1<-y1[y1!=0]
 y2<-(Cl[,j]==2)*Y;y2<-y2[y2!=0]
 y3<-(Cl[,j]==3)*Y;y3<-y3[y3!=0]
 y4<-(Cl[,j]==4)*Y;y4<-y4[y4!=0]
#---------Get MLEs---------------------------------
```

```
#-----Initialize Estimates-------
T<-c(mean(y1),mean(y4),sY,.25)
#-----Start Iteration------------------------------
nit<-1000
for(i in 1:nit){T<-c(MLE(y1,y2,y3,y4,T[1],T[2],T[3],T[4]))}
ML[j,]<-T
LL[j]<-2*(logL(y1,y2,y3,y4,T[1],T[2],T[3],T[4])-LL0)
}
print(cbind(ML,LL))
#-----------------------------------------------------
#---------Define Functions for P1 and P2
P1<-function(y,t1,t2,w,v)
{
temp1<-(1-v)*dnorm(y,mean=t1,sd=w)
temp2<-v*dnorm(y,mean=t2,sd=w)
return(temp1/(temp1+temp2))
}
P2<-function(y,t1,t2,w,v)
{
temp1<-v*dnorm(y,mean=t1,sd=w)
temp2<-(1-v)*dnorm(y,mean=t2,sd=w)
return(temp1/(temp1+temp2))
}
#------------Define log likelihood---------------
logL<-function(y1,y2,y3,y4,t1,t2,w,v)
{
S1<-sum(dnorm(y1,t1,w,log=T))
S2<-sum((1-v)*dnorm(y2,t1,w,log=T)+v*dnorm(y2,t2,w,log=T))
S3<-sum(v*dnorm(y3,t1,w,log=T)+(1-v)*dnorm(y3,t2,w,log=T))
S4<-sum(dnorm(y4,t2,w,log=T))
return(S1+S2+S3+S4)
}
#-------------Define MLE Solver-----------------------
MLE<-function(y1,y2,y3,y4,t1,t2,w,v)
{
n1<-length(y1);n2<-n1+length(y2)
n3<-n2+length(y3);n<-n3+length(y4)
w1<-P1(y2,t1,t2,w,v);w2<-P2(y3,t1,t2,w,v)
t1<-(sum(y1)+sum(w1*y2)+sum(w2*y3))/(n1+sum(w1)+sum(w2))
t2<-(sum((1-w1)*y2)+sum((1-w2)*y3)+sum(y4))/(sum((1-w1))
+sum((1-w2))+n-n3)
s1<-sum((y1-t1)^2)
s2<-sum(w1*(y2-t1)^2+(1-w1)*(y2-t2)^2)
s3<-sum(w2*(y3-t1)^2+(1-w2)*(y3-t2)^2)
s4<-sum((y4-t2)^2)
```

```
w<-sqrt((s1+s2+s3+s4)/n)
v<-(sum(1-w1)+sum(w2))/(n3-n1)
return(c(t1,t2,w,v))
}
```

(4) *R* program for Example 11.2

```
#This gets permutation distribution of MLEs
#for the Poplar data
data<-read.table("PoplarMarker1.txt",
sep=",",header=F)
M<-data[,1:4];nm<-dim(M);
Y<-data[,5];n<-length(Y)
mY<-mean(Y);sY<-sd(Y)
LL0<-sum(dnorm(Y,mY,sY,log=T))
#----------Create Classes-------------------
#----------Class1=11, Class2=10,Class3=01,Class4=00
Cl<-array(1,c(nm[1],(nm[2]-1)));nc<-dim(Cl)
for(i in 1:nc[1])
{
for(j in 1:nc[2])
{
c1<-1*(M[i,j]==2)*(M[i,(j+1)]==2);
c2<-2*(M[i,j]==2)*(M[i,(j+1)]==1)
c3<-3*(M[i,j]==1)*(M[i,(j+1)]==2);
c4<-4*(M[i,j]==1)*(M[i,(j+1)]==1)
Cl[i,j]<-c1+c2+c3+c4
}
}
#---------Start Permutation Loop----------------------
#---------Initialize Estimates----------------------------
nperm<-250;nit<-100;LRT<-1;
for(k in 1:nperm)
{
yp<-sample(Y);LL<-array(0,c(nc[2],1))
#-----Get maximum of likelihood ratio-------------
for(j in 1:nc[2])
{
 y1<-(Cl[,j]==1)*yp;y1<-y1[y1!=0]
 y2<-(Cl[,j]==2)*yp;y2<-y2[y2!=0]
 y3<-(Cl[,j]==3)*yp;y3<-y3[y3!=0]
 y4<-(Cl[,j]==4)*yp;y4<-y4[y4!=0]
T<-c(mean(y1),mean(y4),sY,.25)
for(i in 1:nit){T<-c(MLE(y1,y2,y3,y4,T[1],T[2],T[3],T[4]))}
LL[j]<-2*(logL(y1,y2,y3,y4,T[1],T[2],T[3],T[4])-LL0)
}
```

```
LRT<-c(LRT, max(LL))
}
hist(LRT,freq=F)
#-----------------------------------------------------------
#-----------------------------------------------------------
#----------Define Functions for P1 and P2
P1<-function(y,t1,t2,w,v)
{
temp1<-(1-v)*dnorm(y,mean=t1,sd=w)
temp2<-v*dnorm(y,mean=t2,sd=w)
return(temp1/(temp1+temp2))
}
P2<-function(y,t1,t2,w,v)
{
temp1<-v*dnorm(y,mean=t1,sd=w)
temp2<-(1-v)*dnorm(y,mean=t2,sd=w)
return(temp1/(temp1+temp2))
}
#-------------Define log likelihood---------------
logL<-function(y1,y2,y3,y4,t1,t2,w,v)
{
S1<-sum(dnorm(y1,t1,w,log=T))
S2<-sum((1-v)*dnorm(y2,t1,w,log=T)+v*dnorm(y2,t2,w,log=T))
S3<-sum(v*dnorm(y3,t1,w,log=T)+(1-v)*dnorm(y3,t2,w,log=T))
S4<-sum(dnorm(y4,t2,w,log=T))
return(S1+S2+S3+S4)
}
#--------------Define MLE Solver-----------------------
MLE<-function(y1,y2,y3,y4,t1,t2,w,v)
{
n1<-length(y1);n2<-n1+length(y2)
n3<-n2+length(y3);n<-n3+length(y4)
w1<-P1(y2,t1,t2,w,v);w2<-P2(y3,t1,t2,w,v)
t1<-(sum(y1)+sum(w1*y2)+sum(w2*y3))/(n1+sum(w1)+sum(w2))
t2<-(sum((1-w1)*y2)+sum((1-w2)*y3)+sum(y4))/(sum((1-w1))
+sum((1-w2))+n-n3)
s1<-sum((y1-t1)^2)
s2<-sum(w1*(y2-t1)^2+(1-w1)*(y2-t2)^2)
s3<-sum(w2*(y3-t1)^2+(1-w2)*(y3-t2)^2)
s4<-sum((y4-t2)^2)
w<-sqrt((s1+s2+s3+s4)/n)
v<-(sum(1-w1)+sum(w2))/(n3-n1)
return(c(t1,t2,w,v))}
```

C

References

Akaike, H. (1974). A new look at the statistical model identification. *IEEE Transactions on Automatic Control* **AC-19**, 716–723.

Arunachlum, V. (1977). Heterosis for characters governed by two genes. *Journal of Genetics* **63**, 15–24.

Bailey, N. T. J. (1961). *Introduction to the Mathematical Theory of Genetic Linkage.* Clarendon Press, Oxford.

Bateson, W. (1902). *Mendel's Principles of Heredity: A Defence.* Cambridge University Press, London.

Bliss, S., R. J. Todhunter, R. Quaas, G. Casella, R. L. Wu, G. Lust, A. J. Williams, S. Hamilton, N. L. Dykes, A. Yeager, R. O. Gilbert, N. I. Burton-Wurster, and G. M. Acland (2002). Quantitative genetics of traits associated with canine hip dysplasia in a canine pedigree constructed by mating dysplastic Labrador Retrievers with unaffected Greyhounds. *American Journal of Veterinary Research* **63**, 1029–1035.

Box, G. E. P. and D. R. Cox (1964). An analysis of transformation (with discussion). *Journal of the Royal Statistical Society Series B* **26**, 211–252.

Bradshaw, H. D. Jr. and R. F. Stettler (1995). Molecular genetics of growth and development in *Populus.* IV. mapping QTLs with large effects on growth, form, and phenology traits in a forest tree. *Genetics* **139**, 963–973.

Breen, M., S. Jouquand, C. Renier, C. S. Mellersh, C. Hitte, N. G. Holmes, A. Cheron, N. Suter, F. Vignaux, A. E. Bristow, C. Priat, E. McCann, C. Andre, S. Boundy, P. Gitsham, R. Thomas, W. L. Bridge, H. F. Spriggs, E. J. Ryder, A. Curson, J. Sampson, E. A. Ostrander, M. M. Binns and F. Galibert (2001). Chromosome-specific single-locus FISH probes allow anchorage of an 1800-marker integrated radiation-hybrid/linkage map of the domestic dog genome to all chromosomes. *Mammalian Genome* **11**, 1784–1795.

Broman, K. W. (2005). The genomes of recombinant inbred lines. *Genetics* **169**, 1133–1146.

Buetow, K. N. and A. Chakravarti (1987). Multipoint gene mapping using seriation. *American Journal of Human Genetics* **41**, 189–201.

Carlin, B. P. and T. A. Louis (1998). *Bayes and Emperical Bayes Methods for Data Analysis.* Chapman Hall, New York.

Carter, T. C. and D. S. Falconer (1951). Stocks for detecting linkage in the mouse, and the theory of their design. *Journal of Genetics* **50**, 307–323.

Casella, G. and R. L. Berger (2001). *Statistical Inference*, Second Edition. Brooks-Cole.

Castle, W. E. (1921). An improved method of estimating the number of genetic factors concerned in cases of blending inheritance. *Science* **54**, 223.

Cavalli-Sforza, L. L. and W. F. Bodmer (1971). *The Genetics of Human Populations.* W. H. Freeman and Company, San Francisco.

Celeux, G. and J. Diebolt (1985). The SEM algorithm: a probabilistic teacher algorithm derived from the EM algorithm for the mixture problem. *Computational Statistics Quarterly* **2**, 73–82.

Chen, Z. H. (2005). The full EM algorithm for the MLEs of QTL effects and positions and their estimated variances in multiple-interval mapping. *Biometrics* **61**, 474–480.

Cheverud, J. M. and E. J. Routman(1995). Epistasis and its contribution to genetic variance components. *Genetics* **139**, 1455–1461.

Cheverud, J. M., E. J. Routman, F. A. M. Duarte, B. van Swinderen, K. Cothran and C. Perel (1996). Quantitative trait loci for murine growth. *Genetics* **142**, 1305–1319.

Churchill, G. A. and R. W. Doerge(1994). Empirical threshold values for quantitative trait mapping. *Genetics* **138**, 963–971.

Cockerham, C. C. (1954). An extension of the concept of partitioning hereditary variance for analysis of covariances among relatives when epistasis is present. *Genetics* **39**, 859–882.

Cockerham, C. C. 1963. Estimation of genetic variances. In: *Statistical Genetics and Plant Breeding*, edited by W. D. Hanson and H. F. Robinson. National Academy of Sciences–National Research Council, Washington, D. C. pp. 53–99.

Complex Trait Consortium (2004). The Collaborative Cross, a community resource for the genetic analysis of complex traits. *Nature Genetics* **36**, 1133–1137.

Cox, D. R. and D. V. Hinkley (1974). *Theoretical Statistics.* Chapman and Hall, London.

Daly, M. J., J. D. Rioux, S. F. Schaffner, T. J. Hudson, and E. S. Lander (2001). High-resolution haplotype structure in the human genome. *Nature Genetics* **29**, 229–232.

Darvasi, A., A. Weinreb, V. Minke, J. I. Weller, and M. Soller (1993). Detecting marker-QTL linkage and estimating QTL gene effect and map location using a saturated genetic map. *Genetics* **134**, 943–951.

Davies, R. B. (1977). Hypothesis testing when a nuisance parameter is present only under the alternative. *Biometrika* **64**, 247–254.

Davies, R. B. (1987). Hypothesis testing when a nuisance parameter is present only under the alternative. *Biometrika* **74**, 33–43.

Dawson, E., G. R. Abecasis, S. Bumpstead, Y. Chen, S. Hunt, D. M. Beare, J. Pabial, T. Dibling, E. Tinsley, S. Kirby, D. Carter, M. Papaspyridonos, S. Livingstone,

R. Ganske, E. Lohmussaar, J. Zernant, N. Tonisson, M. Remm, R. Magi, T. Puurand, J. Vilo, A. Kurg, K. Rice, P. Deloukas, R. Mott, A. Metspalu, D. R. Bentley, L. R. Cardon and I. A. Dunham (2002). A first-generation linkage disequilibrium map of human chromosome 22. *Nature* **418**, 544–548.

Dempster, A. P., N. M. Laird and D. B. Rubin (1977). Maximum likelihood from incomplete data via the EM algorithm. *Journal of the Royal Statistical Society Series B* **39**, 1–38.

Doebley, J., A. Stec, and C. Gustus (1995). *Teosinte branched1* and the origin of maize: evidence for epistasis and the evolution of dominance. *Genetics* **141**, 333–346.

Doebley, J., A. Stec, and L. Hubbard (1997). The evolution of apical dominance in maize. *Nature* **386**, 485–488.

Doerge, R. W. and A. Rebai (1996). Significance thresholds for QTL interval mapping tests. *Heredity* **76**, 459–464.

Dupuis, J. and D. Siegmund (1999). Statistical methods for mapping quantitative trait loci from a dense set of markers. *Genetics* **151**, 373–386.

Echt, C., S. Knapp and B.-H. Liu (1992). Genome mapping with non-inbred crosses using GMendel 2.0. *Maize Genet Cooperative Newsletter* **66**, 27–29.

Efron, B. and R. J. Tibshirani (1993). *An Introduction to the Bootstrap.* Chapman and Hall, London.

Ellis, T. H. N. (1997). Neighbor mapping as method for ordering genetic markers. Genetical Research **69**, 35–43.

Elston, R. C. and J. Stewart (1971). A general model for the genetic analysis of pedigree data. *Human Heredity* **21**, 523–542.

Eppig, J. and E. M. Eicher (1983). The mouse linkage map. *Journal of Heredity* **74**, 213–231.

Falconer, D. S. and T. F. C. Mackay (1996). *Introduction to Quantitative Genetics*, Fourth Edition. Longman, New York.

Falk, C. T. (1992). Preliminary ordering of multiple linked loci using pairwise linkage data. *Genetic Epidemiology* **9**, 367–375.

Felsenstein, J. (1979). Alternative methods of phylogenetic inference and their interrelationship. *Systematic Zoology 28*, 49–62.

Ferguson, T. S. (1996). *A Course in Large Sample Theory.* Chapman and Hall, London.

Fisher, R. A. (1918). The correlation between relatives on the supposition of Mendelian inheritance. *Transactions of the Royal Society of Edinburgh* **52**, 399–433.

Frary, A., T. C. Nesbitt, A. Frary, S. Grandillo, E. van der Knaap, B. Cong, J. P. Liu, J. Meller, R. Elber, K. B. Alpert, and S. D. Tanksley (2000). *fw2.2*: A quantitative trait locus key to the evolution of tomato fruit size. *Science* **289**, 85–88.

Fulker, D. W. and L. R. Cardon (1994). A sib-pair approach to interval mapping of quantitative trait loci. *American Journal of Human Genetics* **54**, 1092–1103.

Gallavotti, A., Q. Zhao, J. Kyozuka, R. B. Meeley, M. K. Ritter, J. F. Doebley, M. E. Pe, and R. J. Schmidt (2004). The role of barren stalk1 in the architecture of maize. *Nature* **432**, 630–635.

Gaspin, C. and T. Schiex (1997). Genetic algorithms for genetic mapping. In *Proceedings of the Third European Conference on Evolutionary Computing*, pp. 145–156. Nimes, France.

German, S., and D. German (1984). Stochastic relaxation, Gibbs distributions, and the Bayesian restoration of images, *IEEE Transactions on Pattern Analysis and Machine Intelligence* **6**, 721–741.

Grattapaglia, D. and R. R. Sederoff (1994). Genetic linkage maps of *Eucalyptus grandis* and *Eucalyptus urophylla* using a pseudo-testcross: mapping strategy and RAPD markers. *Genetics* **137**, 1121–1137.

Grattapaglia, D., F. L. G. Bertolucci, R. Penchel and R. R. Sederoff (1996). Genetic mapping of quantitative trait loci controlling growth and wood quality traits in *Eucalyptus grandis* using a maternal half-sib family and RAPD markers. *Genetics* **144**, 1205–1214.

Green, P. J. (1995). Reversible jump MCMC computation and Bayesian model determination. *Biometrika* **82**, 711–732.

Griffiths, A. J. F., W. M. Gelbart, R. C. Lewontin and J. H. Miller (2002). *Modern Genetic Analysis: Integrating Genes and Genomes*, Second Edition. W. H. Freeman, New York.

Groover, A. T., M. Devey, T. Fiddler, J. Lee, R. Megraw, T. Mitchel-Olds, B. Sherman, S. Vujcic, C. Williams and D. Neale (1994). Identification of quantitative trait loci influencing wood specific gravity in an outbred pedigree of loblolly pine. *Genetics* **138**, 1293–1300.

Gutierrez, R. G., R. J. Carroll, N. Wang, G.-H. Lee, and B. H. Taylor (1995). Analysis of tomato root initiation using a normal mixture distribution. *Biometrics* **51**, 1461–1468.

Haldane, J. B. S. (1919). The combination of linkage values and the calculation of distance between the loci of linked factors. *Journal of Genetics* **8**, 299–309.

Haldane, J. B. S. and C. H. Waddington (1931). Inbreeding and linkage. *Genetics* **16**, 357–374.

Haley, C. S. and S. A. Knott (1992). A simple method for mapping quantitative trait loci in line crosses using flanking markers. *Heredity* **69**, 315–324.

Hardy, G. H. (1908). Mendelian proportions in a mixed population. *Science* **28**, 49–50.

Hartl, D. L. and E. W. Jones (2001) *Genetics: Analysis of Genes and Genomes*, Sixth Edition. Jones and Bartlett Publishers, Sudbury, Massachusetts.

Haseman, J. K. and R. C. Elston (1972). The investigation of linkage between a quantitative trait and a marker locus. *Behavioral Genetics* **2**, 3–19.

Hastings, W. K. (1970). Monte Carlo sampling methods using Markov chains and their applications. *Biometrika* **57**, 97–109.

Hill, W. G. and J. Rasbash, (1986). Models of long term artificial selection in finite population. *Genetical Research* **48**, 4150.

Hoeschele, I., P. Ulimari, F. E. Grignola, Q. Zhang and K. M. Gage (1997). Advances in statistical methods to map quantitative trait loci in outbreed populations. *Genetics 147*, 1445–1457.

Huang, N., A. Parco, T. Mew, G. Magpantay, S. McCouch, E. Guiderdoni, J. Xu, P. Subudhi, E. R. Angeles and G. S. Khush (1997). RFLP mapping of isozymes,

RAPD and QTLs for grain shape, brown planthopper resistance in a doubled haploid rice population. *Molecular Breeding* **3**, 105–113.

Jansen, J., A. G. de Jong, and J. W. van Ooijen (2001). Constructing dense genetic linkage maps. *Theoretical and Applied Genetics* **102**, 1113–1122.

Jansen, R. C. (1994). Controlling the type I and type II errors in mapping quantitative trait loci. *Genetics* **138**, 871–881.

Jansen, R. C. and P. Stam. (1994). High resolution mapping of quantitative traits into multiple loci via interval mapping. *Genetics* **136**, 1447–1455.

Jansen, R. C., H. Geerlings, A. J. Van Oeveren and R. C. Van Schaik (2001). A comment on codominant scoring of AFLP markers. *Genetics* **158**, 925–926.

Jeffreys, H. (1961). *Theory of Probability*, Third Edition. Oxford University Press, New York.

Jennings, H. S. (1917). The numerical results of diverse systems of breeding with respect to two characters, etc. *Genetics* **2**, 97–154.

Johnson, D. L., R. C. Jansen and J. A. M. Van Arendonk (1999). Mapping quantitative trait loci in a selectively genotyped outbred population using a mixture model approach. *Genetical Research* **73**, 75–83.

Kao, C.-H. (2000). On the differences between maximum likelihood and regression interval mapping in the analysis of quantitative trait loci. *Genetics* **156**, 855–865.

Kao, C.-H. and Z.-B. Zeng (1997). General formulas for obtaining the MLEs and the asymptotic variance-covariance matrix in mapping quantitative trait loci when using the EM algorithm. *Biometrics* **53**, 653–665.

Kao, C.-H. and Z.-B. Zeng (2002). Modeling epistasis of quantitative trait loci using Cockerham's model. *Genetics* **160**, 1243–1261.

Kao, C.-H., Z.-B. Zeng, and R. D. Teasdale (1999). Multiple interval mapping for quantitative trait loci. *Genetics* **152**, 1203–1216.

Kass, R. E. and A. E. Raftery (1995). Bayes factors. *Journal of the American Statistical Association* **90**, 773–795.

Kimura, M. (1979). The neutral theory of molecular evolution. *Scientific American* **241**, 98–126.

Kosambi, D. D. (1944). The estimation of map distance from recombination values. *Annals of Eugenics* **12**, 172-175.

Lande, R. (1981). The minimum number of genes contributing to quantitative variation between and within populations. *Genetics* **99**, 541–553.

Lande, R. and R. Thompson (1990). Efficiency of marker-assisted selection in the improvement of quantitative traits. *Genetics* **124**, 743–756.

Lander, E. S. and D. Botstein (1989). Mapping Mendelian factors underlying quantitative traits using RFLP linkage maps. *Genetics* **121**, 185–199.

Lander, E. S. and P. Green (1987). Construction of multilocus genetic linkage maps in humans. *Proceedings of the National Academy of Sciences USA* **84**, 2363–2367.

Lander, E. S., P. Green, J. Abrahamson, A. Barlow, M. Daly, S. Lincoln and L. Newburg (1987). MapMaker: an interactive computer package for constructing genetic linkage maps of experimental and natural populations. *Genomics* **1**, 174–181.

Lander, E. S. and N. J. Schork (1994). Genetic dissection of complex traits. *Science* **265**, 2037–2048.

Lange, K. (1997). *Mathematical and Statistical Methods for Genetic Analysis.* Springer Verlag, New York.

Lathrop, G. M., J. M. Lalouel, C. Julier and J. Ott (1984). Strategies for multilocus linkage analysis in human. *Proceedings of the National Academy of Sciences USA* **81**, 3443–3446.

Lathrop, G. M., J. M. Lalouel, C. Julier and J. Ott (1985). Multilocus linkage analysis in humans: detection of linkage and estimation of recombination. *American Journal of Human Genetics* **37**, 482–498.

Lehmann, E. and G. Casella (1998). *Theory of Point Estimation.* Springer-Verlag, New York.

Leips, J. and T. F. C. Mackay (2000). Quantitative trait loci for lifespan in *Drosophila melanogaster*: interactions with genetic background and larval density. *Genetics* **155**, 1773–1788.

Lewin, B. (2005) *Essential Genes.* Prentice-Hall, Engewood Cliff, New Jersey.

Li, B. and R. L. Wu. (1996). Genetic causes of heterosis in juvenile aspen: a quantitative comparison across intra-and interspecific hybrids. *Theoretical and Applied Genetics* **93**, 380–391.

Li, C. B., A. L. Zhou, T. Sang (2006). Rice domestication by reducing shattering. *Science* **311**, 1936–1939.

Liberman, U. and S. Karlin (1984). Theoretical models of genetic map functions. *Theoretical Population Biology* **25**, 331–346.

Lin, M., X.-Y. Lou, M. Chang and R. L. Wu (2003). A general statistical framework for mapping quantitative trait loci in non-model systems: Issue for characterizing linkage phases. *Genetics* **165**, 901–913.

Liu, B.-H. (1998). *Statistical Genomics: Linkage, Mapping, and QTL Analysis.* CRC Press, New York.

Liu, T., R. J. Todhunter, S. Wu, W. Hou, R. Mateescu, Z. W. Zhang, N. I. Burton-Wurster, G. M. Acland, G. Lust and R. L. Wu (2007) A random model for mapping imprinted quantitative trait loci in a structured pedigree: An implication for mapping canine hip dysplasia. *Genomics* (in press).

Lo, Y. T., N. R. Mendell, and D. B. Rubin (2001). Testing the number of components in a normal mixture, *Biometrika* **88**, 767–778.

Louis, T. A. (1982). Finding the observed information matrix when using the EM algorithm. *Journal of the Royal Statistical Society Series B* **44**, 226–233.

Lu, Q., Y. H. Cui and R. L. Wu (2004). A multilocus likelihood approach to joint modeling of linkage, parental diplotype and gene order in a full-sib family. *BMC Genetics* **5**, 20.

Lynch, M. and B. Walsh (1998). *Genetics and Analysis of Quantitative Traits.* Sinauer Associates, Sunderland, MA.

Mackay, T. F. C. (1996). The nature of quantitative genetic variation revisited: lessons from *Drosophila* bristles. *BioEssays* **18**, 113–121.

Mackay, T. F. C. (2001). Quantitative trait loci in *Drosophila. Nature Reviews Genetics* **2**, 11–20.

Mackay, T. F. C., R. F. Lyman, and M. S. Jackson (1992). Effects of P element inserts on quantitative traits in *Drosophila melanogaster*. *Genetics* **130**, 315–332.

Maliepaard, C., J. Jansen and J. W. van Ooijen (1997). Linkage analysis in a fullsib family of an outbreeding plant species: overview and consequences for applications. *Genetical Research* **70**, 237–250.

Mangin, B., B. Goffinet, and A. Rebai (1994). Constructing confidence intervals for QTL location. *Genetics* **138**, 1301–1308.

Martin, O. C. and F. Hospital (2006). Two and three-locus tests for linkage analysis using recombinant inbred lines. *Genetics* **173**, 451–459.

Martinez, O. and R. N. Curnow (1992). Estimating the locations and the sizes of the effects of quantitative trait loci using flanking markers. *Theoretical and Applied Genetics* **85**, 480–488.

Mather, K. (1938). Crossing-over. *Biological Review* **13**, 252–292.

Mather, K. (1943). Polygenic inheritance and natural selection. *Biological Review* **18**, 32–65.

Mather, K. and J. L. Jinks (1982). *Biometrical Genetics*, Third Edition. Chapman and Hall, London.

Mayer, M. (2005). A comparison of regression interval mapping and multiple interval mapping for linked QTL. *Heredity* **94**, 599–605.

McLachlan, G. J. and T. Krishnan (1997). *The EM Algorithm and Extensions.* Wiley, New York.

McLachlan, G. J. and D. Peel (2000). *Finite Mixture Models.* Wiley, New York.

Meng, X.-L. and D. B. Rubin (1993). Maximum likelihood via the ECM algorithm: A general framework. *Biometrika* **80**, 267–278.

Meng, X.-L. and D. B. Rubin (1991). Using EM to obtain asymptotic variance-covariance matrices: The SEM algorithm. *Journal of the American Statistical Association* **86**, 899–909.

Mester, D., Y. Ronin, D. Minkov, E. Nevo and A. Korol (2003). Constructing largescale genetic maps using an evolutionary strategy algorithm. *Genetics* **165**, 2269–2282.

Minvielle, F. (1987). Dominance is not necessary for heterosis: a two-locus model. *Genetical Research* **49**, 245–247.

Morgan, T. H. (1909). What are "factors" in Mendelian explanations? *American Breeders Association Reports* **5**, 365–368.

Morgan, T. H. (1928). *The Theory of Genes.* Yale University Press, New Haven, Connecticut.

Morton, N. E. (1956). The detection and estimation of linkage between the genes for elliptocytosis and the Rh blood type. *American Journal of Human Genetics* **8**, 80–96.

Muller, H. J. (1916). The mechanism of crossing over. *American Naturalist* **50**, 193–207.

Nelder, J. A. and R. Mead (1965). A simplex method for function minimization. *Computer Journal* **7**, 308–313.

Newton, M. A. and A. E. Raftery (1994). Approximate Bayesian inference by the weighted likelihood bootstrap (with discussion). *Journal of the Royal Statistical Society Series B* **56**, 3–48.

Niklas, K. L. (1994). *Plant Allometry: The Scaling of Form and Process.* University of Chicago Press, Chicago, Illinois.

Olson, J. M., and M. Boehnke (1990). Monte Carlo comparison of preliminary methods of ordering multiple genetic loci. *American Journal of Human Genetics* **47**, 470–482.

Ott, J. (1991). *Analysis of Human Genetic Linkage.* 2nd edition. John Hopkins University Press, Baltimore, Maryland.

Otto, S. P. and M. W. Feldman (1997). Deleterious mutations, variable epistatic interactions, and the evolution of recombination. *Theoretical Population Biology* **51**, 134–147.

Piepho, H. P. (2001). A quick method for computing approximate thresholds for quantitative trait loci detection. *Genetics* **157**, 425–432.

Piepho, H. P. and G. Koch (2000). Codominant analysis of banding data from a dominant marker system by normal mixtures. *Genetics* **155**, 1459–1468.

Rebai, A., B. Goffinet, and B. Mangin (1994). Approximate thresholds of interval mapping tests for QTL detection. *Genetics* **138**, 235–240

Richardson, S. and P. J. Green (1997). On Bayesian analysis of mixtures with an unknown number of components (with discussion). *Journal of the Royal Statistical Society Series B* **59**, 731–792.

Ritter, E., C. Gebhardt, and F. Salamini (1990). Estimation of recombination frequencies and construction of RFLP linkage maps in plants from crosses between heterozygous parents. *Genetics* **125**, 645–654.

Ritter, E. and F. Salamini (1996). The calculation of recombination frequencies in crosses of allogamous plant species with applications to linkage mapping. *Genetical Research* **67**, 55–65.

Ridout, M. S., S. Tong, C. J. Vowden, and K. R. Tobutt (1998). Three-point linkage analysis in crosses of allogamous plant species. *Genetical Research* **72**, 111–121.

Robbins, R. B. (1918). Some applications of mathematics to breeding problems. III. *Genetics* **3**, 375–379.

Robert, C. and G. Casella (2004). *Monte Carlo Statistical Methods*, Second Edition. Springer Verlag, New York.

Routman, E. J. and J. M. Cheverud (1997). Gene effects on a quantitative trait: two-locus epistatic effects measured at microsatellite markers and at estimated QTL. *Evolution* **51**, 1654–1662.

Schäfer-Pregl, R., F. Salamini, and C. Gebhardt (1996). Models for mapping quantitative trait loci (QTL) in progeny of non-inbred parents and their behaviour in the presence of distorted segregation ratios. *Genetical Research* **67**, 43–54.

Scheiner, S. M. (1993). Genetics and evolution of phenotypic plasticity. *Annual Review of Ecology and Systematics* **24**, 35–68.

Schlichting, C. D. (1986). The evolution of phenotypic plasticity in plants. *Annual Reviews of Ecology and Systematics* **17**, 667693.

Schlichting, C. D. and M. Pigliucci (1998). *Phenotypic Evolution: A Reaction Norm Perspective.* Sinauer Associates, Sunderland, Massachusetts.

Schnell, F. W. and C. C. Cockerham (1992). Multiplicative vs. arbitrary gene action in heterosis. *Genetics* **131**, 461–469.

Schwarz, G. (1978) Estimating the dimension of a model. *Annals of Statistics* **6**, 461–464.

Sham, P. (1998). *Statistics in Human Genetics.* Wiley, New York.

Shrimpton, A. E., and A. Robertson (1988a). The isolation of polygenic factors controlling bristle score in *Drosophila melanogaster.* I: allocation of third chromosome sterno-pleural bristle effects to chromosome sections. *Genetics* **118**, 437–443.

Shrimpton, A. E., and A. Robertson (1988b). The isolation of polygenic factors controlling bristle score in *Drosophila melanogaster.* II: Distribution of third chromosome bristle effects within chromosome sections. *Genetics* **118**, 445–459.

Sillanpa, M. J. and E. Arjas (1999). Bayesian mapping of multiple quantitative trait loci from incomplete outbred offspring data. *Genetics* **151**, 1605–1619.

Song, J. Z., M. Soller, and A. Genizt (1999). The full-sib intercross line (FSIL): a QTL mapping design for outcrossing species. *Genetical Research* **73**, 61–73.

Speed, T. P., M. S. McPeek, and S. N. Evans (1992). Robustness of the noninterference model for ordering genetic markers. *Proceedings of the National Academy of Sciences USA* **89**, 3103–3106.

Stam, P. (1993). Construction of integrated genetic linkage maps by means of a new computer package: JoinMap. *Plant Journal* **3**, 739–744.

Stephens, D. A. and R. D. Fisch (1998). Bayesian analysis of quantitative trait locus data using reversible jump Markov chain Monte Carlo. *Biometrics* **54**, 1334–1347.

Stuart, A., J. K. Ord, and S. Arnold (1999). *Kendall's Advanced Theory of Statistics* Volume 2A. Oxford University Press, New York.

Suiter, K. A., J. F. Wendel and J. Case (1983). SLINKAGE-1: a PASCAL computer program for the detection and analysis of genetic linkage. *Journal of Heredity* **74**, 203–204.

Tan, Y.-D. and Y.-X. Fu (2006). A novel method for estimating linkage map. *Genetics* **173**, 2383–2390.

Teuscher, F., V. Guiard, P. E. Rudolph, and G. A. Brockmann (2005). The map expansion obtained with recombinant inbred strains and intermated recombinant inbred populations for finite generation designs. *Genetics* **170**, 875–879.

The International HapMap Consortium (2003). The International HapMap Project. *Nature* **426**, 789–794.

Thompson, E. A. (1984). Information gain in joint linkage analysis. *IMA Journal of Mathematics Applied in Medicine and Biology* **1**, 31–49.

Threadgill, D. W., K. W. Hunter, and R. W. Williams (2002). Genetic dissection of complex and quantitative traits: from fantasy to reality via a community effort. *Mammalian Genome* **13**, 175–178.

Tinker, N. A., D. E. Mather, B. G. Rossnagel, K. J. Kasha, A. Kleinhofs, P. M. Hayes, D. E. Falk, T. Ferguson, L. P. Shugar, W. G. Legge, R. B. Irvine, T. M. Choo, K. G. Briggs, S. E. Ullrich, J. D. Franckowiak, T. K. Blake, R. J. Graf, S. M.

Dofing, M. A. Saghai Maroof, G. J. Scoles, D. Hoffman, L. S. Dahleen, A. Kilian, F. Chen, R. M. Biyashev, D. A. Kudrna, and B. J. Steffenson (1996). Regions of the genome that affect agronomic performance in two-row barley. *Crop Science* **36**, 1053–1062.

Todhunter, R. J., R. Mateescu, G. Lust, N. I. Burton-Wurster, N. L. Dykes, S. P. Bliss, A. J. Williams, M. Vernier-Singer, E. Corey, C. Harjes, R. L. Quaas, Z. Zhang, R. O. Gilbert, D. Volkman, G. Casella, R. L. Wu, and G. M. Acland (2005). Quantitative trait loci for hip dysplasia in a crossbreed canine pedigree. *Mammalian Genome* **16**, 720–730.

van Ooijen, J. W. (1999). LOD significance thresholds for QTL analysis in experimental populations of diploid species. *Heredity* **83**, 613–624.

Vaughn, T. T., L. S. Pletscher, A. Peripato, K. King-Ellison, E. Adams, C. Erikson, and J. M. Cheverud (1999). Mapping quantitative trait loci for murine growth–a closer look at genetic architecture. *Genetical Research* **74**, 313–322.

Via, S., R. Gomulkiewicz, G. de Jong, S. E. Scheiner, C. D. Schlichting, and P. van Tienderen (1995). Adaptive phenotypic plasticity: consensus and controversy. *Trends in Ecology and Evolution* **10**, 212–217.

Visscher, P. M., C. S. Haley, and R. Thompson (1996). Marker-assisted introgression in backcross breeding programs. *Genetics* **144**, 1923–1932.

Vuong, Q. H. (1989). Likelihood ratio tests for model selection and non-nested hypotheses. *Econometrica* **57**, 307–333.

Wall, J. D. and J. K. Pritchard (2003). Haplotype blocks and the structure of linkage disequilibrium in the human genome. *Nature Reviews Genetics* **4**, 587–597.

Wang, H., T. Nussbaum-Wagler, B. L. Li, Q. Zhao, Y. Vigouroux, M. Faller, K. Bomblies, L. Lukens, and J. F. Doebley (2005). The origin of the naked grains of maize. *Nature* **436**, 714–719.

Weeks, D. and K. Lange (1987). Preliminary ranking procedures for multilocus ordering. *Genomics* **1**, 236–242.

Weinberg, W. (1908). Uber den Nachweis der Vererbung beim Menschen. *Jh. Ver. vaterl Naturk. Wurttemb.* **64**, 369–382.

Weir, B. S. (1996). *Genetic Data Analysis II.* Sinauer Associates. Sunderland, Massachusetts.

West, G. B., J. H. Brown and B. J. Enquist (1997). A general model for the origin of allometric scaling laws in biology. *Science* **276**, 122–126.

West, G. B., J. H. Brown and B. J. Enquist (1999). The fourth dimension of life: fractal geometry and allometric scaling of organisms. *Science* **284**, 1677–1679.

Wilson, S. (1988). A major simplification in the preliminary ordering of linked loci. *Genetic Epidemiology* **5**, 75–80.

Wu, R. L. (1996). Quantitative genetic dissection of complex traits in a QTL-mapping pedigree. *Theoretical and Applied Genetics* **93**, 447–457.

Wu, R. L. (1998). The detection of plasticity genes in heterogeneous environments. *Evolution* **52**, 967–977.

Wu, R. L., C.-X. Ma, I. Painter and Z.-B. Zeng (2002a). Simultaneous maximum likelihood estimation of linkage and linkage phases in outcrossing species. *Theoretical Population Biology* **61**, 349–363.

Wu, R. L., C.-X. Ma, S. S. Wu, and Z-B. Zeng (2002b). Linkage mapping of sex-specific differences. *Genetical Research* **79**, 85–96.

Wullschleger, S. D., T. M. Yin, S. P. DiFazio, T. J. Tschaplinski, L. E. Gunter, M. F. Davis, and G. A. Tuskan (2005). Genotypic variation in growth and biomass distribution for two advanced-generation (F_2) pedigrees of hybrid poplar (*Populus* spp.). *Canadian Journal of Forest Research* **35**, 1779–1789.

Xu, S. (1995). A comment on the simple regression method for interval mapping. *Genetics* **141**, 1657–1659.

Xu, S. (1998). Mapping quantitative trait loci using multiple families of line crosses. *Genetics* **148**, 517–524.

Xu, S. (2003). Estimating polygenic effects using markers of the entire genome. *Genetics* **163**, 789–801.

Xu, S. and W. R. Atchley (1995). A random model approach to interval mapping of quantitative trait loci. *Genetics* **141**, 1189–1197.

Yan, J. Q., J. Zhu, C. X. He, M. Benmoussa, and P. Wu. (1998). Molecular dissection of developmental behavior of plant height in rice (*Oryza sativa* L.), *Genetics* **150**, 1257–1265.

Yin, T. M., S. P. DiFazio, L. E. Gunter, D. Riemenschneider, and G. A. Tuskan (2004). Large-scale heterospecific segregation distortion in *Populus* revealed by a dense genetic map. *Theoretical and Applied Genetics* **109**, 451–463.

Yin, T. M., M. R. Huang, M. X. Wang, Z.-B. Zeng and R. L. Wu (2001). Interspecific linkage maps of *Populus adenopoda* × *P. alba*. *Genome* **44**, 602–609.

Yin, T. M., X. Y. Zhang, M. R. Huang, M. X. Wang, Q. Zhuge, S. M. Tu, L.-H. Zhu and R. L. Wu (2002). The molecular linkage maps of the Populus genome. *Genome* **45**, 541–555.

Zeng, Z.-B. (1992). Correcting the bias of Wright's estimates of the number of genes affecting a quantitative character—A further improved method. *Genetics* **131**, 987-1001.

Zeng, Z.-B. (1993). Theoretical basis for separation of multiple linked gene effects in mapping quantitative trait loci. *Proceedings of the National Academy of Sciences USA* **90**, 10972-10976.

Zeng, Z.-B. (1994). Precision mapping of quantitative trait loci. *Genetics* **136**, 1457–1468.

Zeng, Z.-B., D. Houle and C. C. Cockerham (1990). How informative is Wright's estimator of the number of genes affecting a quantitative character? *Genetics* **126**, 235–247.

Zeng, Z.-B., J. Liu, L. F. Stam, C.-H. Kao, J. M. Mercer and C.C. Laurie (2000). Genetic architecture of a morphological shape difference between two *Drosophila* species. *Genetics* **154**, 299–310.

Zhang, W.-K., Y.-J. Wang, G.-Z. Luo, J.-S. Zhang, C.-Y. He, X.-L. Wu, J.-Y. Gai, and S.-Y. Chen (2004). QTL mapping of ten agronomic traits on the soybean (*Glycine max* L. Merr.) genetic map and their association with EST markers. *Theoretical and Applied Genetics* **108**, 1131–1139.

Zou, F., J. P. Fine, J. Hu, and D. Y. Lin (2004). An efficient resampling method for assessing genome-wide statistical significance in mapping quantitative trait loci. *Genetics* **168**, 2307–2316.

Author Index

Subject Index

Printed in the United States of America